W9-AXX-206

BRIEF
CALCULUS
AND ITS
APPLICATIONS

Larry J. Goldstein
David C. Lay
David I. Schneider

Department of Mathematics, University of Maryland

Prentice-Hall, Inc., Englewood Cliffs, N.J. 07632

BRIEF
CALCULUS
AND ITS
APPLICATIONS

third edition

Library of Congress Cataloging in Publication Data

GOLDSTEIN, LARRY JOEL.
    Brief calculus and its applications.

    Rev. ed. of: Calculus and its applications. Brief
ed. c1980.
    Includes index.
    1. Calculus.   I. Lay, David C.   II. Schneider,
David I.   III. Title.
QA303.G6252 1984   515   83-3082
ISBN 0-13-111898-6

Editorial/production supervision: Kathryn Gollin Marshak
Interior design: Lee Cohen and Walter A. Behnke
Cover design: Lee Cohen
Cover art: Paul Silverman
Manufacturing buyer: John Hall

BRIEF CALCULUS AND ITS APPLICATIONS, third edition
LARRY J. GOLDSTEIN / DAVID C. LAY / DAVID I. SCHNEIDER

© 1984, 1980 by Prentice-Hall, Inc., Englewood Cliffs, New Jersey 07632

All rights reserved. No part of this book
may be reproduced, in any form
or by any means, without permission in writing
from the publisher.

Printed in the United States of America

10 9 8 7 6 5 4 3 2

Previously published under the title CALCULUS AND ITS APPLICATIONS, BRIEF EDITION

ISBN 0-13-111898-6

PRENTICE-HALL INTERNATIONAL, INC., *London*
PRENTICE-HALL OF AUSTRALIA PTY. LIMITED, *Sydney*
EDITORA PRENTICE-HALL DO BRASIL, LTDA., *Rio de Janeiro*
PRENTICE-HALL CANADA INC., *Toronto*
PRENTICE-HALL OF INDIA PRIVATE LIMITED, *New Delhi*
PRENTICE-HALL OF JAPAN, INC., *Tokyo*
PRENTICE-HALL OF SOUTHEAST ASIA PTE. LTD., *Singapore*
WHITEHALL BOOKS LIMITED, *Wellington, New Zealand*

Chapter Opening Photo Credits

0 George Gerster / Photo Researchers, Inc.

1 Hans Steiner / Photo Researchers, Inc.

2 Irene Springer

3 National Aeronautics and Space Administration

4 Carmichael / Bruce Coleman, Inc.

5 Hubertus Kanus / Rapho/Photo Researchers, Inc.

6 J.E. Pasquier / Rapho/Photo Researchers, Inc.

7 Phyllis Greenberg / Photo Researchers, Inc.

Mathematics and Its Applications

This volume is one of a collection of texts for freshman and sophomore college mathematics courses. Included in this collection are the following.

Calculus and Its Applications, third edition by L. Goldstein, D. Lay, and D. Schneider. A text designed for a two-semester course in calculus for students of business and the social and life sciences. Emphasizes an intuitive approach and integrates applications into the development. Much expanded from the highly successful second edition.

Brief Calculus and Its Applications, third edition by L. Goldstein, D. Lay, and D. Schneider. Consists of the first eight chapters of the above book. Suitable for shorter courses.

Finite Mathematics and Its Applications, second edition by L. Goldstein and D. Schneider. A traditional finite mathematics text for students of business and the social and life sciences. Allows courses to begin with either linear mathematics (linear programming, matrices) or probability and statistics.

Mathematics for the Management, Life, and Social Sciences, second edition by L. Goldstein, D. Lay, and D. Schneider. A text for a two-semester course covering finite mathematics, precalculus, and calculus.

Contents

* Sections preceded by a * are optional in the sense that they are not prerequisites for later material.

Preface

We have been very pleased with the enthusiastic response accorded the Brief Edition of *Calculus and Its Applications,* by teachers and students alike. We have been privileged to hear from a number of them. The present work incorporates many of the suggestions they have put forward.

The current edition, like its predecessor, is designed for a single semester course. Although there are many changes in this edition, we have preserved the approach and the flavor. Our goals remain the same: to begin the calculus as soon as possible; to present calculus in an intuitive yet intellectually satisfying way; and to illustrate the many applications of calculus to the biological, social, and management sciences. We have tried to achieve these goals while paying close attention to students' real and potential problems in learning calculus. Our main concern, as always, is: Will it work for the students? Listed on the following pages are some of the features that illustrate various aspects of this student-oriented approach.

Applications We provide realistic applications that illustrate the uses of calculus in other disciplines. The reader may survey the variety of applications by turning to the Index of Applications on page xv. Wherever possible, we have attempted to use applications to motivate the mathematics. For example, the integral is introduced in Chapter 6 via a discussion of world oil consumption.

Examples We have included many more worked examples than is customary (327). Furthermore, we have included computational details to enhance readability by students whose basic skills are weak.

Exercises The 1901 exercises comprise about one-quarter of the text — the most important part of the text in our opinion. The exercises at the ends of the sections are usually arranged in the order in which the text proceeds, so that the homework assignments may easily be made after only part of a section is discussed. Interesting applications and more challenging problems tend to be located near the ends of the exercise sets. Supplementary exercises at the end of each chapter expand the other exercise sets and provide cumulative exercises that require skills from earlier chapters.

Practice Problems The practice problems introduced in the previous edition have proved to be a popular and useful feature and are included in the present edition. The practice problems are carefully selected exercises that are located at the end of each section, just before the exercise set. Complete solutions are given following the exercise set. The practice problems often focus on points that are potentially confusing or are likely to be overlooked. We recommend that the reader seriously attempt the practice problems and study their solutions before moving on to the exercises. In effect, the practice problems constitute a built-in workbook.

Minimal Prerequisites In Chapter 0, we review those facts that the reader needs to study calculus. A few important topics, such as the laws of exponents, are reviewed again when they are used in a later chapter. A reader familiar with the content of Chapter 0 should begin with Chapter 1 and use Chapter 0 as a reference, whenever needed.

New in This Edition

Among the many changes in this edition, the following are the most significant.

1. *Additional Examples and Exercises.* The already ample stock of examples and exercises has been expanded, the examples by 10% and the exercises by 13%. The new examples introduce new topics and applications, and provide further explanation where needed. Many of the exercise sets have been revised with some "drill" problems added. Among the new exercises are some that will challenge the better students.

2. *The Exponential and Natural Logarithm Functions.* The two basic chapters on these functions have been revised somewhat and combined into one chapter. The applications chapter has also been revised and enlarged to include a section on percentage rates of change and elasticity of demand.

3. *Limits.* We have rewritten Section 1.4 for greater clarity. However, as in the previous edition, the instructor need cover only the first two pages of this section in order to introduce the limit notation. This is the only prerequisite from Section 1.4 for the remainder of the book. In addition, Section 1.5 on Differentiability and Continuity is optional.

4. *Additional Material.* In addition to the new topics mentioned above, we have added a section on related rates and implicit differentiation. We have also added a discussion of level curves to the chapter on calculus of several variables.

The additional topics provide the instructor with considerable flexibility in the design of one-semester or two-quarter courses.

Answers to the odd-numbered exercises are included at the back of the book. Answers to the even-numbered exercises are contained in the Instructor's Manual.

We welcome any comments or suggestions you may have and hope that you enjoy using this text as much as we have enjoyed writing it.

Acknowledgments

While writing this book, we have received assistance from many persons. And our heartfelt thanks goes out to them all. Especially, we should like to thank the following reviewers, who took the time and energy to share their ideas, preferences, and often their enthusiasm, with us.

Reviewers of the first edition: Russell Lee, Allan Hancock College; Donald Hight, Kansas State College of Pittsburgh; Ronald Rose, American River College; W. R. Wilson, Central Piedmant Community College; Bruce Swenson, Foothill College; Samuel Jasper, Ohio University; Carl David Minda, University of Cincinnati; H. Keith Stumpff, Central Missouri State University; Claude Schochet, Wayne State University; and James E. Hunneycutt, North Carolina University.

Reviewers of the second edition: Charles Himmelberg, University of Kansas; James A. Huckaba, University of Missouri; Joyce Longman, Villanova University; T. Y. Lam, University of California, Berkeley; W. T. Kyner, University of New Mexico; Shirley A. Goldman, University of California, Davis; Dennis White, University of Minnesota; Dennis Bertholf, Oklahoma State University; Wallace A. Wood, Bryant College; James L. Heitsch, University of Illinois, Chicago Circle; John H. Mathews, California State University, Fullerton; Arthur J. Schwartz, University of Michigan; Gordon Lukesh, University of Texas, Austin.

Reviewers of the third edition: William McCord, University of Missouri; W.E. Conway, University of Arizona; David W. Pentico, Virginia Commonwealth University; Howard Frisinger, Colorado State University; Robert Brown, University of California, Los Angeles; Carla Wofsy, University of New Mexico; Heath K. Riggs, University of Vermont; James Kaplan, Boston University; Larry Gerstein, University of California, Santa Barbara; Donald E. Myers, University of Arizona, Tempe; Frank Warner, University of Pennsylvania.

The authors would like to thank the many people at Prentice-Hall who have contributed to the success of our books. We appreciate the tremendous efforts of the production, art, manufacturing, marketing, and sales departments. Our sincere thanks go to Kathryn Marshak for courageously and effectively undertaking the mammoth editorial task posed by our series of books and to Lee Cohen for her beautiful design work. An extra special thanks to Robert Sickles, Executive Editor of Prentice-Hall, for his help in planning and executing these new editions. His partnership and friendship have added a warm personal dimension to the writing process.

Larry J. Goldstein
David C. Lay
David I. Schneider

Index of Applications

BRIEF
CALCULUS
AND ITS
APPLICATIONS

Introduction

Often it is possible to give a succinct and revealing description of a situation by drawing a graph. For example, Fig. 1 describes the amount of money in a bank account drawing 5% interest, compounded daily. The graph shows that as time passes, the amount of money in the account grows. In Fig. 2 we have drawn a graph that depicts the weekly sales of a breakfast cereal at various times after advertising has ceased. The graph shows that the longer the time since the last advertisement, the fewer the sales. Figure 3 shows the size of a bacteria culture at various times. The culture grows larger as time passes. But there is a maximum size that the culture cannot exceed. This maximum size reflects the restrictions imposed by food supply, space, and similar factors. The graph in Fig. 4 describes the decay of the radioactive isotope iodine 131. As time passes, less and less of the original radioactive iodine remains.

Each of the graphs in Figs. 1 to 4 describes a change that is taking place. The amount of money in the bank is changing, as are the sales of cereal, the size of the bacteria culture, and the amount of the iodine. Calculus provides mathematical tools to study each of these changes in a quantitative way.

FIGURE 1

FIGURE 2

INTRODUCTION   1

FIGURE 3

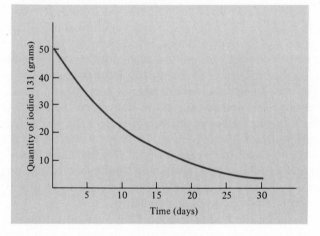

FIGURE 4

Functions

Chapter 0

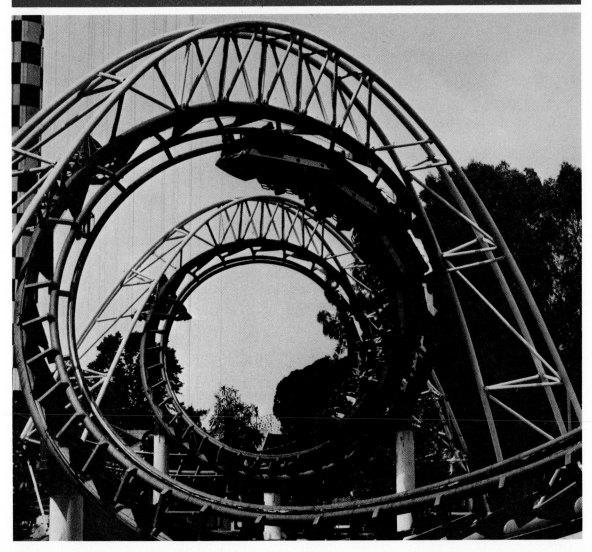

Each of the graphs in Figs. 1 to 4 of the Introduction depicts a relationship between two quantities. For example, Fig. 4 illustrates the relationship between the quantity of iodine (measured in grams) and time (measured in days). In this chapter we develop the concept of a *function*, which is the basic quantitative tool for describing such relationships.

0.1. Functions and Their Graphs

A *function* of a variable x is a *rule f* that assigns to each value of x a unique number $f(x)$, called *the value of the function at x*. [We read "$f(x)$" as "f of x."]

The functions we shall meet in this book will usually be defined by algebraic formulas. For example, the function

$$f(x) = 3x - 1$$

is the rule that takes a number, multiplies it by 3, and then subtracts 1. If we specify a value of x, say $x = 2$, then we find the value of the function at 2 by substituting 2 for x in the formula:

$$f(2) = 3(2) - 1 = 5.$$

EXAMPLE 1   Let f be the function defined by

$$f(x) = 3x^3 - 4x^2 - 3x + 7.$$

Find $f(2)$ and $f(-2)$.

Solution   To find $f(2)$ we substitute 2 for every occurrence of x in the formula for $f(x)$:

$$f(2) = 3(2)^3 - 4(2)^2 - 3(2) + 7$$
$$= 3(8) - 4(4) - 3(2) + 7$$
$$= 24 - 16 - 6 + 7$$
$$= 9.$$

The calculation of $f(-2)$ is similar.

$$f(-2) = 3(-2)^3 - 4(-2)^2 - 3(-2) + 7$$
$$= 3(-8) - 4(4) - 3(-2) + 7$$
$$= -24 - 16 + 6 + 7$$
$$= -27.$$

EXAMPLE 2　If x represents the temperature of an object in degrees Celsius, then the temperature in degrees Fahrenheit is a function of x, given by $f(x) = \frac{9}{5}x + 32$.

(a) Water freezes at 0°C (C = Celsius) and boils at 100°C. What are the corresponding temperatures in degrees Fahrenheit?

(b) Aluminum melts at 660°C. What is its melting point in degrees Fahrenheit?

Solution　(a) $f(0) = \frac{9}{5}(0) + 32 = 32$. Water freezes at 32°F.
$f(100) = \frac{9}{5}(100) + 32 = 180 + 32 = 212$. Water boils at 212°F.

(b) $f(660) = \frac{9}{5}(660) + 32 = 1188 + 32 = 1220$. Aluminum melts at 1220°F.

EXAMPLE 3　(*A Voting Model*) Let x be the proportion of the total popular vote which a Democratic candidate for president receives in a U.S. national election (so x is a number between 0 and 1). Political scientists have observed that a good estimate of the proportion of seats in the House of Representatives going to Democratic candidates is given by

$$f(x) = \frac{x^3}{x^3 + (1 - x)^3}, \qquad 0 \le x \le 1.$$

This formula is called the *cube law*. Compute $f(.6)$ and interpret the result.

Solution　We must substitute .6 for every occurrence of x in $f(x)$:

$$f(.6) = \frac{(.6)^3}{(.6)^3 + (1 - .6)^3} = \frac{(.6)^3}{(.6)^3 + (.4)^3}$$

$$= \frac{.216}{.216 + .064} = \frac{.216}{.280} \approx .77.$$

This calculation shows that the cube law function predicts that if .6 (or 60%) of the total popular vote is for the Democratic candidate for president, then approximately .77 (or 77%) of the seats in the House of Representatives will be won by Democratic candidates; that is, about 335 of the 435 seats will be won by Democrats.

In the preceding examples, the functions involved only a single algebraic formula. However, to define some functions it is necessary to use two or more formulas. Here is an illustration of this phenomenon.

EXAMPLE 4 A leading brokerage firm charges a 6% commission on gold purchases in amounts from \$50 to \$300. For purchases exceeding \$300, the firm charges 2% of the amount purchased plus \$12.00. Let x denote the amount of gold purchased (in dollars) and let $f(x)$ be the commission charge as a function of x.

(a) Describe $f(x)$.

(b) Find $f(100)$ and $f(500)$.

Solution (a) The formula for $f(x)$ depends on whether $50 \leq x \leq 300$ or $300 < x$. When $50 \leq x \leq 300$, the charge is $.06x$ dollars. When $300 < x$, the charge is $.02x + 12$. We write

$$f(x) = \begin{cases} .06x & \text{for } 50 \leq x \leq 300 \\ .02x + 12 & \text{for } 300 < x. \end{cases}$$

(b) Since $x = 100$ satisfies $50 \leq x \leq 300$, we use the first formula for $f(x)$: $f(100) = .06(100) = 6$. Since $x = 500$ satisfies $300 < x$, we use the second formula for $f(x)$: $f(500) = .02(500) + 12 = 22$.

In calculus, it is often necessary to substitute an algebraic expression for x and simplify the result.

EXAMPLE 5 If $f(x) = (4 - x)/(x^2 + 3)$, what is $f(a)$? $f(a + 1)$?

Solution Here a represents some number. To find $f(a)$, we substitute a for x wherever x appears in the formula defining $f(x)$:

$$f(a) = \frac{4 - a}{a^2 + 3}$$

To evaluate $f(a + 1)$, we substitute $a + 1$ for each occurrence of x in the formula for $f(x)$:

$$f(a + 1) = \frac{4 - (a + 1)}{(a + 1)^2 + 3}.$$

The expression for $f(a + 1)$ may be simplified, using the fact that $(a + 1)^2 = (a + 1)(a + 1) = a^2 + 2a + 1$:

$$f(a + 1) = \frac{4 - (a + 1)}{(a + 1)^2 + 3} = \frac{4 - a - 1}{a^2 + 2a + 1 + 3} = \frac{3 - a}{a^2 + 2a + 4}.$$

The Domain of a Function When defining a function, it is necessary to specify the set of acceptable values of the variable. This set is called the *domain* of the function. For instance, the cube law function of Example 3 was given by

$$f(x) = \frac{x^3}{x^3 + (1 - x)^3}, \qquad 0 \leq x \leq 1.$$

The domain of this function consists of those x for which $0 \leq x \leq 1$.

If no domain is specified, we will understand that the intended domain consists of all numbers for which the formula defining the function makes sense. For example, consider the function

$$f(x) = x^2 - x + 1.$$

The expression on the right may be evaluated for any value of x. So in the absence of any explicit restrictions on x, the domain is understood to consist of all numbers. As a second example consider the function

$$f(x) = \frac{1}{x}.$$

Here x may be any number except zero. (Division by zero is not permissible.) So the domain intended is the set of nonzero numbers. Similarly, when we write

$$f(x) = \sqrt{x}$$

we understand the domain of $f(x)$ to be the set of all nonnegative numbers, since the square root of a number x is defined if and only if $x \geq 0$.

Graphs of Functions   Often it is helpful to describe a function f geometrically, using a rectangular xy-coordinate system. Given any x in the domain of f, we can plot the point $(x, f(x))$. This is the point in the xy-plane whose y-coordinate is the value of the function at x. The set of *all* such points $(x, f(x))$ usually forms a curve in the xy-plane and is called the *graph of the function $f(x)$*.

It is possible to approximate the graph of $f(x)$ by plotting the points $(x, f(x))$ for a representative set of values of x and joining them by a smooth curve. (See Fig. 1.) The more closely spaced the values of x, the closer the approximation.

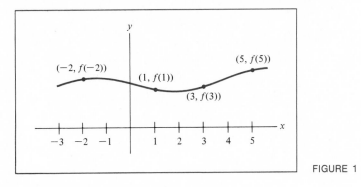

FIGURE 1

EXAMPLE 6   Sketch the graph of the function $f(x) = x^3$.

Solution   The domain consists of all numbers x. We choose some representative values of x and tabulate the corresponding values of $f(x)$. We then plot the points $(x, f(x))$ and sketch the graph indicated. (See Fig. 2.)

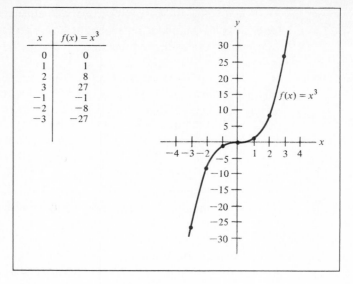

x	$f(x) = x^3$
0	0
1	1
2	8
3	27
−1	−1
−2	−8
−3	−27

FIGURE 2

EXAMPLE 7 Sketch the graph of the function $f(x) = 1/x$.

Solution The domain of the function consists of all numbers except zero. The table in Fig. 3 lists some representative values of x and the corresponding values of $f(x)$. A function often has interesting behavior for x near a number not in the domain. So when we chose representative values of x from the domain, we included some values close to zero. The points $(x, f(x))$ are plotted and the graph sketched in Fig. 3.

FIGURE 3

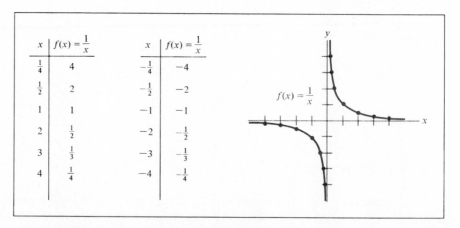

x	$f(x) = \dfrac{1}{x}$
$\frac{1}{4}$	4
$\frac{1}{2}$	2
1	1
2	$\frac{1}{2}$
3	$\frac{1}{3}$
4	$\frac{1}{4}$

x	$f(x) = \dfrac{1}{x}$
$-\frac{1}{4}$	−4
$-\frac{1}{2}$	−2
−1	−1
−2	$-\frac{1}{2}$
−3	$-\frac{1}{3}$
−4	$-\frac{1}{4}$

Graphing functions by plotting points is a tedious procedure. Moreover, we have as yet no way of knowing how many points or which points are sufficient to represent accurately all of the essential features of a graph. In calculus, we learn procedures for sketching graphs of functions that greatly ease the burden of plotting and yet guarantee that the sketch of the graph has the correct shape.

EXAMPLE 8 Suppose that f is the function whose graph is given in Fig. 4. Notice that the point $(x, y) = (3, 2)$ is on the graph of f.

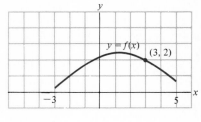

FIGURE 4

(a) What is the value of the function when $x = 3$?

(b) Find $f(-2)$.

(c) What is the domain of f?

Solution (a) Since $(3, 2)$ is on the graph of f, the y-coordinate 2 must be the value of f at the x-coordinate 3. That is, $f(3) = 2$.

(b) To find $f(-2)$ we look at the y-coordinate of the point on the graph where $x = -2$. From Fig. 4 we see that $(-2, 1)$ is on the graph of f. Thus $f(-2) = 1$.

(c) The points on the graph of $f(x)$ all have x-coordinates between -3 and 5 inclusive; and for each value of x between -3 and 5 there is a point $(x, f(x))$ on the graph. So the domain consists of those x for which $-3 \leq x \leq 5$.

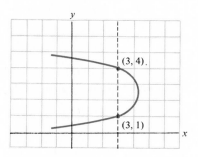

FIGURE 5 A curve that is *not* the graph of a function.

To every x in the domain, a function assigns one and only one value of y, namely, the function value $f(x)$. This implies, among other things, that not every curve is the graph of a function. To see this, refer first to the curve in Fig. 4, which *is* the graph of a function. It has the following important property: For each x between -3 and 5 inclusive there is a *unique* y such that (x, y) is on the curve. Now refer to the curve in Fig. 5. It cannot be the graph of a function, because a function f must assign to each x in its domain a *unique* value $f(x)$. However, for the curve of Fig. 5 there corresponds to $x = 3$ (for example) more than one y-value, namely, $y = 1$ and $y = 4$.

The essential difference between the curves in Figs. 4 and 5 leads us to the following test.

The Vertical Line Test A curve in the xy-plane is the graph of a function if and only if each vertical line cuts or touches the curve at no more than one point.

EXAMPLE 9   Which of the following curves are graphs of functions?

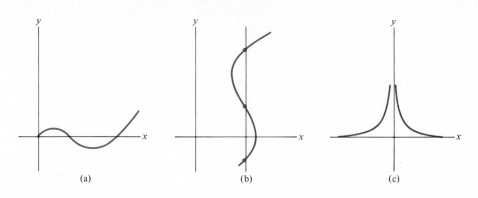

(a)   (b)   (c)

Solution   The curve in (a) is the graph of a function. It appears that vertical lines to the left of the y-axis do not touch the curve at all. This simply means that the function represented in (a) is defined only for $x \geq 0$. The curve in (b) is *not* the graph of a function because some vertical lines cut the curve in three places. The curve in (c) is the graph of a function whose domain is all nonzero x. [There is no point on the curve in (c) whose x-coordinate is 0.]

There is another notation for functions which we will find useful. Suppose that $f(x)$ is a function. When $f(x)$ is graphed on an xy-coordinate system, the values of $f(x)$ give the y-coordinates of points of the graph. For this reason, the function is often abbreviated by the letter y and we find it convenient to speak of "the function $y = f(x)$." For example, the function $y = 2x^2 + 1$ refers to the function $f(x)$ for which $f(x) = 2x^2 + 1$. The graph of a function $f(x)$ is often called *the graph of the equation $y = f(x)$.*

The equations arising in connection with functions are all of the form

$$y = [\text{an expression in } x].$$

However, not all equations connecting the variables x and y are of this sort. For example, consider these equations:

$$2x + 3y = 5$$

$$x = 3$$

$$x^2 + y^2 = 1.$$

It is possible to graph an equation by plotting points just as for functions. The only difference is that the resulting graph may not satisfy the vertical line test. For example, the graphs of the three equations above are shown in Fig. 6. Only the first graph is the graph of a function.

Letters other than f can be used to denote functions and letters other than x and y can be used to denote variables. This is especially common in applied problems where letters are chosen to suggest the quantities they depict. For instance, the revenue of a company as a function of time might be written $R(t)$.

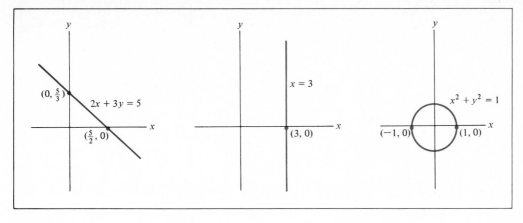

FIGURE 6

PRACTICE PROBLEMS 1

1. Is the point $(3, 12)$ on the graph of the function $g(x) = x^2 + 5x - 10$?

2. Sketch the graph of the function $h(t) = t^2 - 2$.

EXERCISES 1

1. If $f(x) = x^2 - 3x$, find $f(0)$, $f(5)$, $f(3)$, and $f(-7)$.

2. If $f(x) = 9 - 6x + x^2$, find $f(0)$, $f(2)$, $f(3)$, and $f(-13)$.

3. If $f(x) = x^3 + x^2 - x - 1$, find $f(1)$, $f(-1)$, $f(\frac{1}{2})$, and $f(a)$.

4. If $g(t) = t^3 - 3t^2 + t$, find $g(2)$, $g(-\frac{1}{2})$, $g(\frac{2}{3})$, and $g(a)$.

5. If $h(s) = s/(1 + s)$, find $h(\frac{1}{2})$, $h(-\frac{3}{2})$, and $h(a + 1)$.

6. If $f(x) = x^2/(x^2 - 1)$, find $f(\frac{1}{2})$, $f(-\frac{1}{2})$, and $f(a + 1)$.

7. If $f(x) = x^2 - 2x$, find $f(a + 1)$ and $f(a + 2)$.

8. If $f(x) = x^2 + 4x + 3$, find $f(a - 1)$ and $f(a - 2)$.

9. An office supply firm finds that the number of Remington typewriters sold in year x is approximately given by the function $f(x) = 50 + 4x + \frac{1}{2}x^2$, where $x = 0$ corresponds to 1980.

    (a) What does $f(0)$ represent?

    (b) Find the number of Remington typewriters sold in 1982.

10. When a solution of acetylcholine is introduced into the heart muscle of a frog, it diminishes the force with which the muscle contracts. The data from experiments of A. J. Clark are closely approximated by a function of the form

$$R(x) = \frac{100x}{b + x}, \qquad x \geq 0,$$

where x is the concentration of acetylcholine (in appropriate units), b is a positive constant that depends on the particular frog, and $R(x)$ is the response of the muscle to the acetylcholine, expressed as a percentage of the maximum possible effect of the drug.

(a) Suppose that $b = 20$. Find the response of the muscle when $x = 60$.

(b) Determine the value of b if $R(50) = 60$—that is, if a concentration of $x = 50$ units produces a 60% response.

In Exercises 11–14, describe the domain of the function.

11. $f(x) = \dfrac{8x}{(x - 1)(x - 2)}$

12. $f(t) = \dfrac{1}{\sqrt{t}}$

13. $g(x) = \dfrac{1}{\sqrt{3 - x}}$

14. $g(x) = \dfrac{4}{x(x + 2)}$

In Exercises 15–20, decide which curves are graphs of functions.

15.

16.

17.

18.
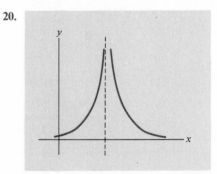

19.

20.

Exercises 21–30 relate to the function whose graph is sketched in Fig. 7.

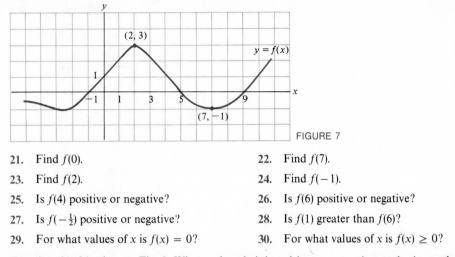

FIGURE 7

21. Find $f(0)$.

22. Find $f(7)$.

23. Find $f(2)$.

24. Find $f(-1)$.

25. Is $f(4)$ positive or negative?

26. Is $f(6)$ positive or negative?

27. Is $f(-\frac{1}{2})$ positive or negative?

28. Is $f(1)$ greater than $f(6)$?

29. For what values of x is $f(x) = 0$?

30. For what values of x is $f(x) \geq 0$?

Exercises 31–34 relate to Fig. 8. When a drug is injected into a person's muscle tissue, the concentration y of the drug in the blood is a function of the time elapsed since the injection. The graph of a typical time-concentration function f is given in Fig. 8, where $t = 0$ corresponds to the time of the injection.

FIGURE 8

31. What is the concentration of the drug when $t = 1$?

32. What is the value of the time-concentration function f when $t = 6$?

33. Find $f(5)$.

34. At what time does $f(t)$ attain its largest value?

35. Is the point $(3, 12)$ on the graph of the function $f(x) = (x - \frac{1}{2})(x + 2)$?

36. Is the point $(-2, 12)$ on the graph of the function $f(x) = x(5 + x)(4 - x)$?

37. Is the point $(\frac{1}{2}, \frac{2}{5})$ on the graph of the function $g(x) = (3x - 1)/(x^2 + 1)$?

38. Is the point $(\frac{2}{3}, \frac{5}{3})$ on the graph of the function $g(x) = (x^2 + 4)/(x + 2)$?

39. Find the y-coordinate of the point $(a + 1, \quad)$ if this point lies on the graph of the function $f(x) = x^3$.

40. Find the y-coordinate of the point $(2 + h, \quad)$ if this point lies on the graph of the function $f(x) = (5/x) - x$.

In Exercises 41–44, compute $f(1)$, $f(2)$, and $f(3)$.

41. $f(x) = \begin{cases} \sqrt{x} & \text{for } 0 \le x < 2 \\ 1 + x & \text{for } 2 \le x \le 5 \end{cases}$

42. $f(x) = \begin{cases} \dfrac{1}{x} & \text{for } 1 \le x \le 2 \\ x^2 & \text{for } 2 < x \end{cases}$

43. $f(x) = \begin{cases} \pi x^2 & \text{for } x < 2 \\ 1 + x & \text{for } 2 \le x \le 2.5 \\ 4x & \text{for } 2.5 < x \end{cases}$

44. $f(x) = \begin{cases} \dfrac{3}{4 - x} & \text{for } x < 2 \\ 2x & \text{for } 2 \le x < 3 \\ \sqrt{x^2 - 5} & \text{for } 3 \le x \end{cases}$

45. Suppose that the brokerage firm in Example 4 decides to keep the commission charges unchanged for purchases up to and including $600, but to charge only 1.5% plus $15 for gold purchases exceeding $600. Express the brokerage commission as a function of the amount x of gold purchased.

SOLUTIONS TO PRACTICE PROBLEMS 1

1. If $(3, 12)$ is on the graph of $g(x) = x^2 + 5x - 10$, then we must have $g(3) = 12$. This is not the case, however, because

$$g(3) = 3^2 + 5(3) - 10$$
$$= 9 + 15 - 10 = 14.$$

Thus $(3, 12)$ is *not* on the graph of $g(x)$.

2. Choose some representative values for t, say $t = 0, \pm 1, \pm 2, \pm 3$. For each value of t, calculate $h(t)$ and plot the point $(t, h(t))$.

t	$h(t) = t^2 - 2$
0	-2
1	-1
2	2
3	7
-1	-1
-2	2
-3	7

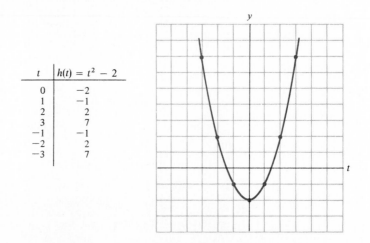

0.2. Some Important Functions

In this section we introduce some of the functions that will play a prominent role in our discussion of calculus.

Linear Functions   As we shall see in the next chapter, a knowledge of the algebraic and geometric properties of straight lines is essential for the study of calculus. Every straight line is the graph of a linear equation of the form

$$cx + dy = e$$

where c, d, and e are given constants, with c and d not both zero. If $d \neq 0$, then we may solve the equation for y to obtain an equation of the form

$$y = mx + b, \tag{1}$$

for appropriate numbers m and b. If $d = 0$, then we may solve the equation for x to obtain an equation of the form

$$x = a, \tag{2}$$

for an appropriate number a. So every straight line is the graph of an equation of type (1) or (2). The graph of an equation of the form (1) is a nonvertical line [Fig. 1(a)], whereas the graph of (2) is a vertical line [Fig. 1(b)].

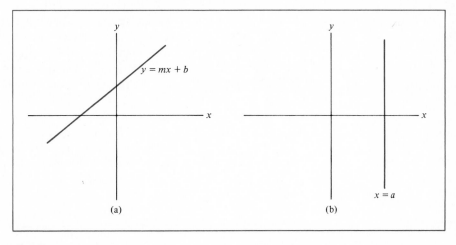

(a)   (b)

FIGURE 1

The straight line of Fig. 1(a) is the graph of the function $f(x) = mx + b$. Such a function, which is defined for all x, is called a *linear function.* Note that

FIGURE 2 Graph of
the constant function
$f(x) = b$.

the straight line of Fig. 1(b) is not the graph of a function since the vertical line test is violated.

An important special case of a linear function occurs if the value of m is zero, that is, if $f(x) = b$ for some number b. In this case, $f(x)$ is called a *constant function* since it assigns the same number, b, to every value of x. Its graph is the horizontal line whose equation is $y = b$. (See Fig. 2.)

Linear functions often arise in real-life situations, as the next example shows.

EXAMPLE 1 When the U.S. Environmental Protection Agency found a certain company dumping sulfuric acid into the Mississippi River, it fined the company \$125,000, plus \$1000 per day until the company complied with federal water pollution regulations. Express the total fine as a function of the number x of days the company continued to violate the federal regulations.

Solution The variable fine for x days of pollution, at \$1000 per day, is $1000x$ dollars. The total fine is therefore given by the function

$$f(x) = 125{,}000 + 1000x.$$

FIGURE 3

Since the graph of a linear function is a line, we may sketch it by locating any two points on the graph and drawing the line through them. For example, to sketch the graph of the function $f(x) = -\frac{1}{2}x + 3$, we may select two convenient values of x, say 0 and 4, and compute $f(0) = -\frac{1}{2}(0) + 3 = 3$ and $f(4) = -\frac{1}{2}(4) + 3 = 1$. The line through the points $(0, 3)$ and $(4, 1)$ is the graph of the function. (See Fig. 3.)

The point at which the graph of a linear function intersects the y-axis is called the *y-intercept* of the graph. The point at which the graph intersects the x-axis is called the *x-intercept*. The next example shows how to determine the intercepts of a linear function.

EXAMPLE 2 Determine the intercepts of the graph of the linear function $f(x) = 2x + 5$.

Solution Since the y-intercept is on the y-axis, its x-coordinate is 0. The point on the line with x-coordinate zero has y-coordinate

$$f(0) = 2(0) + 5 = 5.$$

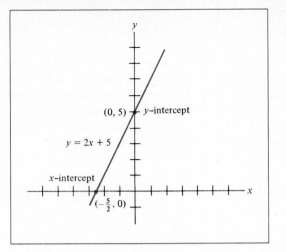

FIGURE 4

So the y-intercept is $(0, 5)$. Since the x-intercept is on the x-axis, its y-coordinate is 0. Since $f(x)$ gives the y-coordinate, we must have

$$2x + 5 = 0$$
$$2x = -5$$
$$x = -\tfrac{5}{2}.$$

So $(-\tfrac{5}{2}, 0)$ is the x-intercept. (See Fig. 4.)

EXAMPLE 3   Sketch the graph of the following function.

$$f(x) = \begin{cases} \tfrac{5}{2}x - \tfrac{1}{2} & \text{for } -1 \le x \le 1 \\ \tfrac{1}{2}x - 2 & \text{for } x > 1. \end{cases}$$

Solution   This function is defined for $x \ge -1$. But it is defined by means of two distinct linear functions. We graph the two linear functions $\tfrac{5}{2}x - \tfrac{1}{2}$ and $\tfrac{1}{2}x - 2$. Then the

FIGURE 5

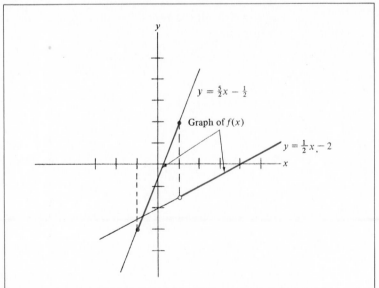

graph of $f(x)$ consists of that part of the graph of $\frac{5}{2}x - \frac{1}{2}$ for which $-1 \le x \le 1$ plus that part of the graph of $\frac{1}{2}x - 2$ for which $x > 1$. (See Fig. 5.)

Quadratic Functions   Economists utilize average cost curves which relate the average unit cost of manufacturing a commodity to the number of units to be produced. (See Fig. 6.) Ecologists use curves that relate the net primary production of nutrients in a plant to the surface area of the foliage. (See Fig. 7.) Each of the curves is bowl-shaped, opening either up or down. The simplest functions whose graphs resemble these curves are the quadratic functions.

FIGURE 6   Average-cost curve.   FIGURE 7   Production of nutrients.

A *quadratic function* is a function of the form

$$f(x) = ax^2 + bx + c,$$

where a, b, and c are constants and $a \ne 0$. The domain of such a function consists of all numbers. The graph of a quadratic function is called a *parabola*. Two typical parabolas are drawn in Figs. 8 and 9. We shall develop techniques for sketching graphs of quadratic functions after we have the properties of the derivative at our disposal.

FIGURE 8   Graph of $f(x) = x^2$.   FIGURE 9   Graph of $f(x) = -x^2 + 4x + 5$.

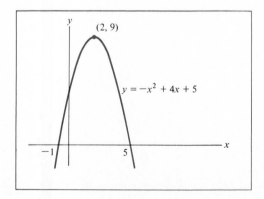

Polynomial and Rational Functions A *polynomial function* $f(x)$ is one of the form

$$f(x) = a_n x^n + a_{n-1} x^{n-1} + \cdots + a_0,$$

where n is a nonnegative integer and $a_0, a_1, \ldots, a_n$ are given numbers. Some examples of polynomial functions are

$$f(x) = 5x^3 - 3x^2 - 2x + 4$$

$$g(x) = x^4 - x + 1.$$

Of course, linear and quadratic functions are special cases of polynomial functions. The domain of a polynomial function consists of all numbers.

A function expressed as the quotient of two polynomials is called a *rational function*. Some examples are

$$h(x) = \frac{x^2 + 1}{x}$$

$$k(x) = \frac{x + 3}{x^2 - 4}.$$

The domain of a rational function excludes all values of x for which the denominator is zero. For example, the domain of $h(x)$ excludes $x = 0$, whereas the domain of $k(x)$ excludes $x = 2$ and $x = -2$. As we shall see, both polynomial and rational functions arise in applications of calculus.

Power Functions Functions of the form $f(x) = x^r$ are called *power functions*. The meaning of x^r is obvious when r is a positive integer. However, the power function $f(x) = x^r$ may be defined for any number r. Power functions are discussed in Section 5.

FIGURE 10

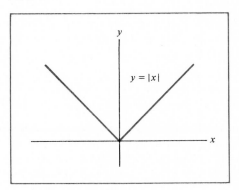

The Absolute Value Function The absolute value of a number x is denoted by $|x|$ and is defined by

$$|x| = \begin{cases} x & \text{if } x \text{ is positive or zero,} \\ -x & \text{if } x \text{ is negative.} \end{cases}$$

For example, $|5| = 5$, $|0| = 0$, and $|-3| = -(-3) = 3$.
The function defined for all numbers x by

$$f(x) = |x|$$

is called the *absolute value function*. Its graph coincides with the graph of the equation $y = x$ for $x \geq 0$ and with the graph of the equation $y = -x$ for $x < 0$. (See Fig. 10.)

PRACTICE PROBLEMS 2

1. A photocopy service has a fixed cost of $2000 per month (for rent, depreciation of equipment, etc.) and variable costs of $.04 for each page it reproduces for customers. Express its total cost as a (linear) function of the number of pages copied per month.

2. Determine the intercepts of the graph of $f(x) = -\frac{3}{8}x + 6$.

EXERCISES 2

Graph the following functions.

1. $f(x) = 2x - 1$
2. $f(x) = 3$
3. $f(x) = 3x + 1$
4. $f(x) = -\frac{1}{2}x - 4$
5. $f(x) = -2x + 3$
6. $f(x) = \frac{1}{4}$

Determine the intercepts of the graphs of the following functions.

7. $f(x) = 9x + 3$
8. $f(x) = -\frac{1}{2}x - 1$
9. $f(x) = 5$
10. $f(x) = 14$
11. $f(x) = -\frac{1}{4}x + 3$
12. $f(x) = 6x - 4$

13. In biochemistry, such as in the study of enzyme kinetics, one encounters a linear function of the form $f(x) = (K/V)x + 1/V$, where K and V are constants.

   (a) If $f(x) = .2x + 50$, find K and V so that $f(x)$ may be written in the form $f(x) = (K/V)x + 1/V$.

   (b) Find the x-intercept and y-intercept of the line $y = (K/V)x + 1/V$ (in terms of K and V).

14. The constants K and V in Exercise 13 are often determined from experimental data. Suppose that a line is drawn through data points and has x-intercept $(-500, 0)$ and y-intercept $(0, 60)$. Determine K and V so that the line is the graph of the function $f(x) = (K/V)x + 1/V$. [*Hint:* Use Exercise 13(b).]

15. In some cities you can still rent a car for $12 per day and $.15 per mile.

   (a) Find the cost of renting the car for one day and driving 200 miles.

   (b) Suppose that the car is to be rented for one day, and express the total rental expense as a function of the number x of miles driven. (Assume that for each fraction of a mile driven, the same fraction of $.15 is charged.)

16. A gas company will pay a property owner $5000 for the right to drill on the land for natural gas and $.10 for each thousand cubic feet of gas extracted from the land. Express the amount of money the landowner will receive as a function of the amount of gas extracted from the land.

17. In 1978, Blue Cross paid $135 per day for a semiprivate hospital room and $150 for an appendectomy operation. Express the total amount Blue Cross paid for an appendectomy as a function of the number of days of hospital confinement.

18. The R-rating of fiberglass home insulation is directly proportional to the thickness of

the insulation. If the R-rating of a 6-inch batt of insulation is 19, express the R-rating of the insulation as a function of the thickness (in inches).

Each of the quadratic functions in Exercises 19–24 has the form $y = ax^2 + bx + c$. Identify a, b, and c.

19. $y = 3x^2 - 4x$

20. $y = \dfrac{x^2 - 6x + 2}{3}$

21. $y = 3x - 2x^2 + 1$

22. $y = 3 - 2x + 4x^2$

23. $y = 1 - x^2$

24. $y = \frac{1}{2}x^2 + \sqrt{3}x - \pi$

Sketch the graphs of the following functions.

25. $f(x) = \begin{cases} 3x & \text{for } 0 \le x \le 1 \\ 5 - 2x & \text{for } x > 1 \end{cases}$

26. $f(x) = \begin{cases} 1 + x & \text{for } x \le 3 \\ 4 & \text{for } x > 3 \end{cases}$

27. $f(x) = \begin{cases} 3 & \text{for } x < 2 \\ 2x + 1 & \text{for } x \ge 2 \end{cases}$

28. $f(x) = \begin{cases} \frac{1}{2}x & \text{for } 0 \le x < 4 \\ 2x - 3 & \text{for } 4 \le x \le 5 \end{cases}$

29. $f(x) = \begin{cases} 4 - x & \text{for } 0 \le x < 2 \\ 2x - 2 & \text{for } 2 \le x < 3 \\ x + 1 & \text{for } x \ge 3 \end{cases}$

30. $f(x) = \begin{cases} 4x & \text{for } 0 \le x < 1 \\ 8 - 4x & \text{for } 1 \le x < 2 \\ 2x - 4 & \text{for } x \ge 2 \end{cases}$

Evaluate each of the functions in Exercises 31–36 at the given value of x.

31. $f(x) = x^{100}, \quad x = -1$

32. $f(x) = x^5, \quad x = \frac{1}{2}$

33. $f(x) = |x|, \quad x = -2.5$

34. $f(x) = |x|, \quad x = \pi$

35. $f(x) = |x|, \quad x = 10^{-2}$

36. $f(x) = |x|, \quad x = -\frac{2}{3}$

SOLUTIONS TO PRACTICE PROBLEMS 2

1. If x represents the number of pages copied per month, then the variable cost is $.04x$ dollars. Now [total cost] = [fixed cost] + [variable cost]. If we define

$$f(x) = 2000 + .04x,$$

then $f(x)$ gives the total cost per month.

2. To find the y-intercept evaluate $f(x)$ at $x = 0$.

$$f(0) = -\tfrac{3}{8}(0) + 6 = 0 + 6 = 6.$$

To find the x-intercept set $f(x) = 0$ and solve for x.

$$-\tfrac{3}{8}x + 6 = 0$$

$$\tfrac{3}{8}x = 6$$

$$x = \tfrac{8}{3} \cdot 6 = 16.$$

Therefore, the y-intercept is $(0, 6)$ and the x-intercept is $(16, 0)$.

0.3. The Algebra of Functions

Many functions we shall encounter later in the text may be viewed as combinations of other functions. For example, let $P(x)$ represent the profit a company makes on the sale of x units of some commodity. If $R(x)$ denotes the revenue received from the sale of x units, and if $C(x)$ is the cost of producing x units, then

$$P(x) \quad = \quad R(x) \quad - \quad C(x).$$

$$[\text{profit}] = [\text{revenue}] - [\text{cost}]$$

Writing the profit function in this way makes it possible to predict the behavior of $P(x)$ from properties of $R(x)$ and $C(x)$. (See Fig. 1.)

The first four examples below review the algebraic techniques needed to combine functions by addition, subtraction, multiplication, and division.

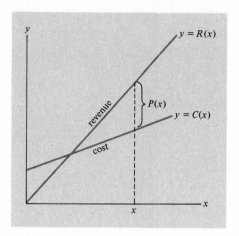

FIGURE 1

EXAMPLE 1   Let $f(x) = 3x + 4$ and $g(x) = 2x - 6$. Find $f(x) + g(x)$, $f(x) - g(x)$, $\dfrac{f(x)}{g(x)}$, and $f(x)g(x)$.

Solution   For $f(x) + g(x)$ and $f(x) - g(x)$ we add or subtract corresponding terms:

$$f(x) + g(x) = (3x + 4) + (2x - 6) = 3x + 4 + 2x - 6 = 5x - 2.$$
$$f(x) - g(x) = (3x + 4) - (2x - 6) = 3x + 4 - 2x + 6 = x + 10.$$

To compute $\dfrac{f(x)}{g(x)}$ and $f(x)g(x)$, we first substitute the formulas for $f(x)$ and $g(x)$.

$$\frac{f(x)}{g(x)} = \frac{3x + 4}{2x - 6}.$$
$$f(x)g(x) = (3x + 4)(2x - 6).$$

The expression for $\dfrac{f(x)}{g(x)}$ is already in simplest form. To simplify the expression for $f(x)g(x)$, we carry out the multiplication indicated in $(3x + 4)(2x - 6)$. We must be careful to multiply each term of $3x + 4$ by each term of $2x - 6$. A common order for multiplying such expressions is: (1) the First terms, (2) the Outer terms,

(3) the Inside terms, and (4) the Last terms. (This procedure may be remembered by the word FOIL.)

$$f(x)g(x) = (3x + 4)(2x - 6) = 6x^2 - 18x + 8x - 24$$
$$= 6x^2 - 10x - 24.$$

EXAMPLE 2 Let $g(x) = \dfrac{2}{x}$, $h(x) = \dfrac{3}{x - 1}$. Express $g(x) + h(x)$ as a rational function.

Solution First we have

$$g(x) + h(x) = \frac{2}{x} + \frac{3}{x - 1}.$$

In order for us to add two fractions, their denominators must be the same. A common denominator for $\dfrac{2}{x}$ and $\dfrac{3}{x - 1}$ is $x(x - 1)$. If we multiply $\dfrac{2}{x}$ by $\dfrac{x - 1}{x - 1}$, we obtain an equivalent expression whose denominator is $x(x - 1)$. Similarly, if we multiply $\dfrac{3}{x - 1}$ by $\dfrac{x}{x}$, we obtain an equivalent expression whose denominator is $x(x - 1)$. Thus

$$\frac{2}{x} + \frac{3}{x - 1} = \frac{2}{x} \cdot \frac{x - 1}{x - 1} + \frac{3}{x - 1} \cdot \frac{x}{x}$$

$$= \frac{2(x - 1)}{x(x - 1)} + \frac{3x}{x(x - 1)}$$

$$= \frac{2(x - 1) + 3x}{x(x - 1)}$$

$$= \frac{5x - 2}{x(x - 1)}.$$

So

$$g(x) + h(x) = \frac{5x - 2}{x(x - 1)}.$$

EXAMPLE 3 Find $f(t)g(t)$, where

$$f(t) = \frac{t}{t - 1} \quad \text{and} \quad g(t) = \frac{t + 2}{t + 1}.$$

Solution   To multiply rational functions, multiply numerator by numerator and denominator by denominator:

$$f(t)g(t) = \frac{t}{t-1} \cdot \frac{t+2}{t+1}$$

$$= \frac{t(t+2)}{(t-1)(t+1)}.$$

An alternative way of expressing $f(t)g(t)$ is obtained by carrying out the indicated multiplications:

$$f(t)g(t) = \frac{t^2+2t}{t^2+t-t-1} = \frac{t^2+2t}{t^2-1}.$$

The choice of which expression to use for $f(t)g(t)$ depends on the particular application involving $f(t)g(t)$.

EXAMPLE 4   Find $\dfrac{f(x)}{g(x)}$, where

$$f(x) = \frac{x}{x-3} \quad \text{and} \quad g(x) = \frac{x+1}{x-5}.$$

Solution   To divide $f(x)$ by $g(x)$, we multiply $f(x)$ by the reciprocal of $g(x)$:

$$\frac{f(x)}{g(x)} = \frac{x}{x-3} \cdot \frac{x-5}{x+1} = \frac{x(x-5)}{(x-3)(x+1)} = \frac{x^2-5x}{x^2-2x-3}.$$

Composition of Functions   Another important way of combining two functions $f(x)$ and $g(x)$ is to substitute the function $g(x)$ for every occurrence of the variable x in $f(x)$. The resulting function is called the *composition* (or *composite*) of $f(x)$ and $g(x)$, and is denoted by $f(g(x))$.

EXAMPLE 5   Let $f(x) = x^2 + 3x + 1$, $g(x) = x - 5$. What is $f(g(x))$?

Solution   We substitute $g(x)$ in place of each x in $f(x)$.

$$f(g(x)) = [g(x)]^2 + 3g(x) + 1$$

$$= (x-5)^2 + 3(x-5) + 1$$

$$= (x^2 - 10x + 25) + (3x - 15) + 1$$

$$= x^2 - 7x + 11.$$

Later in the text we shall need to study expressions of the form $f(x+h)$, where $f(x)$ is a given function and h represents some number. The meaning of $f(x+h)$ is that $x+h$ is to be substituted for each occurrence of x in the formula for $f(x)$. In fact, $f(x+h)$ is just a special case of $f(g(x))$, where $g(x) = x + h$.

EXAMPLE 6   If $f(x) = x^3$, find $f(x+h) - f(x)$.

Solution

$$f(x + h) = (x + h)^3 = x^3 + 3x^2h + 3xh^2 + h^3$$

$$f(x + h) - f(x) = (x^3 + 3x^2h + 3xh^2 + h^3) - x^3$$

$$= 3x^2h + 3xh^2 + h^3.$$

EXAMPLE 7   In a certain lake, the bass feed primarily on minnows and the minnows feed on plankton. Suppose that the size of the bass population is a function $f(n)$ of the number n of minnows in the lake, and the number of minnows is a function $g(x)$ of the amount x of plankton in the lake. Express the size of the bass population as a function of the amount of plankton, if $f(n) = 50 + \sqrt{n/150}$ and $g(x) = 4x + 3$.

Solution   We have $n = g(x)$. Substituting $g(x)$ for n in $f(n)$, we find that the size of the bass population is given by

$$f(g(x)) = 50 + \sqrt{\frac{g(x)}{150}} = 50 + \sqrt{\frac{4x + 3}{150}}.$$

PRACTICE PROBLEMS 3

1.   Let $f(x) = x^5$, $g(x) = x^3 - 4x^2 + x - 8$.

   (a) Find $f(g(x))$.   (b) Find $g(f(x))$.

2.   Let $f(x) = x^2$. Calculate $\dfrac{f(1 + h) - f(1)}{h}$ and simplify.

EXERCISES 3

Let $f(x) = x^2 + 1$, $g(x) = 9x$, $h(x) = 5 - 2x^2$. Calculate the following functions.

1. $f(x) + g(x)$ 2. $f(x) - h(x)$ 3. $f(x)g(x)$

4. $g(x)h(x)$ 5. $\dfrac{f(t)}{g(t)}$ 6. $\dfrac{g(t)}{h(t)}$

In Exercises 7–12, express $f(x) + g(x)$ as a rational function. Carry out all multiplications.

7. $f(x) = \dfrac{2}{x - 3}$, $g(x) = \dfrac{1}{x + 2}$ 8. $f(x) = \dfrac{3}{x - 6}$, $g(x) = \dfrac{-2}{x - 2}$

9. $f(x) = \dfrac{x}{x - 8}$, $g(x) = \dfrac{-x}{x - 4}$ 10. $f(x) = \dfrac{-x}{x + 3}$, $g(x) = \dfrac{x}{x + 5}$

11. $f(x) = \dfrac{x + 5}{x - 10}$, $g(x) = \dfrac{x}{x + 10}$ 12. $f(x) = \dfrac{x + 6}{x - 6}$, $g(x) = \dfrac{x - 6}{x + 6}$

Let $f(x) = \dfrac{x}{x - 2}$, $g(x) = \dfrac{5 - x}{5 + x}$, and $h(x) = \dfrac{x + 1}{3x - 1}$. Express the following as rational functions.

13. $f(x) - g(x)$ 14. $f(t) - h(t)$ 15. $f(x)g(x)$

16. $g(x)h(x)$ 17. $\dfrac{f(x)}{g(x)}$ 18. $\dfrac{h(s)}{f(s)}$

19. $f(x + 1)g(x + 1)$ 20. $f(x + 2) + g(x + 2)$ 21. $\dfrac{g(x + 5)}{f(x + 5)}$

22. $f\left(\dfrac{1}{t}\right)$ 23. $g\left(\dfrac{1}{u}\right)$ 24. $h\left(\dfrac{1}{x^2}\right)$

Let $f(x) = x^6$, $g(x) = \dfrac{x}{1 - x}$, and $h(x) = x^3 - 5x^2 + 1$. Calculate the following functions.

25. $f(g(x))$ 26. $h(f(t))$ 27. $h(g(x))$

28. $g(f(x))$ 29. $g(h(t))$ 30. $f(h(x))$

31. If $f(x) = x^2$, find $f(x + h) - f(x)$ and simplify.

32. If $f(x) = 1/x$, find $f(x + h) - f(x)$ and simplify.

33. If $g(t) = 4t - t^2$, find $\dfrac{g(t + h) - g(t)}{h}$ and simplify.

34. If $g(t) = t^3 + 5$, find $\dfrac{g(t + h) - g(t)}{h}$ and simplify.

35. After t hours of operation, an assembly line has assembled $A(t) = 20t - \frac{1}{2}t^2$ power lawn mowers, $0 \le t \le 10$. Suppose that the factory's cost of manufacturing x units is $C(x)$ dollars, where $C(x) = 3000 + 80x$.

(a) Express the factory's cost as a (composite) function of the number of hours of operation of the assembly line.

(b) What is the cost of the first two hours of operation?

36. During the first half-hour, the employees of a machine shop prepare the work area for the day's work. After that, they turn out 10 precision machine parts per hour, so that the output after t hours is $f(t)$ machine parts, where $f(t) = 10(t - \frac{1}{2}) = 10t - 5$, $\frac{1}{2} \le t \le 8$. The total cost of producing x machine parts is $C(x)$ dollars, where $C(x) = .1x^2 + 25x + 200$.

(a) Express the total cost as a (composite) function of t.

(b) What is the cost of the first four hours of operation?

SOLUTIONS TO PRACTICE PROBLEMS 3

1. (a) $f(g(x)) = [g(x)]^5 = (x^3 - 4x^2 + x - 8)^5$.

   (b) $g(f(x)) = [f(x)]^3 - 4[f(x)]^2 + f(x) - 8$

            $= (x^5)^3 - 4(x^5)^2 + x^5 - 8$

            $= x^{15} - 4x^{10} + x^5 - 8.$

35
a) $C(t) =$
$\quad 3000 + 80\left(20t - \frac{1}{2}t^2\right)$
$\quad 3000 + 1600t - 40t^2$

b) $3000 + 1600(2) - 40(2)^2$
$\quad 3000 + 3200 - 160$
$\quad 1200$
$\quad 6040$

2. $\dfrac{f(1 + h) - f(1)}{h} = \dfrac{(1 + h)^2 - 1^2}{h}$

$$= \dfrac{1 + 2h + h^2 - 1}{h}$$

$$= \dfrac{2h + h^2}{h}$$

$$= 2 + h.$$

0.4 Zeros of Functions—The Quadratic Formula and Factoring

A *zero* of a function $f(x)$ is a value of x for which $f(x) = 0$. For instance, the function $f(x)$ whose graph is shown in Fig. 1 has $x = -3$, $x = 3$, and $x = 7$ as zeros. Throughout this book we shall need to determine zeros of functions or, what amounts to the same thing, to solve the equation $f(x) = 0$.

In Section 2 we found zeros of linear functions. In this section our emphasis is on zeros of quadratic functions.

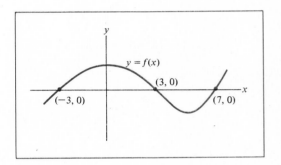

FIGURE 1

The Quadratic Formula Consider the quadratic function $f(x) = ax^2 + bx + c$, $a \neq 0$. The zeros of this function are precisely the solutions of the quadratic equation

$$ax^2 + bx + c = 0.$$

One way of solving such an equation is via the *quadratic formula*.

> The solutions of the equation $ax^2 + bx + c = 0$ are
> $$x = \dfrac{-b \pm \sqrt{b^2 - 4ac}}{2a}.$$

The $\pm$ sign tells us to form two expressions, one with $+$ and one with $-$. The quadratic formula is derived at the end of the section.

EXAMPLE 1 Solve the quadratic equation $3x^2 - 6x + 2 = 0$.

Solution Here $a = 3$, $b = -6$, and $c = 2$. Substituting these values into the quadratic formula, we find that

$$\sqrt{b^2 - 4ac} = \sqrt{(-6)^2 - 4(3)(2)} = \sqrt{36 - 24} = \sqrt{12}$$
$$= \sqrt{4 \cdot 3} = 2\sqrt{3}$$

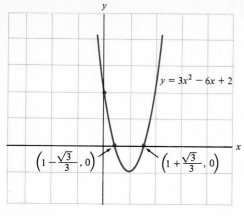

$y = 3x^2 - 6x + 2$

$\left(1 - \frac{\sqrt{3}}{3}, 0\right)$ $\left(1 + \frac{\sqrt{3}}{3}, 0\right)$

FIGURE 2

and

$$x = \frac{-b \pm \sqrt{b^2 - 4ac}}{2a}$$

$$= \frac{-(-6) \pm 2\sqrt{3}}{2(3)}$$

$$= \frac{6 \pm 2\sqrt{3}}{6}$$

$$= 1 \pm \frac{\sqrt{3}}{3}.$$

Therefore, the solutions of the equation are $1 + \sqrt{3}/3$ and $1 - \sqrt{3}/3$. (See Fig. 2.)

EXAMPLE 2 Find the zeros of the following quadratic functions.

(a) $f(x) = 4x^2 - 4x + 1$ (b) $f(x) = \frac{1}{2}x^2 - 3x + 5$

Solution (a) We must solve $4x^2 - 4x + 1 = 0$. Here $a = 4$, $b = -4$, and $c = 1$, so that

$$\sqrt{b^2 - 4ac} = \sqrt{(-4)^2 - 4(4)(1)} = \sqrt{0} = 0.$$

Thus there is only one zero, namely,

$$x = \frac{-(-4) \pm 0}{2(4)} = \frac{4}{8} = \frac{1}{2}.$$

The graph of $f(x)$ is sketched in Fig. 3.

(b) We must solve $\frac{1}{2}x^2 - 3x + 5 = 0$. Here $a = \frac{1}{2}$, $b = -3$, and $c = 5$, so that

$$\sqrt{b^2 - 4ac} = \sqrt{(-3)^2 - 4(\tfrac{1}{2})(5)} = \sqrt{9 - 10} = \sqrt{-1}.$$

The square root of a negative number is undefined, so we conclude that $f(x)$ has no zeros. The reason for this is clear from Fig. 4. The graph of $f(x)$ lies entirely above the x-axis and has no x-intercepts.

FIGURE 3   FIGURE 4

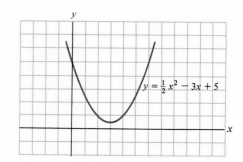

$y = 4x^2 - 4x + 1$

$(\frac{1}{2}, 0)$

$y = \frac{1}{2}x^2 - 3x + 5$

In the following example the quadratic formula is used to solve a type of problem that will arise in another context in our discussion of the definite integral.

EXAMPLE 3 Find the points of intersection of the graphs of the functions $y = x^2 + 1$ and $y = 4x$. (See Fig. 5.)

Solution If a point (x, y) is on both graphs, then its coordinates must satisfy both equations. That is, x and y must satisfy $y = x^2 + 1$ and $y = 4x$. Equating the two expressions for y, we have

$$x^2 + 1 = 4x.$$

To use the quadratic formula, we rewrite the equation in the form

$$x^2 - 4x + 1 = 0.$$

From the quadratic formula,

$$x = \frac{4 \pm \sqrt{16 - 4}}{2} = \frac{4 \pm \sqrt{12}}{2} = \frac{4 \pm 2\sqrt{3}}{2} = 2 \pm \sqrt{3}.$$

Thus the x-coordinates of the points of intersection are $2 + \sqrt{3}$ and $2 - \sqrt{3}$. To find the y-coordinates, we substitute these values of x into either equation, $y = x^2 + 1$ or $y = 4x$. The second equation is simpler. We obtain $y = 4(2 + \sqrt{3}) = 8 + 4\sqrt{3}$ and $y = 4(2 - \sqrt{3}) = 8 - 4\sqrt{3}$. Thus the points of intersection are $(2 + \sqrt{3}, 8 + 4\sqrt{3})$ and $(2 - \sqrt{3}, 8 - 4\sqrt{3})$.

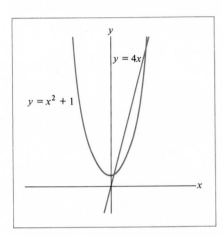

FIGURE 5

Factoring   If $f(x)$ is a polynomial, we can often write $f(x)$ as a product of linear factors (i.e., factors of the form $ax + b$). If this can be done, then the zeros of $f(x)$ can be determined by setting each of the linear factors equal to zero and solving for x. (The reason is that the product of numbers can be zero only when one of the numbers is zero.)

EXAMPLE 4 Factor the following quadratic functions.

(a) $x^2 + 7x + 12$ (b) $x^2 - 13x + 12$ (c) $x^2 - 4x - 12$ (d) $x^2 + 4x - 12$

Solution Note first that for any numbers c and d,

$$(x + c)(x + d) = x^2 + (c + d)x + cd.$$

In the quadratic on the right, the constant term is the product cd, while the coefficient of x is the sum $c + d$.

(a) Think of all integers c and d such that $cd = 12$. Then choose the pair that satisfies $c + d = 7$; that is, take $c = 3, d = 4$. Thus

$$x^2 + 7x + 12 = (x + 3)(x + 4).$$

(b) We want $cd = 12$. Since 12 is positive, c and d must be both positive or both negative. We must also have $c + d = -13$. These facts lead us to

$$x^2 - 13x + 12 = (x - 12)(x - 1).$$

(c) We want $cd = -12$. Since -12 is negative, c and d must have opposite signs. Also, they must sum to give -4. We find that

$$x^2 - 4x - 12 = (x - 6)(x + 2).$$

(d) This is almost the same as part (c).

$$x^2 + 4x - 12 = (x + 6)(x - 2).$$

EXAMPLE 5   Factor the following polynomials.

     (a) $x^2 - 6x + 9$ (b) $x^2 - 25$ (c) $3x^2 - 21x + 30$ (d) $20 + 8x - x^2$

Solution   (a) We look for $cd = 9$ and $c + d = -6$. The solution is $c = d = -3$, and

$$x^2 - 6x + 9 = (x - 3)(x - 3) = (x - 3)^2.$$

In general,

$$x^2 - 2cx + c^2 = (x - c)(x - c) = (x - c)^2.$$

(b) We use the identity

$$x^2 - c^2 = (x + c)(x - c).$$

Hence

$$x^2 - 25 = (x + 5)(x - 5).$$

(c) We first factor out a common factor of 3, and then use the method of Example 4.

$$3x^2 - 21x + 30 = 3(x^2 - 7x + 10)$$
$$= 3(x - 5)(x - 2).$$

(d) We first factor out a -1 in order to make the coefficient of x^2 equal to $+1$.

$$20 + 8x - x^2 = (-1)(x^2 - 8x - 20)$$
$$= (-1)(x - 10)(x + 2).$$

EXAMPLE 6   Factor the following polynomials.

     (a) $x^2 - 8x$ (b) $x^3 + 3x^2 - 18x$ (c) $x^3 - 10x$

Solution   In each case we first factor out a common factor of x.

     (a) $x^2 - 8x = x(x - 8)$

     (b) $x^3 + 3x^2 - 18x = x(x^2 + 3x - 18) = x(x + 6)(x - 3)$

(c) $x^3 - 10x = x(x^2 - 10)$. To factor $x^2 - 10$, we use the identity $x^2 - c^2 = (x + c)(x - c)$, where $c^2 = 10$ and $c = \sqrt{10}$. Thus $x^3 - 10x = x(x^2 - 10) = x(x + \sqrt{10})(x - \sqrt{10})$.

The main use of factoring in this text will be to solve equations.

EXAMPLE 7 Solve the following equations.

(a) $x^2 - 2x - 15 = 0$ (b) $x^2 - 20 = x$ (c) $\dfrac{x^2 + 10x + 25}{x + 1} = 0$

Solution (a) The equation $x^2 - 2x - 15 = 0$ may be written in the form

$$(x - 5)(x + 3) = 0.$$

The product of two numbers is zero only if one or the other of the numbers (or both) is zero. Hence

$$x - 5 = 0 \quad \text{or} \quad x + 3 = 0.$$

That is,

$$x = 5 \quad \text{or} \quad x = -3.$$

(b) First we must rewrite the equation $x^2 - 20 = x$ in the form $ax^2 + bx + c = 0$; that is,

$$x^2 - x - 20 = 0$$
$$(x - 5)(x + 4) = 0.$$

We conclude that

$$x - 5 = 0 \quad \text{or} \quad x + 4 = 0;$$

that is,

$$x = 5 \quad \text{or} \quad x = -4.$$

(c) A rational function will be zero only if the numerator is zero. Thus

$$x^2 + 10x + 25 = 0,$$
$$(x + 5)^2 = 0,$$
$$x + 5 = 0; \quad \text{that is, } x = -5.$$

Since the denominator is not 0 at $x = -5$, we conclude that $x = -5$ is the solution.

Derivation of the Quadratic Formula

$$ax^2 + bx + c = 0$$
$$ax^2 + bx = -c$$
$$4a^2x^2 + 4abx = -4ac \qquad \text{(both sides were multiplied by } 4a\text{),}$$
$$4a^2x^2 + 4abx + b^2 = b^2 - 4ac \qquad (b^2 \text{ added to both sides).}$$

Now note that $4a^2x^2 + 4abx + b^2 = (2ax + b)^2$. To check this, simply multiply out the right-hand side! Therefore,

$$(2ax + b)^2 = b^2 - 4ac$$

$$2ax + b = \pm\sqrt{b^2 - 4ac}$$

$$2ax = -b \pm \sqrt{b^2 - 4ac}$$

$$x = \frac{-b \pm \sqrt{b^2 - 4ac}}{2a}.$$

PRACTICE PROBLEMS 4

1. Solve the equation $x - 14/x = 5$.

2. Use the quadratic formula to solve $7x^2 - 35x + 35 = 0$.

EXERCISES 4

Use the quadratic formula to find the zeros of the functions in Exercises 1–6.

1. $f(x) = 2x^2 - 7x + 6$
2. $f(x) = 3x^2 + 2x - 1$
3. $f(x) = 4x^2 - 12x + 9$
4. $f(x) = \frac{1}{4}x^2 + x + 1$
5. $f(x) = -2x^2 + 3x - 4$
6. $f(x) = 11x^2 - 7x + 1$

Use the quadratic formula to solve the equations in Exercises 7–12.

7. $5x^2 - 4x - 1 = 0$
8. $x^2 - 4x + 5 = 0$
9. $15x^2 - 135x + 300 = 0$
10. $x^2 - \sqrt{2}x - \frac{5}{4} = 0$
11. $\frac{3}{2}x^2 - 6x + 5 = 0$
12. $9x^2 - 12x + 4 = 0$

Factor the polynomials in Exercises 13–24.

13. $x^2 + 8x + 15$
14. $x^2 - 10x + 16$
15. $x^2 - 16$
16. $x^2 - 1$
17. $3x^2 + 12x + 12$
18. $2x^2 - 12x + 18$
19. $30 - 4x - 2x^2$
20. $15 + 12x - 3x^2$
21. $3x - x^2$
22. $4x^2 - 1$
23. $6x - 2x^3$
24. $16x + 6x^2 - x^3$

Find the points of intersection of the pairs of curves in Exercises 25–32.

25. $y = 2x^2 - 5x - 6, y = 3x + 4$
26. $y = x^2 - 10x + 9, y = x - 9$
27. $y = x^2 - 4x + 4, y = 12 + 2x - x^2$
28. $y = 3x^2 + 9, y = 2x^2 - 5x + 3$
29. $y = x^3 - 3x^2 + x, y = x^2 - 3x$
30. $y = \frac{1}{2}x^3 - 2x^2, y = 2x$
31. $y = \frac{1}{2}x^3 + x^2 + 5, y = 3x^2 - \frac{1}{2}x + 5$
32. $y = 30x^3 - 3x^2, y = 16x^3 + 25x^2$

Solve the equations in Exercises 33–38.

33. $\dfrac{21}{x} - x = 4$

34. $x + \dfrac{2}{x - 6} = 3$

35. $x + \dfrac{14}{x + 4} = 5$

36. $1 = \dfrac{5}{x} + \dfrac{6}{x^2}$

37. $\dfrac{x^2 + 14x + 49}{x^2 + 1} = 0$

38. $\dfrac{x^2 - 8x + 16}{1 + \sqrt{x}} = 0$

SOLUTIONS TO PRACTICE PROBLEMS 4

1. Multiply both sides of the equation by x. Then

$$x^2 - 14 = 5x.$$

Now, take the $5x$ term to the left side of the equation and solve by factoring.

$$x^2 - 5x - 14 = 0$$

$$(x - 7)(x + 2) = 0$$

$$x = 7 \quad \text{or} \quad x = -2.$$

2. In this case, each coefficient is a multiple of 7. To simplify the arithmetic, we divide both sides of the equation by 7 before using the quadratic formula.

$$x^2 - 5x + 5 = 0$$

$$\sqrt{b^2 - 4ac} = \sqrt{(-5)^2 - 4(1)(5)} = \sqrt{5}$$

$$x = \frac{-b \pm \sqrt{b^2 - 4ac}}{2a} = \frac{5 \pm \sqrt{5}}{2 \cdot 1} = \frac{5}{2} \pm \frac{1}{2}\sqrt{5}.$$

0.5 Exponents and Power Functions

In this section we review the operations with exponents that occur frequently throughout the text. We begin with the definition of b^r for various types of numbers b and r.

For any nonzero number b and any positive integer n, we have by definition that

$$b^n = \underbrace{b \cdot b \cdot \ldots \cdot b,}_{n \text{ times}}$$

$$b^{-n} = \frac{1}{b^n},$$

and

$$b^0 = 1.$$

For example, $2^4 = 2 \cdot 2 \cdot 2 \cdot 2 = 16$, $2^{-4} = \dfrac{1}{2^4} = \frac{1}{16}$, and $2^0 = 1$.

Next, we consider numbers of the form $b^{1/n}$, where n is a positive integer. For instance,

$2^{1/2}$ is the positive number whose square is 2: $2^{1/2} = \sqrt{2}$;

$2^{1/3}$ is the positive number whose cube is 2: $2^{1/3} = \sqrt[3]{2}$;

$2^{1/4}$ is the positive number whose fourth power is 2: $2^{1/4} = \sqrt[4]{2}$;

and so on. In general, when b is zero or positive, $b^{1/n}$ is zero or the positive number whose nth power is b.

In the special case when n is odd, we may permit b to be negative as well as positive. For example, $(-8)^{1/3}$ is the number whose cube is -8; that is,

$$(-8)^{1/3} = -2.$$

Thus, when b is negative and n is odd, we again define $b^{1/n}$ to be the number whose nth power is b.

Finally, let us consider numbers of the form $b^{m/n}$ and $b^{-m/n}$, where m and n are positive integers. We may assume that the fraction m/n is in lowest terms (so that m and n have no common factor). Then we define

$$b^{m/n} = (b^{1/n})^m$$

whenever $b^{1/n}$ is defined, and

$$b^{-m/n} = \frac{1}{b^{m/n}}$$

whenever $b^{m/n}$ is defined and is not zero. For example,

$$8^{5/3} = (8^{1/3})^5 = (2)^5 = 32,$$

$$8^{-5/3} = \frac{1}{8^{5/3}} = \frac{1}{32},$$

$$(-8)^{5/3} = [(-8)^{1/3}]^5 = [-2]^5 = -32.$$

Exponents may be manipulated algebraically according to the following rules:

Laws of Exponents

1. $b^r b^s = b^{r+s}$ 4. $(b^r)^s = b^{rs}$

2. $b^{-r} = \dfrac{1}{b^r}$ 5. $(ab)^r = a^r b^r$

3. $\dfrac{b^r}{b^s} = b^r \cdot b^{-s} = b^{r-s}$ 6. $\left(\dfrac{a}{b}\right)^r = \dfrac{a^r}{b^r}$

EXAMPLE 1 Use the laws of exponents to calculate the following quantities.

(a) $2^{1/2}50^{1/2}$ (b) $(2^{1/2}2^{1/3})^6$ (c) $\dfrac{5^{3/2}}{\sqrt{5}}$

Solution (a) $2^{1/2}50^{1/2} = (2 \cdot 50)^{1/2}$ (Law 5)

$$= \sqrt{100}$$

$$= 10.$$

(b) $(2^{1/2}2^{1/3})^6 = (2^{(1/2)+(1/3)})^6$ (Law 1)

$$= (2^{5/6})^6$$

$$= 2^{(5/6)6}$$ (Law 4)

$$= 2^5$$

$$= 32.$$

(c) $\dfrac{5^{3/2}}{\sqrt{5}} = \dfrac{5^{3/2}}{5^{1/2}}$

$$= 5^{(3/2)-(1/2)}$$ (Law 3)

$$= 5^1$$

$$= 5.$$

EXAMPLE 2 Simplify the following expressions.

(a) $\dfrac{1}{x^{-4}}$ (b) $\dfrac{x^2}{x^5}$ (c) $\sqrt{x}(x^{3/2} + 3\sqrt{x})$

Solution (a) $\dfrac{1}{x^{-4}} = x^{-(-4)}$ (Law 2 with $r = -4$)

$$= x^4.$$

(b) $\dfrac{x^2}{x^5} = x^{2-5}$ (Law 3)

$$= x^{-3}.$$

It is also correct to write this answer as $\dfrac{1}{x^3}$.

(c) $\sqrt{x}(x^{3/2} + 3\sqrt{x}) = x^{1/2}(x^{3/2} + 3x^{1/2})$

$$= x^{1/2}x^{3/2} + 3x^{1/2}x^{1/2}$$

$$= x^{(1/2)+(3/2)} + 3x^{(1/2)+(1/2)}$$ (Law 1)

$$= x^2 + 3x.$$

A *power function* is a function of the form

$$f(x) = x^r,$$

for some number r.

EXAMPLE 3   Let $f(x)$ and $g(x)$ be the power functions

$$f(x) = x^{-1} \quad \text{and} \quad g(x) = x^{1/2}.$$

Determine the following functions.

(a) $\dfrac{f(x)}{g(x)}$ (b) $f(x)g(x)$ (c) $\dfrac{g(x)}{f(x)}$

Solution   (a) $\dfrac{f(x)}{g(x)} = \dfrac{x^{-1}}{x^{1/2}}$ (b) $f(x)g(x) = x^{-1}x^{1/2}$ (c) $\dfrac{g(x)}{f(x)} = \dfrac{x^{1/2}}{x^{-1}}$

$$= x^{-1-(1/2)} \qquad\qquad = x^{-1+(1/2)} \qquad\qquad = x^{(1/2)-(-1)}$$

$$= x^{-3/2} \qquad\qquad\qquad = x^{-1/2} \qquad\qquad\qquad = x^{3/2}.$$

$$= \dfrac{1}{x^{3/2}}. \qquad\qquad\qquad = \dfrac{1}{x^{1/2}}$$

$$\qquad\qquad\qquad\qquad = \dfrac{1}{\sqrt{x}}.$$

PRACTICE PROBLEMS 5

1. Compute. (a) -5^2 (b) $16^{.75}$

2. Simplify. (a) $(4x^3)^2$ (b) $\dfrac{\sqrt[3]{x}}{x^3}$ (c) $\dfrac{2 \cdot (x + 5)^6}{x^2 + 10x + 25}$

EXERCISES 5

In Exercises 1–28, compute the numbers.

1. 3^2 2. $(-2)^3$ 3. 1^{100} 4. 0^{25}

5. $(.1)^4$ 6. $(100)^4$ 7. -4^2 8. $(.01)^3$

9. $(16)^{1/2}$ 10. $(27)^{1/3}$ 11. $(.000001)^{1/3}$ 12. $\left(\dfrac{1}{125}\right)^{1/3}$

13. 6^{-1} 14. $\left(\dfrac{1}{2}\right)^{-1}$ 15. $(.01)^{-1}$ 16. $(-5)^{-1}$

17. $8^{4/3}$ 18. $16^{3/4}$ 19. $(25)^{3/2}$ 20. $(27)^{2/3}$

21. $(1.8)^0$ 22. $9^{1.5}$ 23. $16^{.5}$ 24. $(81)^{.75}$

25. $4^{-1/2}$ **26.** $\left(\dfrac{1}{8}\right)^{-2/3}$ **27.** $(.01)^{-1.5}$ **28.** $1^{-1.2}$

In Exercises 29–40, use the laws of exponents to compute the numbers.

29. $5^{1/3} \cdot 200^{1/3}$ **30.** $(3^{1/3} \cdot 3^{1/6})^6$ **31.** $6^{1/3} \cdot 6^{2/3}$ **32.** $(9^{4/5})^{5/8}$

33. $\dfrac{10^4}{5^4}$ **34.** $\dfrac{3^{5/2}}{3^{1/2}}$ **35.** $(2^{1/3} \cdot 3^{2/3})^3$ **36.** $20^{.5} \cdot 5^{.5}$

37. $\left(\dfrac{8}{27}\right)^{2/3}$ **38.** $(125 \cdot 27)^{1/3}$ **39.** $\dfrac{7^{4/3}}{7^{1/3}}$ **40.** $(6^{1/2})^0$

In Exercises 41–70, use the laws of exponents to simplify the following algebraic expressions. Your answer should not involve parentheses or negative exponents.

41. $(xy)^6$ **42.** $(x^{1/3})^6$ **43.** $\dfrac{x^4 \cdot y^5}{xy^2}$ **44.** $\dfrac{1}{x^{-3}}$

45. $x^{-1/2}$ **46.** $(x^3 \cdot y^6)^{1/3}$ **47.** $\left(\dfrac{x^4}{y^2}\right)^3$ **48.** $\left(\dfrac{x}{y}\right)^{-2}$

49. $(x^3 y^5)^4$ **50.** $\sqrt{1+x}(1+x)^{3/2}$ **51.** $x^5 \cdot \left(\dfrac{y^2}{x}\right)^3$ **52.** $x^{-3} \cdot x^7$

53. $(2x)^4$ **54.** $\dfrac{-3x}{15x^4}$ **55.** $\dfrac{-x^3 y}{-xy}$ **56.** $\dfrac{x^3}{y^{-2}}$

57. $\dfrac{x^{-4}}{x^3}$ **58.** $(-3x)^3$ **59.** $\sqrt[3]{x} \cdot \sqrt[3]{x^2}$ **60.** $(9x)^{-1/2}$

61. $\left(\dfrac{3x^2}{2y}\right)^3$ **62.** $\dfrac{x^2}{x^5 y}$ **63.** $\dfrac{2x}{\sqrt{x}}$ **64.** $\dfrac{1}{yx^{-5}}$

65. $(16x^8)^{-3/4}$ **66.** $(-8y^9)^{2/3}$ **67.** $\sqrt{x}\left(\dfrac{1}{4x}\right)^{5/2}$ **68.** $\dfrac{(25xy)^{3/2}}{x^2 y}$

69. $\dfrac{(-27x^5)^{2/3}}{\sqrt[3]{x}}$ **70.** $(-32y^{-5})^{3/5}$

The expressions in Exercises 71–74 may be factored as shown. Find the missing factors.

71. $\sqrt{x} - \dfrac{1}{\sqrt{x}} = \dfrac{1}{\sqrt{x}}(\quad)$ **72.** $2x^{2/3} - x^{-1/3} = x^{-1/3}(\quad)$

73. $x^{-1/4} + 6x^{1/4} = x^{-1/4}(\quad)$ **74.** $\sqrt{\dfrac{x}{y}} - \sqrt{\dfrac{y}{x}} = \sqrt{xy}\,(\quad)$

75. Explain why $\sqrt{a} \cdot \sqrt{b} = \sqrt{ab}$.   **76.** Explain why $\sqrt{a}/\sqrt{b} = \sqrt{a/b}$.

In Exercises 77–84, evaluate $f(4)$.

77. $f(x) = x^2$ **78.** $f(x) = x^3$ **79.** $f(x) = x^{-1}$ **80.** $f(x) = x^{1/2}$

81. $f(x) = x^{3/2}$ **82.** $f(x) = x^{-1/2}$ **83.** $f(x) = x^{-5/2}$ **84.** $f(x) = x^0$

SOLUTIONS TO PRACTICE PROBLEMS 5

1. (a) $-5^2 = -25$. [Note that -5^2 is the same as $-(5^2)$. This number is different from $(-5)^2$, which equals 25. Whenever there are no parentheses, apply the exponent first and then apply the other operations.]

   (b) Since $.75 = \frac{3}{4}$, $16^{.75} = 16^{3/4} = (\sqrt[4]{16})^3 = 2^3 = 8$.

2. (a) Apply Law 5 with $a = 4$ and $b = x^3$. Then use Law 4.

   $$(4x^3)^2 = 4^2 \cdot (x^3)^2 = 16 \cdot x^6.$$

   [A common error is to forget to square the 4. If that had been our intent, we would have asked for $4(x^3)^2$.]

   (b) $\dfrac{\sqrt[3]{x}}{x^3} = \dfrac{x^{1/3}}{x^3} = x^{(1/3)-3} = x^{-8/3}$. [The answer can also be given as $1/x^{8/3}$.] When simplifying expressions involving radicals, it is usually a good idea to first convert the radical to an exponent.

   (c) $\dfrac{2(x+5)^6}{x^2+10x+25} = \dfrac{2\cdot(x+5)^6}{(x+5)^2} = 2(x+5)^{6-2} = 2(x+5)^4$. [Here the third law of exponents was applied to $(x+5)$. The laws of exponents apply to any algebraic expression.]

Chapter 0: CHECKLIST

- □ Function
- □ Value of a function at x
- □ Domain of a function
- □ Graph of a function
- □ Vertical line test
- □ Graph of an equation
- □ Linear function
- □ Constant function
- □ x- and y-intercepts
- □ Quadratic function
- □ Polynomial and rational functions
- □ Power function
- □ Absolute value function
- □ Addition, subtraction, multiplication, and division of functions
- □ Composition of functions
- □ Zero of a function
- □ Quadratic formula
- □ Factoring polynomials
- □ Laws of exponents

Chapter 0: SUPPLEMENTARY EXERCISES

1. Let $f(x) = x^3 + 1/x$. Evaluate $f(1)$, $f(3)$, $f(-1)$, $f(-\frac{1}{2})$, and $f(\sqrt{2})$.

2. Let $f(x) = 2x + 3x^2$. Evaluate $f(0)$, $f(-\frac{1}{4})$, and $f(1/\sqrt{2})$.

3. Let $f(x) = x^2 - 2$. Evaluate $f(a - 2)$.

4. Let $f(x) = [1/(x + 1)] - x^2$. Evaluate $f(a + 1)$.

Determine the domains of the following functions.

5. $f(x) = \dfrac{1}{x(x + 3)}$

6. $f(x) = \sqrt{x - 1}$

7. $f(x) = \sqrt{x^2 + 1}$

8. $f(x) = \dfrac{1}{\sqrt{3x}}$

9. Is the point $(\frac{1}{2}, -\frac{3}{5})$ on the graph of the function $h(x) = (x^2 - 1)/(x^2 + 1)$?

10. Is the point $(1, -2)$ on the graph of the function $k(x) = x^2 + (2/x)$?

Factor the polynomials in Exercises 11–14.

11. $5x^3 + 15x^2 - 20x$

12. $3x^2 - 3x - 60$

13. $18 + 3x - x^2$

14. $x^5 - x^4 - 2x^3$

15. Find the zeros of the quadratic function $y = 5x^2 - 3x - 2$.

16. Find the zeros of the quadratic function $y = -2x^2 - x + 2$.

17. Find the points of intersection of the curves $y = 5x^2 - 3x - 2$ and $y = 2x - 1$.

18. Find the points of intersection of the curves $y = -x^2 + x + 1$ and $y = x - 5$.

Let $f(x) = x^2 - 2x$, $g(x) = 3x - 1$, and $h(x) = \sqrt{x}$. Find the following functions.

19. $f(x) + g(x)$

20. $f(x) - g(x)$

21. $f(x)h(x)$

22. $f(x)g(x)$

23. $f(x)/h(x)$

24. $g(x)h(x)$

Let $f(x) = x/(x^2 - 1)$, $g(x) = (1 - x)/(1 + x)$, and $h(x) = 2/(3x + 1)$. Express the following as rational functions.

25. $f(x) - g(x)$

26. $f(x) - g(x + 1)$

27. $g(x) - h(x)$

28. $f(x) + h(x)$

29. $g(x) - h(x - 3)$

30. $f(x) + g(x)$

Let $f(x) = x^2 - 2x + 4$, $g(x) = 1/x^2$, and $h(x) = 1/(\sqrt{x} - 1)$. Determine the following functions.

31. $f(g(x))$

32. $g(f(x))$

33. $g(h(x))$

34. $h(g(x))$

35. $f(h(x))$

36. $h(f(x))$

37. Simplify $(81)^{3/4}$, $8^{5/3}$, and $(.25)^{-1}$.

38. Simplify 5^{-2}, $(100)^{3/2}$, and $(.001)^{1/3}$.

39. The population of a city is estimated to be $750 + 25t + .1t^2$ thousand people t years from the present. Ecologists estimate that the average level of carbon monoxide in the air above the city will be $1 + .4x$ ppm (parts per million) when the population is x thousand people. Express the carbon monoxide level as a function of the time t.

40. The revenue $R(x)$ (in thousands of dollars) a company receives from the sale of x thousand units is given by $R(x) = 5x - x^2$. The sales level x is in turn a function $f(d)$ of the number d of dollars spent on advertising, where

$$f(d) = 6\left(1 - \frac{200}{d + 200}\right).$$

Express the revenue as a function of the amount spent on advertising.

In Exercises 41–44, use the laws of exponents to simplify the algebraic expressions.

41. $(\sqrt{x + 1})^4$ **42.** $\dfrac{xy^3}{x^{-5}y^6}$ **43.** $\dfrac{x^{3/2}}{\sqrt{x}}$ **44.** $\sqrt[3]{x}(8x^{2/3})$

The Derivative

Chapter 1

The *derivative* is a mathematical tool used to measure rate of change. To illustrate the sort of change we have in mind, consider the following example. Suppose that a rumor spreads through a town of 10,000 people.* Denote by $N(t)$ the number of people who have heard the rumor after t days. We have tabulated the values for $N(t)$ corresponding to $t = 0, 1, \ldots, 10$ in the chart in Fig. 1. For example, the rumor starts at $t = 0$ with one person $[N(0) = 1]$; after 1 day, the number having heard the rumor increases to 6; after 2 days, to 40; and so forth.

FIGURE 1

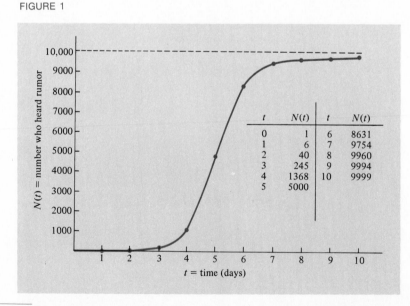

t	$N(t)$	t	$N(t)$
0	1	6	8631
1	6	7	9754
2	40	8	9960
3	245	9	9994
4	1368	10	9999
5	5000		

---

* The graph used in this example is derived from a mathematical model used by sociologists. See J. Coleman, *An Introduction to Mathematical Sociology* (London: Collier-Macmillan Ltd., 1964), p. 43.

Each day the number of people who have heard the rumor increases. We may describe the increase geometrically by plotting the points corresponding to $t = 0, 1, \ldots, 10$ and drawing a smooth curve through them. (See Fig. 1.)

Let us describe the change that is taking place in this example. Initially, the number of *new* people to hear the rumor each day is rather small (5 the first day, 34 the second, 205 the third). As we shall see later, this situation is reflected in the fact that the graph is not very steep at $t = 1, 2, 3$. (It is almost flat.) On the fourth day, 1123 additional people hear the rumor; on the fifth day, 3632. The more rapid spread of the rumor on the fourth and fifth days is reflected in the increasing steepness of the graph at $t = 4$ and $t = 5$. After the fifth day the rumor spreads less rapidly, since there are fewer people to whom it can spread. Correspondingly, after $t = 5$, the graph becomes less steep. Finally, at $t = 9$ and $t = 10$, the graph is practically flat, indicating little change in the number who heard the rumor.

In the example above, there is a correlation between the rate at which $N(t)$ is changing and the steepness of the graph. This illustrates one of the fundamental ideas of calculus, which may be put roughly as follows: Measure rates of change in terms of steepness of graphs. This chapter is devoted to the *derivative*, which provides a numerical measure of the steepness of a curve at a particular point. By studying the derivative we will be able to deal numerically with the rates of change which occur in applied problems.

1.1. The Slope of a Straight Line

As we shall see later, the study of straight lines is crucial for the study of the steepness of curves. So this section is devoted to a discussion of the geometric and algebraic properties of straight lines.

A nonvertical line* L has an equation of the form $y = mx + b$. The number m is called the slope of L and the point $(0, b)$ is called the *y-intercept*.

FIGURE 1

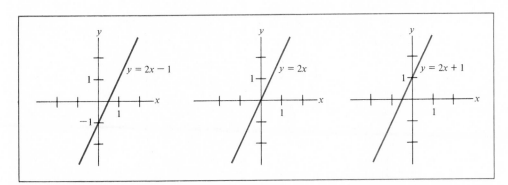

___
* Throughout we confine our discussion of straight lines to those that are not vertical.

FIGURE 2

If we set $x = 0$, we see that $y = b$, so that $(0, b)$ is on the line L. Thus the y-intercept tells us where the line L crosses the y-axis. The slope measures the steepness of the line. In Fig. 1 we give three examples of lines with slope $m = 2$. In Fig. 2 we give three examples of lines with slope $m = -2$.

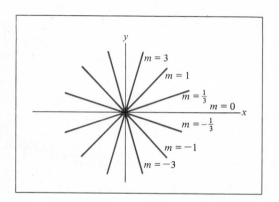

FIGURE 3

To conceptualize the meaning of slope, think of walking along a line from left to right. On lines of positive slope we will be walking uphill; the greater the slope, the steeper the ascent. On lines of negative slope we will be walking down-hill; the more negative the slope, the steeper the descent. Walking on lines of zero slope corresponds to walking on level ground. In Fig. 3 we have graphed lines with $m = 3, 1, \frac{1}{3}, 0, -\frac{1}{3}, -1, -3$, all having $b = 0$. The reader can readily verify our conceptualization of slope for these lines.

The slope and y-intercept of a straight line often have physical interpretations, as the following three examples illustrate.

EXAMPLE 1 A manufacturer finds that the total cost of producing x units of a commodity is $2x + 1000$ dollars. What is the economic significance of the y-intercept and the slope of the line $y = 2x + 1000$? (See Fig. 4.)

FIGURE 4

Solution The y-intercept is $(0, 1000)$. The number 1000 represents the *fixed costs* of the manufacturer—those overhead costs, such as rent and insurance, that must be paid no matter how many items are produced. Thus when $x = 0$ (no units produced), the cost is still $y = 1000$ dollars.

The slope of the line is 2. This number represents the cost of producing each additional unit. To see this, we can calculate some typical costs.

Quantity produced	Total cost
$x = 1500$	$y = 2(1500) + 1000 = 4000$
$x = 1501$	$y = 2(1501) + 1000 = 4002$
$x = 1502$	$y = 2(1502) + 1000 = 4004$

Each time x is increased by 1, the value of y increases by 2. The number 2 is called the marginal cost.

EXAMPLE 2   An apartment complex has a storage tank to hold its heating oil. The tank was filled on January 1, but no more deliveries of oil will be made until some time in March. Let t denote the number of days after January 1 and let y denote the number of gallons of fuel oil in the tank. Current records of the apartment complex show that y and t are related approximately by the equation

$$y = 30,000 - 400t. \qquad (1)$$

What interpretation can be given to the y-intercept and slope of this line? (See Fig. 5.)

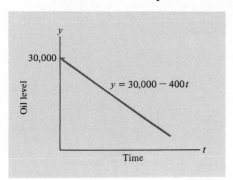

FIGURE 5

Solution   The y-intercept is (0, 30,000). This value of y corresponds to $t = 0$, so there were 30,000 gallons of oil in the tank on January 1. Let us examine how fast the oil is removed from the tank.

Days after January 1	Gallons of oil in the tank
$t = 0$	$y = 30,000 - 400(0) = 30,000$
$t = 1$	$y = 30,000 - 400(1) = 29,600$
$t = 2$	$y = 30,000 - 400(2) = 29,200$
$t = 3$	$y = 30,000 - 400(3) = 28,800$
$\vdots$	$\vdots$

The oil level in the tank drops by 400 gallons each day; that is, the oil is being used at the rate of 400 gallons per day. The slope of the line (1) is -400. Thus the slope gives the rate at which the level of oil in the tank is changing. The negative sign on the -400 indicates that the oil level is decreasing rather than increasing.

EXAMPLE 3   For tax purposes, businesses are allowed to regard equipment as decreasing in value (or depreciating) each year. The amount of depreciation may be taken as an income tax deduction.

Suppose that the value y of a piece of equipment, x years after its purchase, is given by

$$y = 500,000 - 50,000x.$$

Interpret the y-intercept and the slope of the graph.

Solution The y-intercept is $(0, 500{,}000)$ and corresponds to the value of y when $x = 0$; that is, the y-intercept gives the original value, \$500,000, of the equipment. The slope indicates the rate at which the equipment is changing in value. That is, the value of the equipment is decreasing at a rate of 50,000 dollars per year.

Properties of the Slope of a Line   Let us now examine several useful properties of the slope of a straight line. At the end of the section we shall explain why these properties are valid.

Slope Property 1   Suppose that we start at a point on a line of slope m and move one unit to the right. Then we must move m units in the y-direction in order to return to the line.

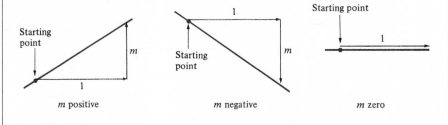

m positive m negative m zero

Slope Property 2   We can compute the slope of a line by knowing two points on the line. If (x_1, y_1) and (x_2, y_2) are on the line, then the slope of the line is $\dfrac{y_2 - y_1}{x_2 - x_1}$.

As we move from (x_1, y_1) to (x_2, y_2), the change in the y-coordinates is $y_2 - y_1$ and the change in the x-coordinates is $x_2 - x_1$. Thus the slope of the line is simply the ratio of the change in y to the change in x. This interpretation of the slope is also illustrated by the diagrams given above for Slope Property 1. The slope of a line equals the change in y per unit change in x. We say that the slope gives the *rate of change of y with respect to x*.

Slope Property 3   The equation of a line can be obtained if we know the slope and one point on the line. If the slope is m and if (x_1, y_1) is on the line, then the equation of the line is

$$y - y_1 = m(x - x_1).$$

This equation is called the *point-slope form* of the equation of the line.

Slope Property 4   Distinct lines of the same slope are parallel. Conversely, if two lines are parallel, they have the same slope.

Slope Property 5   When two lines are perpendicular, the product of their slopes is -1.

Calculations Involving Slope of a Line

EXAMPLE 4   Find the slope and the y-intercept of the line whose equation is $2x + 3y = 6$.

Solution   We solve for y in terms of x.

$$3y = -2x + 6$$
$$y = -\tfrac{2}{3}x + 2.$$

The slope is $-\tfrac{2}{3}$ and the y-intercept is $(0, 2)$.

EXAMPLE 5   Sketch the graph of the line

(a) passing through $(2, -1)$ with slope 3,

(b) passing through $(2, 3)$ with slope $-\tfrac{1}{2}$.

Solution   We use Slope Property 1. (See Fig. 6.) In each case, we begin at the given point, move one unit to the right, and then m units in the y-direction (upward for positive m, downward for negative m). The new point reached will also be on the line. Draw the straight line through these two points.

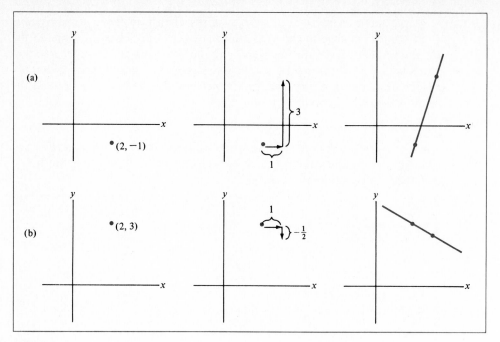

FIGURE 6

EXAMPLE 6   Find the slope of the line passing through the points $(6, -2)$ and $(9, 4)$.

Solution   We apply Slope Property 2 with $(x_1, y_1) = (6, -2)$ and $(x_2, y_2) = (9, 4)$. Then

$$\frac{y_2 - y_1}{x_2 - x_1} = \frac{4 - (-2)}{9 - 6} = \frac{6}{3} = 2.$$

Thus the slope is 2. [We would have reached the same answer if we had let $(x_1, y_1) = (9, 4)$ and $(x_2, y_2) = (6, -2)$.] The slope is just the difference of the y-coordinates divided by the difference of the x-coordinates, with each difference formed in the same order.

EXAMPLE 7   Find an equation of the line passing through $(-1, 2)$ with slope 3.

Solution   We let $(x_1, y_1) = (-1, 2)$ and $m = 3$, and we use Slope Property 3. The equation of the line is

$$y - 2 = 3[x - (-1)]$$

or

$$y - 2 = 3(x + 1).$$

If desired, this equation can be put into the form $y = mx + b$,

$$y - 2 = 3(x + 1) = 3x + 3$$

$$y = 3x + 5.$$

EXAMPLE 8 Find an equation of the line passing through the points $(1, -2)$ and $(2, -3)$.

Solution By Slope Property 2, the slope of the line is

$$\frac{-3 - (-2)}{2 - 1} = \frac{-3 + 2}{1} = -1.$$

Since $(1, -2)$ is on the line, we can use Property 3 to get the equation of the line:

$$y - (-2) = (-1)(x - 1)$$
$$y + 2 = -x + 1$$
$$y = -x - 1.$$

EXAMPLE 9 Find an equation of the line passing through $(5, 3)$ parallel to the line $2x + 5y = 7$.

Solution We first find the slope of the line $2x + 5y = 7$.

$$2x + 5y = 7$$
$$5y = 7 - 2x$$
$$y = -\tfrac{2}{5}x + \tfrac{7}{5}.$$

The slope of this line is $-\tfrac{2}{5}$. By Slope Property 4, any line parallel to this line will also have slope $-\tfrac{2}{5}$. Using the given point $(5, 3)$ and Slope Property 3, we get the desired equation

$$y - 3 = -\tfrac{2}{5}(x - 5).$$

This equation can also be written as

$$y = -\tfrac{2}{5}x + 5.$$

Verification of the Properties of Slope It is convenient to verify the properties of slope in the order 2, 3, 1. Verifications of Properties 4 and 5 are outlined in Exercises 40 and 41.

Verification of Property 2 Suppose that the equation of the line is $y = mx + b$. Then, since (x_2, y_2) is on the line, we have $y_2 = mx_2 + b$. Similarly, since (x_1, y_1) is on the line, we have $y_1 = mx_1 + b$. Subtracting, we see that

$$y_2 - y_1 = mx_2 - mx_1$$
$$y_2 - y_1 = m(x_2 - x_1),$$

so that

$$\frac{y_2 - y_1}{x_2 - x_1} = m.$$

This is the formula for the slope m stated in Property 2.

Verification of Property 3 The equation $y - y_1 = m(x - x_1)$ may be put into the form

$$y = mx + \underbrace{(y_1 - mx_1)}_{b}. \tag{2}$$

This is the equation of a line with slope m. Furthermore, the point (x_1, y_1) is on this line, because the equation (2) remains true when we plug in x_1 for x and y_1 for y. Thus the equation $y - y_1 = m(x - x_1)$ corresponds to the line of slope m passing through (x_1, y_1).

Verification of Property 1 Let $P = (x_1, y_1)$ be on a line l and let y_2 be the coordinate of the point on l obtained by moving P one unit to the right and then moving vertically to return to the line. (See Fig. 7.) By Property 2, the slope m of this line l satisfies

$$m = \frac{y_2 - y_1}{(x_1 + 1) - x_1} = \frac{y_2 - y_1}{1} = y_2 - y_1.$$

Thus the difference of the y-coordinates of R and Q is m. This is Property 1.

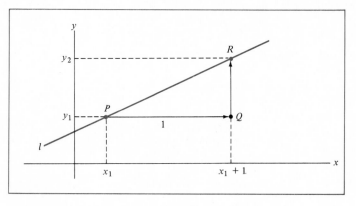

FIGURE 7

PRACTICE PROBLEMS 1

Find the slopes of the following lines.

1. The line whose equation is $x = 3y - 7$.

2. The line passing through the points $(2, 5)$ and $(2, 8)$.

EXERCISES 1

Find the slopes of the following lines.

1. $y = 2 - 5x$
2. $y = -5x$
3. $y = 2$
4. $y = \frac{1}{3}(x + 2)$
5. $y = (2x - 1)/7$
6. $y = \frac{1}{4}$
7. $2x + 3y = 6$
8. $x - y = 2$

Find the equations of the following lines.

9. Slope is 3; y-intercept is $(0, -1)$
10. Slope is $-\frac{1}{2}$; y-intercept is $(0, 0)$
11. Slope is 1; $(1, 2)$ on line
12. Slope is $-\frac{1}{3}$; $(6, -2)$ on line

13. Slope is -7; $(5, 0)$ on line

14. Slope is $\frac{1}{2}$; $(2, -3)$ on line

15. Slope is 0; $(7, 4)$ on line

16. Slope is $-\frac{2}{3}$; $(0, 5)$ on line

17. $(2, 1)$ and $(4, 2)$ on line

18. $(5, -3)$ and $(-1, 3)$ on line

19. $(0, 0)$ and $(1, -2)$ on line

20. $(2, -1)$ and $(3, -1)$ on line

21. Parallel to $y = -2x + 1$; $(\frac{1}{2}, 5)$ on line

22. Parallel to $3x + y = 7$; $(-1, -1)$ on line

23. Parallel to $3x - 6y = 1$; $(1, 0)$ on line

24. Parallel to $5x + 2y = -4$; $(0, 17)$ on line

In each of Exercises 25–28, we specify a line by giving the slope and one point on the line. We give the first coordinate of some points on the line. Without deriving the equation of the line, find the second coordinate of each of the points.

25. Slope is 2, $(1, 3)$ on line; $(2, \quad)$; $(0, \quad)$; $(-1, \quad)$

26. Slope is -3, $(2, 2)$ on line; $(3, \quad)$; $(4, \quad)$; $(1, \quad)$

27. Slope is $-\frac{1}{4}$, $(-1, -1)$ on line; $(0, \quad)$; $(1, \quad)$; $(-2, \quad)$

28. Slope is $\frac{1}{3}$, $(-5, 2)$ on line; $(-4, \quad)$; $(-3, \quad)$; $(-2, \quad)$

For each pair of lines in the figures below, determine the one with the greater slope.

29.

30.

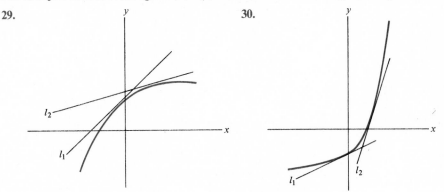

Find the equation and sketch the graph of the following lines.

31. With slope -2 and y-intercept $(0, -1)$

32. With slope $\frac{1}{3}$ and y-intercept $(0, 1)$

33. Through $(2, 0)$ with slope $\frac{4}{5}$

34. Through $(-1, 3)$ with slope 0

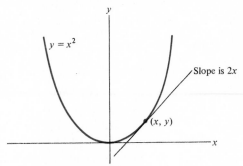

Later in the chapter we shall show that the tangent line to the parabola $y = x^2$ passing through the point with coordinates (x, y) has slope $2x$. Thus the slopes of the tangent lines at the points $(1, 1)$, $(0, 0)$, $(-\frac{1}{2}, \frac{1}{4})$ are, respectively, 2, 0, and -1. Find the equation of the tangent lines through each of the following points.

35. $(1, 1)$ 36. $(0, 0)$ 37. $(-\frac{1}{2}, \frac{1}{4})$

38. A salesperson's weekly pay depends on the volume of sales. If she sells x units of goods, then her pay is $y = 5x + 60$ dollars. Give an interpretation to the slope and the y-intercept of this straight line.

39. The demand equation for a monopolist is $y = -.02x + 7$, where x is the number of units produced and y is the price. That is, in order to sell x units of goods, the price must be $y = -.02x + 7$ dollars. Interpret the slope and y-intercept of this line.

40. Prove Property 4 of straight lines. [*Hint:* If $y = mx + b$ and $y = m'x + b'$ are two lines, then they have a point in common if and only if the equation $mx + b = m'x + b'$ has a solution x.]

41. Prove Property 5 of straight lines. [*Hint:* Without loss of generality, assume that both lines pass through the origin. Use Slope Property 1 and the Pythagorean theorem. See the accompanying figure.]

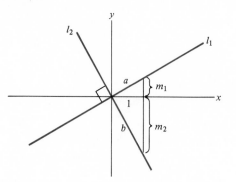

SOLUTIONS TO PRACTICE PROBLEMS 1

1. We solve for y in terms of x.

$$y = \tfrac{1}{3}x + \tfrac{7}{3}.$$

The slope of the line is the coefficient of x, that is, $\frac{1}{3}$.

2. The line passing through these two points is a vertical line; therefore, its slope is undefined.

1.2. The Slope of a Curve at a Point

In order to extend the concept of slope from straight lines to more general curves, we must first discuss the notion of the tangent line to a curve at a point.

We have a clear idea of what is meant by the tangent line to a circle at a point P. It is the straight line that touches the circle at just the one point P. Let us focus on the region near P, designated by the dashed rectangle shown in Fig. 1. The enlarged portion of the circle looks almost straight, and the straight line that it resembles is the tangent line. Further enlargements would make the circle near P look even straighter and have an even closer resemblance to the tangent line. In this sense, the tangent line to the circle at the point P is the straight line through P that best approximates the circle near P. In particular, the tangent line at P reflects the steepness of the circle at P. Thus it seems reasonable to define the *slope* of the circle at P to be the slope of the tangent line at P.

Similar reasoning leads us to a suitable definition of slope for an arbitrary curve at a point P. Consider the three curves drawn in Fig. 2. We have drawn an enlarged version of the dashed box around each point P. Notice that the portion of each curve lying in the boxed region looks almost straight. If we further magnify the curve near P, it would appear even straighter. Indeed, if we applied higher and higher magnification, the portion of the curve near P approaches a certain straight line ever more exactly. (See Fig. 3.) This straight line is called the *tangent line to*

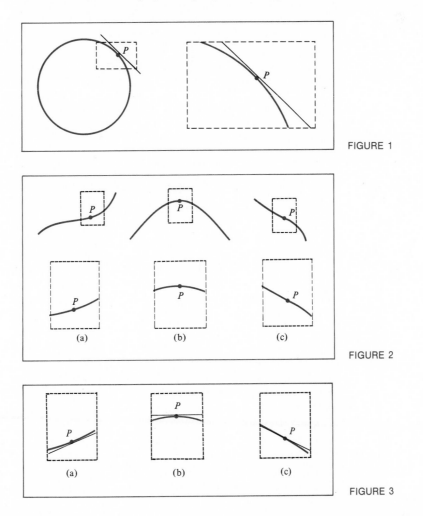

FIGURE 1

FIGURE 2

FIGURE 3

the curve at P. This line best approximates the curve near *P*. We define the *slope of a curve at a point P* to be the slope of the tangent line to the curve at *P*.

The portion of the curve near *P* can be, at least to within an approximation, replaced by the tangent line. Therefore, the slope of the curve at *P*—that is, the slope of the tangent line at *P*—measures the rate of increase or decrease of the curve as it passes through *P*.

EXAMPLE 1 In Example 2 of Section 1 the apartment complex used approximately 400 gallons of oil per day. Suppose that we keep a continuous record of the oil level in the storage tank. The graph for a typical 2-day period appears in Fig. 4. What is the physical significance of the slope of the graph at the point *P*?

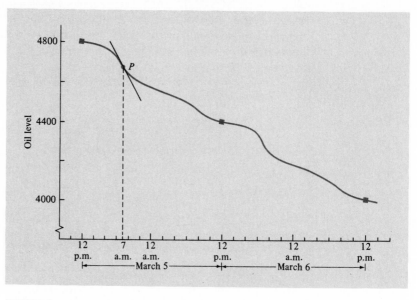

FIGURE 4

Solution The curve near *P* is closely approximated by its tangent line. So think of the curve as replaced by its tangent line near *P*. Then the slope at *P* is just the rate of decrease of the oil level at 7 A.M. on March 5.

Notice that during the entire day of March 5 the graph in Fig. 4 seems to be the steepest at 7 A.M. That is, the oil level is falling the fastest at that time. This corresponds to the fact that most people awake around 7 A.M., turn up their thermostats, take showers, and so on. Example 1 provides a typical illustration of the manner in which slopes can be interpreted as rates of change. We shall return to this idea in Section 8.

In calculus, we can usually compute slopes by using formulas. For instance, in Section 3 we shall show that the tangent line to the graph of $y = x^2$ at the point $(1, 1)$ has slope 2; the tangent line at $(3, 9)$ has slope 6; and the tangent line at $(-\frac{5}{2}, \frac{25}{4})$ has slope -5. The various tangent lines are shown in Fig. 5. Notice that

FIGURE 5

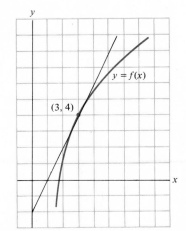

FIGURE 6

the slope at each point is two times the x-coordinate of the point. This is a general fact. In Section 3 we shall derive this simple formula (see Fig. 6.):

$$[\text{slope of the graph of } y = x^2 \text{ at the point } (x, y)] = 2x.$$

EXAMPLE 2 (a) What is the slope of the graph of $y = x^2$ at the point $(\frac{3}{4}, \frac{9}{16})$?

(b) Write the equation of the tangent line to the graph of $y = x^2$ at the point $(\frac{3}{4}, \frac{9}{16})$.

Solution (a) The x-coordinate of $(\frac{3}{4}, \frac{9}{16})$ is $\frac{3}{4}$, so the slope of $y = x^2$ at this point is $2(\frac{3}{4}) = \frac{3}{2}$.

(b) We shall write the equation of the tangent line in the point-slope form. The point is $(\frac{3}{4}, \frac{9}{16})$, and the slope is $\frac{3}{2}$ by part (a). Hence the equation is

$$y - \tfrac{9}{16} = \tfrac{3}{2}(x - \tfrac{3}{4}).$$

PRACTICE PROBLEMS 2

1. Refer to the accompanying graph.

    (a) What is the slope of the curve at $(3, 4)$?

    (b) What is the equation of the tangent line at the point where $x = 3$?

2. What is the equation of the tangent line to the graph of $y = \frac{1}{2}x + 1$ at the point $(4, 3)$?

EXERCISES 2

Trace the curves in Exercises 1–6 onto another piece of paper and sketch the tangent line in each case at the designated point P.

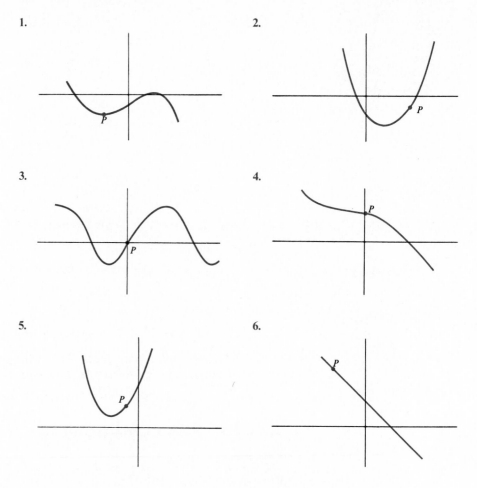

1.

2.

3.

4.

5.

6.

Estimate the slope of each of the following curves at the designated point P.

7.

8.

9.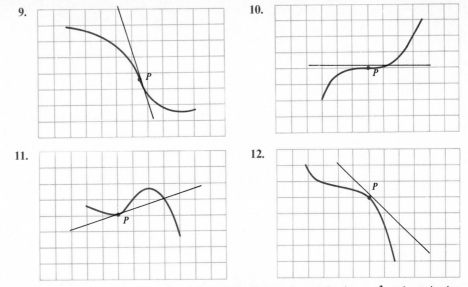

10.

11.

12.

In Exercises 13–15, find the slope of the tangent line to the graph of $y = x^2$ at the point indicated, and then write the corresponding equation of the tangent line.

13. $(-2, 4)$ **14.** $(-.4, .16)$ **15.** $\left(\frac{4}{3}, \frac{16}{9}\right)$

16. Find the slope of the tangent line to the graph of $y = x^2$ at the point where $x = -\frac{1}{2}$.

17. Write the equation of the tangent line to the graph of $y = x^2$ at the point where $x = 1.5$.

18. Write the equation of the tangent line to the graph of $y = x^2$ at the point where $x = .6$.

19. Find the point on the graph of $y = x^2$ where the curve has slope $\frac{5}{3}$.

20. Find the point on the graph of $y = x^2$ where the curve has slope -4.

21. Find the point on the graph of $y = x^2$ where the tangent line is parallel to the line $x + 2y = 4$.

22. Find the point on the graph of $y = x^2$ where the tangent line is parallel to the line $3x - y = 2$.

In the next section we shall see that the tangent line to the graph of $y = x^3$ at the point (x, y) has slope $3x^2$. Using this result, find the slope of the curve at the following points.

23. $(2, 8)$

24. $\left(\frac{3}{2}, \frac{27}{8}\right)$

25. $\left(-\frac{1}{2}, -\frac{1}{8}\right)$

26. Find the slope of the curve $y = x^3$ at the point where $x = \frac{1}{4}$.

27. Write the equation of the line tangent to the graph of $y = x^3$ at the point where $x = -1$.

28. Write the equation of the line tangent to the graph of $y = x^3$ at the point where $x = \frac{1}{2}$.

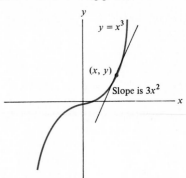

1. (a) The slope of the curve at the point (3, 4) is, by definition, the slope of the tangent line at (3, 4). Note that the point (4, 6) is also on the line. Therefore, the slope is

$$\frac{6 - 4}{4 - 3} = \frac{2}{1} = 2.$$

   (b) Use the point-slope formula. The equation of the line passing through the point (3, 4) and having slope 2 is

$$y - 4 = 2(x - 3)$$

   or

$$y = 2x - 2.$$

2. The tangent line at (4, 3) is, by definition, the line that best approximates the curve at (4, 3). Since the "curve" in this case is itself a line, the curve and its tangent line at (4, 3) (and at every other point) must be the same. Therefore, the equation is $y = \frac{1}{2}x + 1$.

1.3. The Derivative

by definition
$$f'(x) = \frac{f(x + \Delta x) - f(x)}{\Delta x}$$
first lots of practice, then "easy way"

Suppose that a curve is the graph of a function $f(x)$. It is usually possible to obtain a formula that gives the slope of the curve $y = f(x)$ at any point. This slope formula is called the *derivative* of $f(x)$ and is written $f'(x)$. For each value of x, $f'(x)$ gives the slope of the curve $y = f(x)$ at the point with first coordinate x.* (See Fig. 1.)

FIGURE 1

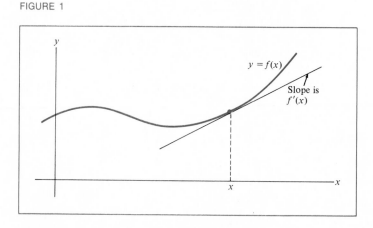

y

$y = f(x)$

Slope is $f'(x)$

x

x

----

* As we shall see, there are are curves which do not have tangent lines at every point. At values of x corresponding to such points, the derivative $f'(x)$ is not defined. For the sake of the current discussion, which is designed to develop an intuitive feeling for the derivative, let us assume that the graph of $f(x)$ has a tangent line for each x in the domain of f.

The process of computing $f'(x)$ for a given function $f(x)$ is called *differentiation*.

As we shall see later, the concept of a derivative occurs frequently in applications. In economics, the derivatives are often described by the adjective "marginal." For instance, if $C(x)$ is a cost function (the cost of producing x units of a commodity), then the derivative $C'(x)$ is called the *marginal cost function*. The derivative $P'(x)$ of a profit function $P(x)$ is called the *marginal profit function*; the derivative of a revenue function is called the *marginal revenue function*; and so on. We shall discuss the economic meaning of these marginal concepts in Section 8.

For the remainder of this section as well as the next few sections we will concentrate on calculating derivatives.

The case of a linear function $f(x) = mx + b$ is particularly simple. The graph of $y = mx + b$ is a straight line L of slope m. The tangent line to L (at any point) is just L itself, and so the slope of the graph is m at every point. (See Fig. 2.) In other words, the value of the derivative $f'(x)$ is always equal to m. We summarize this fact as follows:

$$\text{If } f(x) = mx + b, \text{ then } f'(x) = m. \tag{1}$$

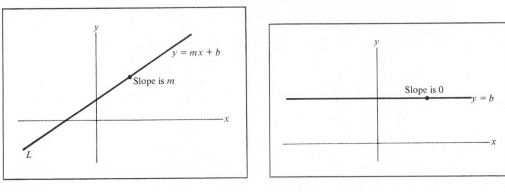

FIGURE 2

FIGURE 3

Set $m = 0$ in equation (1). Then the function becomes $f(x) = b$, which has the value b for each value of x. The graph is a horizontal line of slope 0, so $f'(x) = 0$ for all x. (See Fig. 3.) Thus we have:

$$\text{The derivative of a constant function } f(x) = b \text{ is } f'(x) = 0. \tag{2}$$

Next, consider the function $f(x) = x^2$. As we stated in Section 2 (and will prove at the end of this section), the slope of the graph of $y = x^2$ at the point

(x, y) is equal to $2x$. That is, the value of the derivative $f'(x)$ is $2x$:

$$\text{If } f(x) = x^2, \text{ then } f'(x) = 2x. \tag{3}$$

In Exercises 23–28 of Section 2 we made use of the fact that the slope of the graph of $y = x^3$ at the point (x, y) is $3x^2$. This can be restated in terms of derivatives as follows:

$$\text{If } f(x) = x^3, \text{ then } f'(x) = 3x^2. \tag{4}$$

We should, at this stage at least, keep the geometric meaning of these formulas clearly in mind. Figure 4 shows the graphs of x^2 and x^3 together with the interpretations of formulas (3) and (4) in terms of slope.

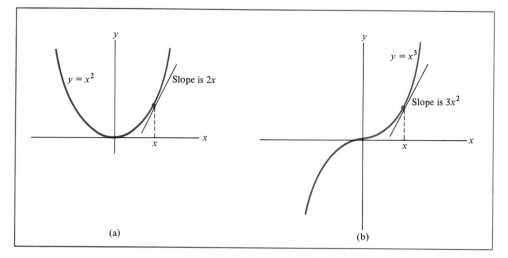

(a)   (b)

FIGURE 4

One of the reasons calculus is so useful is that it provides general techniques which can be easily used to determine derivatives. One such general rule, which contains formulas (3) and (4) as special cases, is the so-called power rule.

Power Rule   Let r be any number and let $f(x) = x^r$. Then $f'(x) = rx^{r-1}$.

Indeed, if $r = 2$, then $f(x) = x^2$ and $f'(x) = 2x^{2-1} = 2x$, which is formula (3). If $r = 3$, then $f(x) = x^3$ and $f'(x) = 3x^{3-1} = 3x^2$, which is (4). We shall prove the power rule in Chapter 4. Until then, we shall use it to calculate derivatives.

EXAMPLE 1   Let $f(x) = \sqrt{x}$. What is $f'(x)$?

Solution   Recall that $\sqrt{x} = x^{1/2}$. We may apply the power rule with $r = \tfrac{1}{2}$.

$$f(x) = x^{1/2}$$

$$f'(x) = \tfrac{1}{2}x^{1/2-1} = \tfrac{1}{2}x^{-1/2}$$

$$= \frac{1}{2} \cdot \frac{1}{x^{1/2}} = \frac{1}{2\sqrt{x}}.$$

Another important special case of the power rule occurs for $r = -1$, corresponding to $f(x) = x^{-1}$. In this case, $f'(x) = (-1)x^{-1-1} = -x^{-2}$. However, since $x^{-1} = 1/x$ and $x^{-2} = 1/x^2$, the power rule for $r = -1$ may also be written as follows:*

$$\text{If } f(x) = \frac{1}{x}, \text{ then } f'(x) = -\frac{1}{x^2} \; (x \neq 0). \qquad (5)$$

EXAMPLE 2   Find the slope of the curve $y = 1/x$ at $(2, \tfrac{1}{2})$.

Solution   Set $f(x) = 1/x$. The point $(2, \tfrac{1}{2})$ corresponds to $x = 2$, so in order to find the slope at this point, we compute $f'(2)$. From formula (5) we find that

$$f'(2) = -\frac{1}{2^2} = -\frac{1}{4}.$$

Thus the slope of $y = 1/x$ at the point $(2, \tfrac{1}{2})$ is $-\tfrac{1}{4}$. (See Fig. 5.)

FIGURE 5

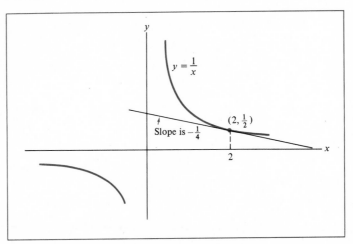

---

* The formula gives $f'(x)$ for $x \neq 0$. The derivative of $f(x)$ is not defined at $x = 0$ since $f(x)$ itself is not defined there.

Warning Do not confuse $f'(2)$, the value of the derivative at 2, with $f(2)$, the value of the y-coordinate at the point on the graph at which $x = 2$. In Example 2 we have $f'(2) = -\frac{1}{4}$, whereas $f(2) = \frac{1}{2}$. The number $f'(2)$ gives the *slope* of the graph at $x = 2$; the number $f(2)$ gives the *height* of the graph at $x = 2$.

Notation   The operation of forming a derivative $f'(x)$ from a function $f(x)$ is also indicated by the symbol $\dfrac{d}{dx}$ (read: the derivative with respect to x). Thus

$$\frac{d}{dx} f(x) = f'(x).$$

For example,

$$\frac{d}{dx}(x^6) = 6x^5, \qquad \frac{d}{dx}(x^{5/3}) = \frac{5}{3} x^{2/3}, \qquad \frac{d}{dx}\left(\frac{1}{x}\right) = -\frac{1}{x^2}.$$

When working with an equation of the form $y = f(x)$, we often write $\dfrac{dy}{dx}$ as a symbol for the derivative $f'(x)$. For example, if $y = x^6$, we may write

$$\frac{dy}{dx} = 6x^5.$$

The Secant-Line Calculation of the Derivative   So far, we have said nothing about how to derive differentiation formulas, such as (3), (4), or (5). Let us remedy that omission now. The derivative gives the slope of the tangent line, so we must describe a procedure for calculating that slope.*

The fundamental idea for calculating the slope of the tangent line at a point P is to approximate the tangent line very closely by *secant lines*. A secant line at P is a straight line passing through P and a nearby point Q on the curve. (See Fig. 6.) By taking Q very close to P, we can make the slope of the secant line approximate the slope of the tangent line to any desired degree of accuracy. Let us see what this amounts to in terms of calculations.

Suppose that the point P is $(x, f(x))$. Suppose also that Q is h horizontal units away from P. Then Q has x-coordinate $x + h$ and y-coordinate $f(x + h)$. The slope of the secant line through the points $P = (x, f(x))$ and $Q = (x + h, f(x + h))$ is simply

$$[\text{slope of secant line}] = \frac{f(x + h) - f(x)}{(x + h) - x} = \frac{f(x + h) - f(x)}{h}.$$

(See Fig. 7.)

In order to move Q close to P along the curve, we let h approach zero. Then the secant line approaches the tangent line, and so

$$[\text{slope of secant line}] \quad \text{approaches} \quad [\text{slope of tangent line}];$$

---

* The following discussion is designed to develop geometric intuition for the derivative. It will also provide the basis for a formal definition of the derivative in terms of limits in Section 4.

FIGURE 6

FIGURE 7

that is,

$$\frac{f(x + h) - f(x)}{h} \quad \text{approaches} \quad f'(x).$$

Change in Y over change in X.

Since we can make the secant line as close to the tangent line as we wish, by taking h sufficiently small, the quantity $[f(x + h) - f(x)]/h$ can be made to approximate $f'(x)$ to any desired degree of accuracy. And we arrive at the following method to compute the derivative $f'(x)$.

To calculate $f'(x)$:

1. First calculate $\dfrac{f(x + h) - f(x)}{h}$ for $h \neq 0$.

2. Then let h approach zero.

3. The quantity $\dfrac{f(x + h) - f(x)}{h}$ will approach $f'(x)$.

Let us use this method to verify the differentiation formulas (3) and (5), which as we saw were special cases of the power rule for $r = 2$ and $r = -1$, respectively.

Verification of the Power Rule for $r = 2$

Here $f(x) = x^2$, so that the slope of the secant line is

$$\frac{f(x + h) - f(x)}{h} = \frac{(x + h)^2 - x^2}{h}.$$

However, by multiplying out, we have $(x + h)^2 = x^2 + 2xh + h^2$, so that

$$\frac{f(x + h) - f(x)}{h} = \frac{x^2 + 2xh + h^2 - x^2}{h} = \frac{(2x + h)h}{h}$$

$$= 2x + h.$$

As h approaches zero (i.e., as the secant line approaches the tangent line), the quantity $2x + h$ approaches $2x$. Thus we have

$$f'(x) = 2x,$$

which is formula (3).

Verification of the Power Rule for $r = -1$

Here $f(x) = x^{-1} = 1/x$, so that the slope of the secant line is

$$\frac{f(x + h) - f(x)}{h} = \frac{1}{h}\left[\frac{1}{x + h} - \frac{1}{x}\right] = \frac{1}{h}\left[\frac{x - (x + h)}{(x + h)x}\right]$$

$$= \frac{1}{h}\left[\frac{-h}{(x + h)x}\right] = -\frac{1}{x(x + h)}.$$

As h approaches zero, $-\dfrac{1}{x(x + h)}$ approaches $-\dfrac{1}{x^2}$. Hence

$$f'(x) = -\frac{1}{x^2},$$

which is formula (5).

Similar arguments can be used to verify the power rule for other values of r. The cases $r = 3$ and $r = \frac{1}{2}$ are outlined in Exercises 47 and 48, respectively.

PRACTICE PROBLEMS 3

1. Consider the curve $y = f(x)$ in the accompanying sketch.

   (a) Find $f(5)$.

   (b) Find $f'(5)$.

2. Let $f(x) = 1/x^4$.

   (a) Find its derivative.

   (b) Find $f'(2)$.

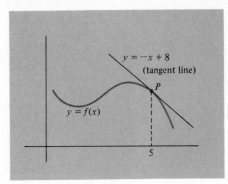

$y = -x + 8$
(tangent line)

P

$y = f(x)$

5

EXERCISES 3

Find the derivatives of the following functions.

1. $f(x) = 2x - 5$

2. $f(x) = 3 - \frac{1}{2}x$

3. $f(x) = x^8$

4. $f(x) = x^{75}$

5. $f(x) = x^{5/2}$

6. $f(x) = x^{4/3}$

7. $f(x) = \sqrt[3]{x}$

8. $f(x) = x^{3/4}$

9. $f(x) = x^{-2}$

10. $f(x) = 5$

11. $f(x) = x^{-1/4}$

12. $f(x) = x^{-3}$

13. $f(x) = \frac{3}{4}$

14. $f(x) = 1/\sqrt[3]{x}$

15. $f(x) = 1/x^3$

16. $f(x) = 1/x^5$

In Exercises 17–24, find the derivative of $f(x)$ at the designated value of x.

17. $f(x) = x^6$, at $x = -2$

18. $f(x) = x^3$, at $x = \frac{1}{4}$

19. $f(x) = 1/x$, at $x = 3$

20. $f(x) = 5x$, at $x = 2$

21. $f(x) = 4 - x$, at $x = 5$

22. $f(x) = x^{2/3}$, at $x = 1$

23. $f(x) = x^{3/2}$, at $x = 9$

24. $f(x) = 1/x^2$, at $x = 2$

25. Find the slope of the curve $y = x^4$ at $x = 3$.

26. Find the slope of the curve $y = x^5$ at $x = -2$.

27. Find the slope of the curve $y = \sqrt{x}$ at $x = 9$.

28. Find the slope of the curve $y = x^{-3}$ at $x = 3$.

29. If $f(x) = x^2$, compute $f(-5)$ and $f'(-5)$.

30. If $f(x) = x + 6$, compute $f(3)$ and $f'(3)$.

31. If $f(x) = 1/x^5$, compute $f(2)$ and $f'(2)$.

32. If $f(x) = 1/x^2$, compute $f(5)$ and $f'(5)$.

33. If $f(x) = x^{4/3}$, compute $f(8)$ and $f'(8)$.

34. If $f(x) = x^{3/2}$, compute $f(16)$ and $f'(16)$.

35. Find the slope of the tangent line to the curve $y = x^3$ at the point $(4, 64)$, and write the equation of this line.

36. Find the slope of the tangent line to the curve $y = \sqrt{x}$ at the point $(25, 5)$, and write the equation of this line.

In Exercises 37–44, find the indicated derivative.

37. $\dfrac{d}{dx}(x^8)$

38. $\dfrac{d}{dx}(x^{-3})$

39. $\dfrac{d}{dx}(x^{3/4})$

40. $\dfrac{d}{dx}(x^{-1/3})$

41. $\dfrac{dy}{dx}$ if $y = 1$

42. $\dfrac{dy}{dx}$ if $y = x^{-4}$

43. $\dfrac{dy}{dx}$ if $y = x^{1/5}$

44. $\dfrac{dy}{dx}$ if $y = \dfrac{x-1}{3}$

45. Consider the curve $y = f(x)$ in Fig. 8. Find $f(6)$ and $f'(6)$.

46. Consider the curve $y = f(x)$ in Fig. 9. Find $f(1)$ and $f'(1)$.

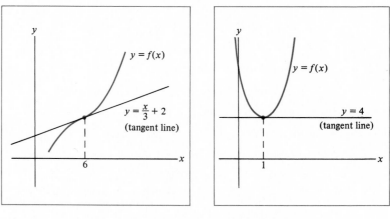

FIGURE 8   FIGURE 9

47. Use an argument like those used to verify formulas (3) and (5) to show that the derivative of $f(x) = x^3$ is $3x^2$. [*Hint:* Recall that $(x + h)^3 = x^3 + 3x^2h + 3xh^2 + h^3$.]

48. Use an argument like those used to verify formulas (3) and (5) to show that the derivative of $f(x) = \sqrt{x}$ is $1/(2\sqrt{x})$. [*Hint:* After forming the expression for the slope of a secant line, eliminate the square roots from the numerator by multiplying both the numerator and denominator by the quantity $\sqrt{x + h} + \sqrt{x}$.]

SOLUTIONS TO PRACTICE PROBLEMS 3

1. (a) The number $f(5)$ is the y-coordinate of the point P. Since the tangent line passes through P, the coordinates of P satisfy the equation $y = -x + 8$. Since its x-coordinate is 5, its y-coordinate is $-5 + 8 = 3$. Therefore, $f(5) = 3$.

   (b) The number $f'(5)$ is the slope of the tangent line at P, which is readily seen to be -1.

2. (a) The function $1/x^4$ can be written as the power function x^{-4}. Here $r = -4$. Therefore,

$$f'(x) = (-4)x^{(-4)-1} = -4x^{-5} = \frac{-4}{x^5}.$$

(b) $f'(2) = -4/2^5 = -4/32 = -\frac{1}{8}.$

1.4. Limits and the Derivative

The notion of a limit is one of the fundamental ideas of calculus. Indeed, any "theoretical" development of calculus rests on an extensive use of the theory of limits. Even in this book, where we have adopted an intuitive viewpoint, limit arguments are used occasionally (although in an informal way). In this section we give a brief introduction to limits and their role in calculus. As we shall see, the limit concept will allow us to define the notion of a derivative independently of our geometric reasoning.

Actually, we have already considered a limit in our discussion of the derivative, although we did not use the term "limit." Using the geometric reasoning of the previous section, we have the following procedure for calculating the derivative of a function $f(x)$ at $x = a$. First calculate the *difference quotient*

$$\frac{f(a + h) - f(a)}{h},$$

where h is a nonzero number. Next, allow h to approach zero by allowing it to assume both positive and negative numbers arbitrarily close to zero but different from zero. In symbols, we write $h \to 0$. The values of the difference quotient then approach the value of the derivative $f'(a)$. We say that the number $f'(a)$ is the *limit* of the difference quotient as h approaches zero, and in symbols we write

$$f'(a) = \lim_{h \to 0} \frac{f(a + h) - f(a)}{h}. \tag{1}$$

As a numerical example of formula (1), consider the case of the derivative of $f(x) = x^2$ at $x = 2$. The difference quotient in this case has the form

$$\frac{f(2 + h) - f(2)}{h} = \frac{(2 + h)^2 - 2^2}{h}.$$

Table 1 gives some typical values of this difference quotient for progressively smaller values of h, both positive and negative. It is clear that the values of the difference quotient are approaching 4 as $h \to 0$. In other words, 4 *is the limit* of the difference quotient as $h \to 0$. Thus

$$\lim_{h \to 0} \frac{(2 + h)^2 - 2^2}{h} = 4.$$

Since the values of the difference quotient approach the derivative $f'(2)$, we conclude that $f'(2) = 4$.

TABLE 1

h	$\dfrac{(2+h)^2 - 2^2}{h}$	h	$\dfrac{(2+h)^2 - 2^2}{h}$
1	$\dfrac{(2+1)^2 - 2^2}{1} = 5$	-1	$\dfrac{(2+(+1))^2 - 2^2}{-1} = 3$
.1	$\dfrac{(2+.1)^2 - 2^2}{.1} = 4.10$	$-.1$	$\dfrac{(2+(-.1))^2 - 2^2}{-.1} = 3.9$
.01	$\dfrac{(2+.01)^2 - 2^2}{.01} = 4.01$	$-.01$	$\dfrac{(2+(-.01))^2 - 2^2}{-.01} = 3.99$
.001	$\dfrac{(2+.001)^2 - 2^2}{.001} = 4.001$	$-.001$	$\dfrac{(2+(-.001))^2 - 2^2}{-.001} = 3.999$
.0001	$\dfrac{(2+.0001)^2 - 2^2}{.0001} = 4.0001$	$-.0001$	$\dfrac{(2+(-.0001))^2 - 2^2}{-.0001} = 3.9999$

Our discussion of the derivative has been based on an intuitive geometric concept of the tangent line. However, the limit on the right in (1) may be considered independently of its geometric interpretation. In fact, we may use (1) to define $f'(a)$. We say that f is *differentiable* at $x = a$ if $\dfrac{f(a+h) - f(a)}{h}$ approaches some number as $h \to 0$, and we denote this limiting number by $f'(a)$. If the difference quotient $\dfrac{f(a+h) - f(a)}{h}$ does not approach any specific number as $h \to 0$, we say that f is *nondifferentiable* at $x = a$. Essentially all of the functions in this text are differentiable at all points in their domain. A few exceptions are described in Section 5.

To better understand the limit concept used to define the derivative, it will be helpful to look at limits in a more general setting. If we let $g(h) = \dfrac{(2+h)^2 - 2^2}{h}$, then Table 1 gives values of $g(h)$ as $h \to 0$. These values obviously approach 4. We may express this by writing

$$\lim_{h \to 0} g(h) = 4.$$

The discussion above suggests the following definition: Let $g(x)$ be a function, a a number. We say that the number L is *the limit of $g(x)$ as x approaches a* provided that, as x gets arbitrarily close (but not equal) to a, the values of $g(x)$ approach L. In this case we write

$$\lim_{x \to a} g(x) = L.$$

If, as x approaches a, the values $g(x)$ do *not* approach a specific number, then we say that the limit of $g(x)$ as x approaches a *does not exist*. Let us give some further examples of limits.

EXAMPLE 1   Determine $\lim_{x \to 2} (3x - 5)$.

Solution   Let us make a table of values of x approaching 2 and the corresponding values of $3x - 5$:

x	$3x - 5$	x	$3x - 5$
2.1	1.3	1.9	.7
2.01	1.03	1.99	.97
2.001	1.003	1.999	.997
2.0001	1.0003	1.9999	.9997

As x approaches 2, we see that $3x - 5$ approaches 1. In terms of our notation,

$$\lim_{x \to 2} (3x - 5) = 1.$$

EXAMPLE 2   For each of the following functions, determine if $\lim_{x \to 2} g(x)$ exists. (The circles drawn on the graphs are meant to represent breaks in the graph, indicating that the functions under consideration are not defined at $x = 2$.)

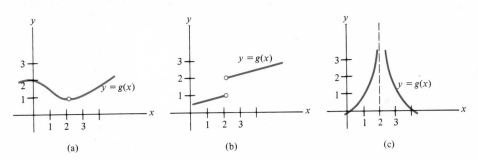

(a)   (b)   (c)

Solution   (a) $\lim_{x \to 2} g(x) = 1$. We can see that as x gets closer and closer to 2, the values of $g(x)$ get closer and closer to 1. This is true for values of x to both the right and the left of 2.

(b) $\lim_{x \to 2} g(x)$ does not exist. As x approaches 2 from the right, $g(x)$ approaches 2. However, as x approaches 2 from the left, $g(x)$ approaches 1. In order for a limit to exist, the function must approach the *same* value from each direction.

(c) $\lim_{x \to 2} g(x)$ does not exist. As x approaches 2, the values of $g(x)$ become larger and larger and do not approach a fixed number.

The following limit theorems, which we cite without proof, allow us to reduce the computation of limits for combinations of functions to computations of limits involving the constituent functions.

Limit Theorems Suppose that $\lim\limits_{x \to a} f(x)$ and $\lim\limits_{x \to a} g(x)$ both exist. Then we have the following results.

I. If k is a constant, then $\lim\limits_{x \to a} k \cdot f(x) = k \cdot \lim\limits_{x \to a} f(x)$.

II. If r is a positive constant, then $\lim\limits_{x \to a} [f(x)]^r = \left[\lim\limits_{x \to a} f(x) \right]^r$.

III. $\lim\limits_{x \to a} [f(x) + g(x)] = \lim\limits_{x \to a} f(x) + \lim\limits_{x \to a} g(x)$.

IV. $\lim\limits_{x \to a} [f(x) - g(x)] = \lim\limits_{x \to a} f(x) - \lim\limits_{x \to a} g(x)$.

V. $\lim\limits_{x \to a} [f(x) \cdot g(x)] = \left[\lim\limits_{x \to a} f(x) \right]\left[\lim\limits_{x \to a} g(x) \right]$.

VI. If $\lim\limits_{x \to a} g(x) \neq 0$, then $\lim\limits_{x \to a} \dfrac{f(x)}{g(x)} = \dfrac{\lim\limits_{x \to a} f(x)}{\lim\limits_{x \to a} g(x)}$.

EXAMPLE 3 Use the limit theorems to compute the following limits.

(a) $\lim\limits_{x \to 2} x^3$ (b) $\lim\limits_{x \to 2} 5x^3$ (c) $\lim\limits_{x \to 2} (5x^3 - 15)$ (d) $\lim\limits_{x \to 2} \sqrt{5x^3 - 15}$

(e) $\lim\limits_{x \to 2} (\sqrt{5x^3 - 15/x^5})$

Solution (a) Since $\lim\limits_{x \to 2} x = 2$, we have by Limit Theorem II that

$$\lim_{x \to 2} x^3 = \left(\lim_{x \to 2} x \right)^3 = 2^3 = 8.$$

(b) $\lim\limits_{x \to 2} 5x^3 = 5 \lim\limits_{x \to 2} x^3$ (Limit Theorem I with $k = 5$)

$$= 5 \cdot 8 \qquad \text{[by part (a)]}$$

$$= 40.$$

(c) $\lim\limits_{x \to 2} (5x^3 - 15) = \lim\limits_{x \to 2} 5x^3 - \lim\limits_{x \to 2} 15$ (Limit Theorem IV).

Note that $\lim\limits_{x \to 2} 15 = 15$. This is because the constant function $g(x) = 15$ always has the value 15, and so its limit as x approaches *any* number is 15. By part (b), $\lim\limits_{x \to 2} 5x^3 = 40$. Thus

$$\lim_{x \to 2} (5x^3 - 15) = 40 - 15 = 25.$$

(d)　$\lim\limits_{x \to 2} \sqrt{5x^3 - 15} = \lim\limits_{x \to 2} (5x^3 - 15)^{1/2}$

$$= \left[\lim\limits_{x \to 2} (5x^3 - 15)\right]^{1/2} \quad \text{[Limit Theorem II with } r = \tfrac{1}{2},$$
$$\qquad\qquad\qquad\qquad\qquad\qquad f(x) = 5x^3 - 15]$$

$$= 25^{1/2} \qquad\qquad \text{[by part (c)]}$$

$$= 5.$$

(e) The limit of the denominator is $\lim\limits_{x \to 2} x^5$, which is $2^5 = 32$, a nonzero number.
So by Limit Theorem VI we have

$$\lim\limits_{x \to 2} \frac{\sqrt{5x^3 - 15}}{x^5} = \frac{\lim\limits_{x \to 2} \sqrt{5x^3 - 15}}{\lim\limits_{x \to 2} x^5}$$

$$= \frac{5}{32} \quad \text{[by part (d)]}.$$

The following facts, which may be deduced by repeated applications of the various limit theorems, are extremely handy in evaluating limits.

Limit of a Polynomial Function　Let $p(x)$ be a polynomial function, a any number. Then

$$\lim\limits_{x \to a} p(x) = p(a).$$

Limit of a Rational Function　Let $r(x) = p(x)/q(x)$ be a rational function, where $p(x)$ and $q(x)$ are polynomials. Let a be a number such that $q(a) \neq 0$. Then

$$\lim\limits_{x \to a} r(x) = r(a).$$

In other words, to determine a limit for a polynomial or a rational function, simply evaluate the function at $x = a$, provided, of course, that the function is defined at $x = a$. For instance, we can rework the solution to Example 3(c) as follows:

$$\lim\limits_{x \to 2} (5x^3 - 15) = 5(2)^3 - 15 = 25.$$

Many situations require algebraic simplifications before the limit theorems can be applied.

EXAMPLE 4 Compute the following limits.

(a) $\lim\limits_{x \to 3} \dfrac{x^2 - 9}{x - 3}$

(b) $\lim\limits_{x \to 0} \dfrac{\sqrt{x + 4} - 2}{x}$

Solution (a) The function $\dfrac{x^2 - 9}{x - 3}$ is not defined when $x = 3$, since $\dfrac{3^2 - 9}{3 - 3} = \dfrac{0}{0}$, which is undefined. That causes no difficulty, since the limit as x approaches 3 depends only on the values of x *near* 3 and excludes consideration of the value at $x = 3$ itself. To evaluate the limit, note that $x^2 - 9 = (x - 3)(x + 3)$. So for $x \neq 3$,

$$\frac{x^2 - 9}{x - 3} = \frac{(x - 3)(x + 3)}{x - 3} = x + 3.$$

As x approaches 3, $x + 3$ approaches 6. Therefore,

$$\lim\limits_{x \to 3} \frac{x^2 - 9}{x - 3} = 6.$$

(b) Since the denominator approaches zero when taking the limit, we may not apply Limit Theorem VI directly. However, if we first apply an algebraic trick, the limit may be evaluated. Multiply numerator and denominator by $\sqrt{x + 4} + 2$.

$$\frac{\sqrt{x + 4} - 2}{x} \cdot \frac{\sqrt{x + 4} + 2}{\sqrt{x + 4} + 2} = \frac{(x + 4) - 4}{x(\sqrt{x + 4} + 2)}$$

$$= \frac{x}{x(\sqrt{x + 4} + 2)}$$

$$= \frac{1}{\sqrt{x + 4} + 2}.$$

Thus

$$\lim\limits_{x \to 0} \frac{\sqrt{x + 4} - 2}{x} = \lim\limits_{x \to 0} \frac{1}{\sqrt{x + 4} + 2}$$

$$= \frac{\lim\limits_{x \to 0} 1}{\lim\limits_{x \to 0} (\sqrt{x + 4} + 2)} \qquad \text{(Limit Theorem VI)}$$

$$= \frac{1}{4}.$$

EXAMPLE 5 Use limits to compute the derivative $f'(5)$ for the following functions.

(a) $f(x) = 15 - x^2$

(b) $f(x) = \dfrac{1}{2x - 3}$

Solution   In each case, we must calculate $\lim_{h \to 0} \dfrac{f(5 + h) - f(5)}{h}$.

(a)
$$\frac{f(5 + h) - f(5)}{h} = \frac{[15 - (5 + h)^2] - (15 - 5^2)}{h}$$

$$= \frac{15 - (25 + 10h + h^2) - (15 - 25)}{h}$$

$$= \frac{-10h - h^2}{h} = -10 - h.$$

Therefore, $f'(5) = \lim_{h \to 0} (-10 - h) = -10.$

(b)
$$\frac{f(5 + h) - f(5)}{h} = \frac{\dfrac{1}{2(5 + h) - 3} - \dfrac{1}{2(5) - 3}}{h}$$

$$= \frac{\dfrac{1}{7 + 2h} - \dfrac{1}{7}}{h} = \frac{\dfrac{7 - (7 + 2h)}{(7 + 2h)7}}{h}$$

$$= \frac{-2h}{(7 + 2h)7 \cdot h} = \frac{-2}{(7 + 2h)7} = \frac{-2}{49 + 14h}.$$

$$f'(5) = \lim_{h \to 0} \frac{-2}{49 + 14h} = -\frac{2}{49}.$$

Remark   When computing the limit in the example above, we considered only values of h near zero (and not $h = 0$ itself). Therefore, we were able to freely divide both numerator and denominator by h.

Infinity and Limits   Consider the function $f(x)$ whose graph is sketched in Fig. 1. As x grows large the value of $f(x)$ approaches 2. In this circumstance, we say that 2 is *the limit of $f(x)$ as x approaches infinity*. Infinity is denoted by the symbol ∞.

FIGURE 1

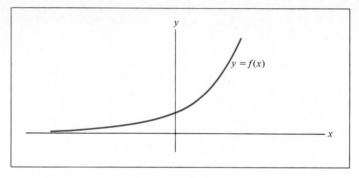

FIGURE 2

The limit statement above is expressed in the following notation:

$$\lim_{x \to \infty} f(x) = 2.$$

In a similar vein, consider the function whose graph is sketched in Fig. 2. As x grows large in the negative direction, the value of $f(x)$ approaches 0. In this circumstance, we say that 0 is *the limit of $f(x)$ as x approaches minus infinity*. In symbols,

$$\lim_{x \to -\infty} f(x) = 0.$$

EXAMPLE 6   Calculate the following limits.

(a) $\displaystyle\lim_{x \to \infty} \frac{1}{x^2 + 1}$

(b) $\displaystyle\lim_{x \to \infty} \frac{x + 1}{x - 1}$

Solution   (a) As x increases without bound, so does $x^2 + 1$. Therefore, $1/(x^2 + 1)$ approaches zero as x approaches ∞.

 * Multiply numerator and denominator by reciprocal of highest power of x.

(b) Both $x + 1$ and $x - 1$ increase without bound as x does. To determine the limit of their quotient, we employ an algebraic trick. Divide both numerator and denominator by x to obtain

$$\lim_{x \to \infty} \frac{x + 1}{x - 1} = \lim_{x \to \infty} \frac{1 + \dfrac{1}{x}}{1 - \dfrac{1}{x}}.$$

As x increases without bound, $1/x$ approaches zero, so that both $1 + (1/x)$ and $1 - (1/x)$ approach 1. Thus the desired limit is $1/1 = 1$.

PRACTICE PROBLEMS 4

Determine which of the following limits exist. Compute the limits that exist.

1. $\displaystyle\lim_{x \to 6} \frac{x^2 - 4x - 12}{x - 6}$

2. $\displaystyle\lim_{x \to 6} \frac{4x + 12}{x - 6}$

EXERCISES 4

For each of the following functions, $g(x)$, determine whether or not $\lim_{x \to 3} g(x)$ exists. If so, give the limit.

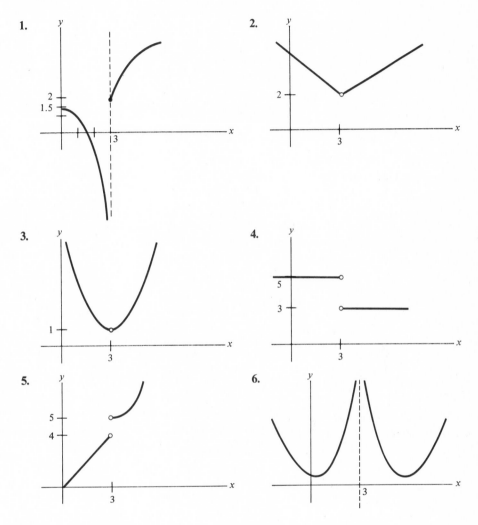

1.

2.

3.

4.

5.

6.

Determine which of the following limits exist. Compute the limits that exist.

7. $\lim_{x \to 1} (1 - 6x)$

8. $\lim_{x \to 2} \dfrac{x}{x - 2}$

9. $\lim_{x \to 3} \sqrt{x^2 + 16}$

10. $\lim_{x \to 4} (x^3 - 7)$

11. $\lim_{x \to 5} \dfrac{x^2 + 1}{5 - x}$

12. $\lim_{x \to 6} \left(\sqrt{6x} + 3x - \dfrac{1}{x} \right)(x^2 - 4)$

13. $\lim\limits_{x \to 7} (x + \sqrt{x - 6})(x^2 - 2x + 1)$

14. $\lim\limits_{x \to 8} \dfrac{\sqrt{5x - 4} - 1}{3x^2 + 2}$

15. $\lim\limits_{x \to 9} \dfrac{\sqrt{x^2 - 5x - 36}}{8 - 3x}$

16. $\lim\limits_{x \to 10} (2x^2 - 15x - 50)^{20}$

17. $\lim\limits_{x \to 0} \dfrac{x^2 + 3x}{x}$

18. $\lim\limits_{x \to 1} \dfrac{x^2 - 1}{x - 1}$

19. $\lim\limits_{x \to 2} \dfrac{-2x^2 + 4x}{x - 2}$

20. $\lim\limits_{x \to 3} \dfrac{x^2 - x - 6}{x - 3}$

21. $\lim\limits_{x \to 4} \dfrac{x^2 - 16}{4 - x}$

22. $\lim\limits_{x \to 5} \dfrac{2x - 10}{x^2 - 25}$

23. $\lim\limits_{x \to 6} \dfrac{x^2 - 6x}{x^2 - 5x - 6}$

24. $\lim\limits_{x \to 7} \dfrac{x^3 - 2x^2 + 3x}{x^2}$

25. $\lim\limits_{x \to 8} \dfrac{x^2 + 64}{x - 8}$

26. $\lim\limits_{x \to 9} \dfrac{1}{(x - 9)^2}$

27. $\lim\limits_{x \to 0} \dfrac{-2}{\sqrt{x + 16} + 7}$

28. $\lim\limits_{x \to 0} \dfrac{4x}{x(x^2 + 3x + 5)}$

Use limits to compute the following derivatives.

29. $f'(3)$ where $f(x) = x^2 + 1$

30. $f'(2)$ where $f(x) = x^3$

31. $f'(0)$ where $f(x) = x^3 + 3x + 1$

32. $f'(0)$ where $f(x) = x^2 + 2x + 2$

33. $f'(3)$ where $f(x) = \dfrac{1}{2x + 5}$

34. $f'(4)$ where $f(x) = \sqrt{2x - 1}$

35. $f'(2)$ where $f(x) = \sqrt{5 - x}$

36. $f'(3)$ where $f(x) = \dfrac{1}{7 - 2x}$

37. $f'(0)$ where $f(x) = \sqrt{1 - x^2}$

38. $f'(2)$ where $f(x) = (5x - 4)^2$

39. $f'(0)$ where $f(x) = (x + 1)^3$

40. $f'(0)$ where $f(x) = \sqrt{x^2 + x + 1}$

If degree of num < degree of denominator then limit = 0.

Compute the following limits.

41. $\lim\limits_{x \to \infty} \dfrac{1}{x^2}$ ○

42. $\lim\limits_{x \to -\infty} \dfrac{1}{x^2}$ ○

43. $\lim\limits_{x \to \infty} \dfrac{1}{x - 8}$ ○

44. $\lim\limits_{x \to \infty} \dfrac{1}{3x + 5}$ ○

45. $\lim\limits_{x \to \infty} \dfrac{2x + 1}{x + 2}$ 2

46. $\lim\limits_{x \to \infty} \dfrac{x^2 + x}{x^2 - 1}$ 1

SOLUTIONS TO PRACTICE PROBLEMS 4

1. The function under consideration is a rational function. Since the denominator has value 0 at $x = 6$, we cannot immediately determine the limit by just evaluating the function at $x = 6$. Now also, $\lim\limits_{x \to 6} (x - 6) = 0$. Since the function in the denominator has limit 0, we cannot apply Limit Theorem VI. However, since the definition of limit considers only values of x different from 6, the quotient can be simplified by factoring and canceling.

$$\frac{x^2 - 4x - 12}{x - 6} = \frac{(x + 2)(x - 6)}{(x - 6)} = x + 2 \qquad \text{for } x \neq 6.$$

Now $\lim_{x \to 6} (x + 2) = 8$. Therefore, $\lim_{x \to 6} \dfrac{x^2 - 4x - 12}{x - 6} = 8$.

2. No limit exists. It is easily seen that $\lim_{x \to 6} (4x + 12) = 36$ and $\lim_{x \to 6} (x - 6) = 0$. As x approaches 6, the denominator gets very small and the numerator approaches 36. For example, if $x = 6.00001$, then the numerator is 36.00004 and the denominator is .00001. The quotient is 3,600,004. As x approaches 6 even more closely, the quotient gets arbitrarily large and cannot possibly approach a limit.

1.5. Differentiability and Continuity

In the preceding section we defined differentiability of $f(x)$ at $x = a$ in terms of a limit. If this limit does not exist, then we say that $f(x)$ is *nondifferentiable* at $x = a$. Geometrically, the nondifferentiability of $f(x)$ at $x = a$ can manifest itself in several different ways. First of all, the graph of $f(x)$ could have no tangent line at

FIGURE 1

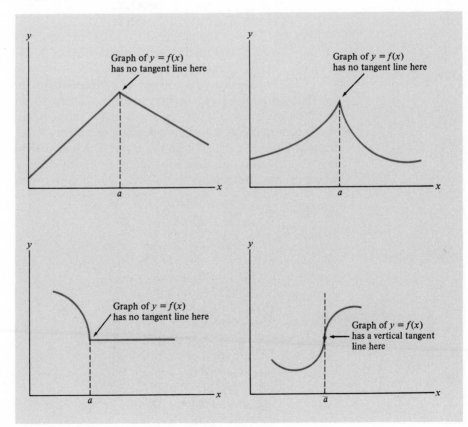

$x = a$. Second, the graph could have a vertical tangent line at $x = a$. (Recall that slope is not defined for vertical lines.) Some of the various geometric possibilities are illustrated in Fig. 1.

The following example illustrates how nondifferentiable functions can arise in practice.

EXAMPLE 1   A railroad company charges $10 per mile to haul a boxcar up to 200 miles and $8 per mile for each mile exceeding 200. In addition, the railroad charges a $1000 handling charge per boxcar. Graph the cost of sending a boxcar x miles.

Solution   If x is at most 200 miles, then the cost $C(x)$ is given by $C(x) = 1000 + 10x$ dollars. The cost for 200 miles is $C(200) = 1000 + 2000 = 3000$ dollars. If x exceeds 200 miles, then the total cost will be

$$C(x) = \underbrace{3000}_{\substack{\text{cost of first} \\ \text{200 miles}}} + \underbrace{8(x - 200)}_{\substack{\text{cost of miles in} \\ \text{excess of 200}}} = 1400 + 8x.$$

Thus

$$C(x) = \begin{cases} 1000 + 10x, & 0 < x \le 200, \\ 1400 + 8x, & x > 200. \end{cases}$$

The graph of $C(x)$ is sketched in Fig. 2. Note that $C(x)$ is not differentiable at $x = 200$.

FIGURE 2

Closely related to the concept of differentiability is that of continuity. We say that a function $f(x)$ is *continuous* at $x = a$ provided that, roughly speaking, its graph has no breaks (or gaps) as it passes through the point $(a, f(a))$. That is, $f(x)$ is continuous at $x = a$ provided that we can draw the graph through $(a, f(a))$ without lifting our pencil from the paper. The functions whose graphs are drawn in Figs. 1 and 2 are continuous for all values of x. By contrast, however, the function whose graph is drawn in Fig. 3(a) is not continuous (we say it is *discontinuous*) at

FIGURE 3

(a)   (b)

$x = 1$ and $x = 2$, since the graph has breaks there. Similarly, the function whose graph is drawn in Fig. 3(b) is discontinuous at $x = 2$.

Discontinuous functions can occur in applications, as the following example shows.

EXAMPLE 2 Suppose that a manufacturing plant is capable of producing 15,000 units in one shift of 8 hours. For each shift worked, there is a fixed cost of $2000 (for light, heat, etc.). Suppose that the variable cost (the cost of labor and raw materials) is $2 per unit. Graph the cost $C(x)$ of manufacturing x units.

Solution If $x \leq 15,000$, a single shift will suffice, so that

$$C(x) = 2000 + 2x, \qquad 0 \leq x \leq 15,000.$$

If x is between 15,000 and 30,000, one extra shift will be required, and

$$C(x) = 4000 + 2x, \qquad 15,000 < x \leq 30,000.$$

If x is between 30,000 and 45,000, the plant will need to work three shifts, and

$$C(x) = 6000 + 2x, \qquad 30,000 < x \leq 45,000.$$

The graph of $C(x)$ for $0 \leq x \leq 45,000$ is drawn in Fig. 4. Note that the graph has breaks at two points.

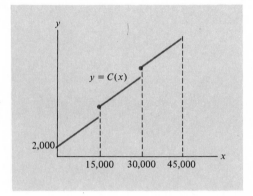

FIGURE 4

The relationship between differentiability and continuity is this:

□ **Theorem 1** If $f(x)$ is differentiable at $x = a$, then $f(x)$ is continuous at $x = a$.

Note, however, that the converse statement is definitely false: A function may be continuous at $x = a$ but still not be differentiable there. The functions whose graphs are drawn in Fig. 1 provide examples of this phenomenon.

Just as with differentiability, the notion of continuity can be phrased in terms of limits. In order for $f(x)$ to be continuous at $x = a$, the values of $f(x)$ for all x near a must be close to $f(a)$ (otherwise, the graph would have a break at $x = a$). In fact, the closer x is to a, the closer $f(x)$ must be to $f(a)$ (again, in order to avoid a break in the graph). In terms of limits, we must therefore have

$$\lim_{x \to a} f(x) = f(a).$$

Conversely, an intuitive argument shows that if the limit relation above holds, then the graph of $y = f(x)$ has no break at $x = a$. So we may state the following result.

Limit Criterion for Continuity   A function $f(x)$ is continuous at $x = a$ provided the following limit relation holds:

$$\lim_{x \to a} f(x) = f(a). \tag{1}$$

In order for (1) to hold, three conditions must be fulfilled.

1. $f(x)$ must be defined at $x = a$.
2. $\lim\limits_{x \to a} f(x)$ must exist.
3. The limit $\lim\limits_{x \to a} f(x)$ must have the value $f(a)$.

A function will fail to be continuous at $x = a$ when any one of these conditions fails to hold. The various possibilities are illustrated in the next example.

EXAMPLE 3 Determine whether the functions whose graphs are drawn in Fig. 5 are continuous at $x = 3$. Use the limit criterion.

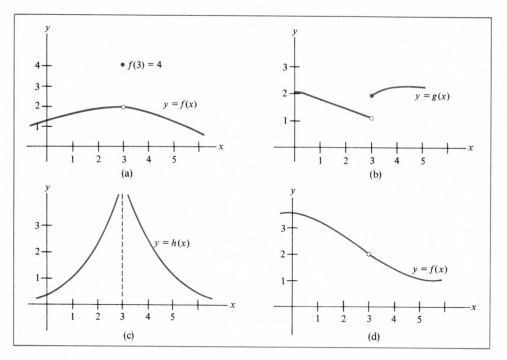

FIGURE 5

Solution (a) Here $\lim\limits_{x \to 3} f(x) = 2$. However, $f(3) = 4$. So

$$\lim\limits_{x \to 3} f(x) \neq f(3)$$

and $f(x)$ is not continuous at $x = 3$. (Geometrically, this is clear. The graph has a break at $x = 3$.)

(b) $\lim\limits_{x \to 3} g(x)$ does not exist, so $g(x)$ is not continuous at $x = 3$.

(c) $\lim\limits_{x \to 3} h(x)$ does not exist, so $h(x)$ is not continuous at $x = 3$.

(d) $f(x)$ is not defined at $x = 3$, so $f(x)$ is not continuous at $x = 3$.

Using our result on the limit of a polynomial function (Section 4), we see that

$$p(x) = a_0 + a_1x + \cdots + a_nx^n, \qquad a_0, \ldots, a_n \text{ constants,}$$

is continuous at all x. Similarly, a rational function

$$\frac{p(x)}{q(x)}, \qquad p(x), q(x) \text{ polynomials,}$$

is continuous at all x for which $q(x) \neq 0$.

PRACTICE PROBLEMS 5

Let

$$f(x) = \begin{cases} \dfrac{x^2 - x - 6}{x - 3} & \text{for } x \neq 3 \\ 4 & \text{for } x = 3. \end{cases}$$

$\lim\limits_{x \to 3} \dfrac{x^2-x-6}{x-3}$

$\lim\limits_{x \to 3} \dfrac{(x-3)(x+2)}{x-3} = 5$

1. Is $f(x)$ continuous at $x = 3$? No
2. Is $f(x)$ differentiable at $x = 3$? No

EXERCISES 5

Is the function whose graph is drawn in Fig. 6 continuous at the following values of x?

1. $x = 0$ No   2. $x = -3$ YES   3. $x = 3$ YES

4. $x = .001$ YES   5. $x = -2$ No   6. $x = 2$ No

Is the function whose graph is drawn in Fig. 6 differentiable at the following values of x?

7. $x = 0$ No   8. $x = -3$ No   9. $x = 3$ YES

10. $x = .001$ YES   11. $x = -2$ No   12. $x = 2$ No

Determine whether each of the following functions is continuous and/or differentiable at $x = 1$.

13. $f(x) = x^2 + 8x$ Both

14. $f(x) = x - \dfrac{1}{x}$

$|-1| = 0$

$|-.9| = .1$

$|-1.1| = .1$

FIGURE 6

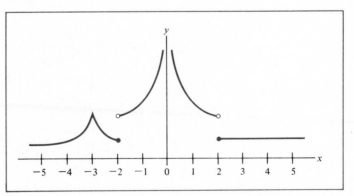

15. $f(x) = \begin{cases} x + 2 & \text{for } -1 \le x \le 1 \\ 3x & \text{for } 1 < x < 5 \end{cases}$

16. $f(x) = \begin{cases} x^2 & \text{for } 1 \le x \le 2 \\ x^3 & \text{for } 0 \le x < 1 \end{cases}$

17. $f(x) = \begin{cases} 2x - 1 & \text{for } 0 \le x \le 1 \\ 1 & \text{for } 1 < x \end{cases}$

18. $f(x) = \begin{cases} x & \text{for } x \ne 1 \\ 2 & \text{for } x = 1 \end{cases}$

19. $f(x) = \begin{cases} \dfrac{1}{x - 1} & \text{for } x \ne 1 \\ 0 & \text{for } x = 1 \end{cases}$

20. $f(x) = \begin{cases} x - 1 & \text{for } 0 \le x < 1 \\ 1 & \text{for } x = 1 \\ 2x - 2 & \text{for } x > 1 \end{cases}$

SOLUTIONS TO PRACTICE PROBLEMS 5

1. The function $f(x)$ is defined at $x = 3$, namely, $f(3) = 4$. When computing $\lim_{x \to 3} f(x)$ we exclude consideration of $x = 3$; therefore, we can simplify the expression for $f(x)$ as follows:

$$f(x) = \frac{x^2 - x - 6}{x - 3} = \frac{(x - 3)(x + 2)}{x - 3} = x + 2.$$

Clearly,

$$\lim_{x \to 3} f(x) = \lim_{x \to 3} (x + 2) = 5.$$

Since $\lim_{x \to 3} f(x) = 5 \ne 4 = f(3)$, $f(x)$ is not continuous at $x = 3$.

2. There is no need to compute any limits in order to answer this question. By Theorem 1, since $f(x)$ is not continuous at $x = 3$, it cannot possibly be differentiable there.

1.6. Some Rules for Differentiation

Three additional rules of differentiation greatly extend the number of functions that we can differentiate.

1. Constant-Multiple Rule

$$\frac{d}{dx}[k \cdot f(x)] = k \cdot \frac{d}{dx}[f(x)], \qquad k \text{ a constant.}$$

2. Sum Rule

$$\frac{d}{dx}[f(x) + g(x)] = \frac{d}{dx}[f(x)] + \frac{d}{dx}[g(x)].$$

3. General Power Rule

$$\frac{d}{dx}([g(x)]^r) = r \cdot [g(x)]^{r-1} \cdot \frac{d}{dx}[g(x)].$$

We shall discuss these rules and then prove the first two.

The Constant-Multiple Rule   Starting with a function $f(x)$, we can multiply it by a constant number k in order to obtain a new function $k \cdot f(x)$. For instance, if $f(x) = x^2 - 4x + 1$ and $k = 2$, then

$$2f(x) = 2(x^2 - 4x + 1) = 2x^2 - 8x + 2.$$

The constant-multiple rule says that the derivative of the new function $k \cdot f(x)$ is just k times the derivative of the original function.* In other words, when faced with the differentiation of a constant times a function, simply carry along the constant and differentiate the function.

EXAMPLE 1   Calculate.

(a) $\dfrac{d}{dx}(2x^5)$ (b) $\dfrac{d}{dx}\left(\dfrac{x^3}{4}\right)$ (c) $\dfrac{d}{dx}\left(-\dfrac{3}{x}\right)$ (d) $\dfrac{d}{dx}(5\sqrt{x})$.

Solution   (a) With $k = 2$ and $f(x) = x^5$, we have

$$\frac{d}{dx}(2 \cdot x^5) = 2 \cdot \frac{d}{dx}(x^5) = 2(5x^4) = 10x^4.$$

(b) Write $\dfrac{x^3}{4}$ in the form $\dfrac{1}{4} \cdot x^3$. Then

$$\frac{d}{dx}\left(\frac{x^3}{4}\right) = \frac{1}{4} \cdot \frac{d}{dx}(x^3) = \frac{1}{4}(3x^2) = \frac{3}{4}x^2.$$

(c) Write $-\dfrac{3}{x}$ in the form $(-3) \cdot \dfrac{1}{x}$. Then

$$\frac{d}{dx}\left(-\frac{3}{x}\right) = (-3) \cdot \frac{d}{dx}\left(\frac{1}{x}\right) = (-3) \cdot \frac{-1}{x^2} = \frac{3}{x^2}.$$

(d) $\dfrac{d}{dx}(5\sqrt{x}) = 5\dfrac{d}{dx}(\sqrt{x}) = 5\dfrac{d}{dx}(x^{1/2}) = \dfrac{5}{2}x^{-1/2}.$

This answer may also be written in the form $\dfrac{5}{2\sqrt{x}}$.

The Sum Rule   To differentiate a sum of functions, differentiate each function individually and add the derivatives together.† Another way of saying this is "the derivative of a sum of functions is the sum of the derivatives."

---

* More precisely, the constant-multiple rule asserts that if $f(x)$ is differentiable at $x = a$, then so is the function $k \cdot f(x)$, and the derivative of $k \cdot f(x)$ at $x = a$ may be computed using the given formula.

† More precisely, the sum rule asserts that if both $f(x)$ and $g(x)$ are differentiable at $x = a$, then so is $f(x) + g(x)$, and the derivative (at $x = a$) of the sum is then the sum of the derivatives (at $x = a$).

EXAMPLE 2 Find.

(a) $\dfrac{d}{dx}(x^3 + 5x)$ (b) $\dfrac{d}{dx}\left(x^4 - \dfrac{3}{x^2}\right)$ (c) $\dfrac{d}{dx}(2x^7 - x^5 + 8)$

Solution (a) Let $f(x) = x^3$ and $g(x) = 5x$. Then

$$\frac{d}{dx}(x^3 + 5x) = \frac{d}{dx}(x^3) + \frac{d}{dx}(5x) = 3x^2 + 5.$$

(b) The sum rule applies to differences as well as sums (see Exercise 36). Indeed, by the sum rule,

$$\frac{d}{dx}\left(x^4 - \frac{3}{x^2}\right) = \frac{d}{dx}(x^4) + \frac{d}{dx}\left(-\frac{3}{x^2}\right) \qquad \text{(sum rule)}$$

$$= \frac{d}{dx}(x^4) - 3\frac{d}{dx}(x^{-2}) \qquad \text{(constant-multiple rule)}$$

$$= 4x^3 - 3(-2x^{-3})$$

$$= 4x^3 + 6x^{-3}.$$

After some practice, one usually omits most or all of the intermediate steps and simply writes

$$\frac{d}{dx}\left(x^4 - \frac{3}{x^2}\right) = 4x^3 + 6x^{-3}.$$

(c) We apply the sum rule repeatedly and use the fact that the derivative of a constant function is 0:

$$\frac{d}{dx}(2x^7 - x^5 + 8) = \frac{d}{dx}(2x^7) - \frac{d}{dx}(x^5) + \frac{d}{dx}(8)$$

$$= 2(7x^6) - 5x^4 + 0$$

$$= 14x^6 - 5x^4.$$

The General Power Rule Frequently, we will encounter expressions of the form $[g(x)]^r$—for instance, $(x^3 + 5)^2$, where $g(x) = x^3 + 5$ and $r = 2$. The general power rule says that, to differentiate $[g(x)]^r$, we must first treat $g(x)$ as if it were simply an x, form $r[g(x)]^{r-1}$, and then multiply it by a "correction factor" $g'(x)$.* Thus

$$\frac{d}{dx}(x^3 + 5)^2 = 2(x^3 + 5)^1 \cdot \frac{d}{dx}(x^3 + 5)$$

$$= 2(x^3 + 5) \cdot (3x^2)$$

$$= 6x^2(x^3 + 5).$$

---

  * More precisely, the general power rule asserts that if $g(x)$ is differentiable at $x = a$, and if $[g(x)]^r$ and $[g(x)]^{r-1}$ are both defined at $x = a$, then $[g(x)]^r$ is also differentiable at $x = a$ and its derivative is given by the formula stated.

In this special case it is easy to verify that the general power rule gives the correct answer. We first expand $(x^3 + 5)^2$ and then differentiate.

$$(x^3 + 5)^2 = (x^3 + 5)(x^3 + 5) = x^6 + 10x^3 + 25.$$

From the constant-multiple rule and the sum rule, we have

$$\frac{d}{dx}(x^3 + 5)^2 = \frac{d}{dx}(x^6 + 10x^3 + 25)$$

$$= 6x^5 + 30x^2 + 0$$

$$= 6x^2(x^3 + 5).$$

The two methods give the same answer.

Note that if we set $g(x) = x$ in the general power rule, we recover the power rule. So the general power rule contains the power rule as a special case.

EXAMPLE 3 Differentiate $\sqrt{1 - x^2}$.

Solution

$$\frac{d}{dx}(\sqrt{1 - x^2}) = \frac{d}{dx}((1 - x^2)^{1/2}) = \frac{1}{2}(1 - x^2)^{-1/2} \cdot \frac{d}{dx}(1 - x^2)$$

$$= \frac{1}{2}(1 - x^2)^{-1/2} \cdot (-2x)$$

$$= \frac{-x}{(1 - x^2)^{1/2}} = \frac{-x}{\sqrt{1 - x^2}}.$$

EXAMPLE 4 Differentiate $y = \dfrac{1}{x^3 + 4x}$.

Solution

$$y = \frac{1}{x^3 + 4x} = (x^3 + 4x)^{-1}.$$

$$\frac{dy}{dx} = (-1)(x^3 + 4x)^{-2} \cdot \frac{d}{dx}(x^3 + 4x)$$

$$= \frac{-1}{(x^3 + 4x)^2}(3x^2 + 4)$$

$$= -\frac{3x^2 + 4}{(x^3 + 4x)^2}.$$

Proofs of the Constant-Multiple and Sum Rules Let us verify both rules when x has the value a. Recall that if $f(x)$ is differentiable at $x = a$, then its derivative is the limit

$$\lim_{h \to 0} \frac{f(a + h) - f(a)}{h}.$$

Constant-Multiple Rule   We assume that $f(x)$ is differentiable at $x = a$. We must prove that $k \cdot f(x)$ is differentiable at $x = a$ and that its derivative there is $k \cdot f'(a)$. This amounts to showing that the limit

$$\lim_{h \to 0} \frac{k \cdot f(a + h) - k \cdot f(a)}{h}$$

exists and has the value $k \cdot f'(a)$. However,

$$\lim_{h \to 0} \frac{k \cdot f(a + h) - k \cdot f(a)}{h} = \lim_{h \to 0} k\left[\frac{f(a + h) - f(a)}{h}\right]$$

$$= k \cdot \lim_{h \to 0} \frac{f(a + h) - f(a)}{h} \qquad \text{(by Limit Theorem I)}$$

$$= k \cdot f'(a) \qquad \text{(since } f(x) \text{ is differentiable at } x = a),$$

which is what we desired to show.

Sum Rule   We assume that both $f(x)$ and $g(x)$ are differentiable at $x = a$. We must prove that $f(x) + g(x)$ is differentiable at $x = a$ and that its derivative is $f'(a) + g'(a)$. That is, we must show that the limit

$$\lim_{h \to 0} \frac{[f(a + h) + g(a + h)] - [f(a) + g(a)]}{h}$$

exists and equals $f'(a) + g'(a)$. Using Limit Theorem III and the fact that $f(x)$ and $g(x)$ are differentiable at $x = a$, we have

$$\lim_{h \to 0} \frac{[f(a + h) + g(a + h)] - [f(a) + g(a)]}{h}$$

$$= \lim_{h \to 0} \left[\frac{f(a + h) - f(a)}{h} + \frac{g(a + h) - g(a)}{h}\right]$$

$$= \lim_{h \to 0} \frac{f(a + h) - f(a)}{h} + \lim_{h \to 0} \frac{g(a + h) - g(a)}{h}$$

$$= f'(a) + g'(a).$$

The general power rule will be proven as a special case of the chain rule in Chapter 3.

PRACTICE PROBLEMS 6

1. Find $\dfrac{d}{dx}(x)$.

2. Differentiate $y = \dfrac{x + (x^5 + 1)^{10}}{3}$.

EXERCISES 6

Differentiate.

1. $y = x^3 + x^2$

2. $y = x^2 + \dfrac{1}{x}$

3. $y = x^2 + 3x - 1$

4. $y = x^3 + 2x + 5$

5. $f(x) = x^5 + \dfrac{1}{x}$

6. $f(x) = x^8 - x$

7. $f(x) = x^4 + x^3 + x$

8. $f(x) = x^5 + x^2 - x$

9. $y = 3x^2$

10. $y = 2x^3$

11. $y = x^3 + 7x^2$

12. $y = -2x$

13. $y = \dfrac{4}{x^2}$

14. $y = 2\sqrt{x}$

15. $y = 3x - \dfrac{1}{x}$

16. $y = -x^2 + 3x + 1$

17. $f(x) = \tfrac{1}{3}x^3 - \tfrac{1}{2}x^2$

18. $f(x) = 100x^{100}$

19. $f(x) = -\dfrac{1}{5x^5}$

20. $f(x) = x^2 - \dfrac{1}{x^2}$

21. $f(x) = 1 - \sqrt{x}$

22. $f(x) = -3x^2 + 7$

23. $f(x) = (3x + 1)^{10}$

24. $f(x) = \dfrac{1}{x^2 + x + 1}$

25. $f(x) = \sqrt{3x^3 + x}$

26. $y = \dfrac{1}{(x^2 - 7)^5}$

27. $y = (2x^2 - x + 4)^6$

28. $y = \sqrt{-2x + 1}$

29. $y = \dfrac{x}{3} + \dfrac{3}{x}$

30. $y = \dfrac{2x - 1}{5}$

31. $y = \dfrac{2}{1 - 5x}$

32. $y = \dfrac{4}{3\sqrt{x}}$

33. $y = \dfrac{1}{1 - x^4}$

34. $y = \left(x^3 + \dfrac{x}{2} + 1\right)^5$

35. $f(x) = \dfrac{4}{\sqrt{x^2 + x}}$

36. $f(x) = \dfrac{6}{x^2 + 2x + 5}$

37. $f(x) = \left(\dfrac{\sqrt{x}}{2} + 1\right)^{3/2}$

38. $f(x) = \left(4 - \dfrac{2}{x}\right)^3$

Find the slope of the graph of $y = f(x)$ at the designated point.

39. $f(x) = 3x^2 - 2x + 1, (1, 2)$

40. $f(x) = x^{10} + 1 + \sqrt{1 - x}, (0, 2)$

41. Find the slope of the tangent line to the curve $y = x^3 + 3x - 8$ at $(2, 6)$.

42. Write the equation of the tangent line to the curve $y = x^3 + 3x - 8$ at $(2, 6)$.

43. Find the slope of the tangent line to the curve $y = (x^2 - 15)^6$ at $x = 4$. Then write the equation of this tangent line.

44. Find the equation of the tangent line to the curve $y = \dfrac{8}{x^2 + x + 2}$ at $x = 2$.

45. Differentiate the function $f(x) = (3x^2 + x - 2)^2$ in two ways.

(a) Use the general power rule.

(b) Multiply $3x^2 + x - 2$ by itself and then differentiate the resulting polynomial.

46. Using the sum rule and the constant-multiple rule, show that for any functions $f(x)$ and $g(x)$

$$\frac{d}{dx}[f(x) - g(x)] = \frac{d}{dx}f(x) - \frac{d}{dx}g(x).$$

47. Find m such that the line $y = mx$ is tangent to the curve $y = (3x - 5)^2 + 11$. [*Hint:* The line and the curve must both pass through some point (x, y) and have the same slope at that point.]

48. There are two points on the curve $y = x^2 + 3$ where the tangent line to the curve passes through the point $(1, 0)$. Find these two points.

SOLUTIONS TO PRACTICE PROBLEMS 6

1. The problem asks for the derivative of the function $y = x$, a straight line of slope 1. Therefore, $\frac{d}{dx}(x) = 1$. The result can also be obtained from the power rule with $r = 1$. If $f(x) = x^1$, then $\frac{d}{dx}(f(x)) = 1 \cdot x \quad = x^0 = 1$.

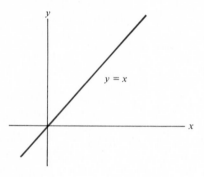

2. All three rules are required to differentiate this function.

$$\frac{dy}{dx} = \frac{d}{dx}\frac{1}{3} \cdot [x + (x^5 + 1)^{10}]$$

$$= \frac{1}{3}\frac{d}{dx}[x + (x^5 + 1)^{10}] \qquad \text{(constant-multiple rule)}$$

$$= \frac{1}{3}\left[\frac{d}{dx}(x) + \frac{d}{dx}(x^5 + 1)^{10}\right] \qquad \text{(sum rule)}$$

$$= \tfrac{1}{3}[1 + 10(x^5 + 1)^9 \cdot (5x^4)] \qquad \text{(general power rule)}$$

$$= \tfrac{1}{3}[1 + 50x^4(x^5 + 1)^9].$$

1.7. More About Derivatives

In many applications it is convenient to use variables other than x and y. One might, for instance, study the function $f(t) = t^2$ instead of writing $f(x) = x^2$. In this case, the notation for the derivative involves t rather than x, but the concept of the derivative as a slope formula is unaffected. (See Fig. 1.) When the independent variable is t instead of x, we write $\dfrac{d}{dt}$ in place of $\dfrac{d}{dx}$. For instance,

$$\frac{d}{dt}(t^3) = 3t^2, \qquad \frac{d}{dt}(2t^2 + 3t) = 4t + 3.$$

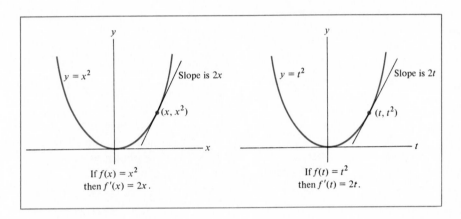

FIGURE 1

Recall that if y is a function of x, say $y = f(x)$, then we may write $\dfrac{dy}{dx}$ in place of $f'(x)$. We sometimes call $\dfrac{dy}{dx}$ "the derivative of y with respect to x." Similarly, if v is a function of t, then the derivative of v with respect to t is written as $\dfrac{dv}{dt}$. For example, if $v = 4t^2$, then $\dfrac{dv}{dt} = 8t$.

Of course, other letters can be used to denote variables. The formulas

$$\frac{d}{dP}(P^3) = 3P^2, \qquad \frac{d}{ds}(s^3) = 3s^2, \qquad \frac{d}{dz}(z^3) = 3z^2$$

are all expressing the same basic fact that the slope formula for the cubic curve $y = x^3$ is given by $3x^2$.

EXAMPLE 1   Compute.

(a) $\dfrac{ds}{dp}$ if $s = 3(p^2 + 5p + 1)^{10}$ (b) $\dfrac{d}{dt}(at^2 + St^{-1} + S^2)$

Solution   (a)
$$\frac{d}{dp}3(p^2 + 5p + 1)^{10} = 30(p^2 + 5p + 1)^9 \cdot \frac{d}{dp}(p^2 + 5p + 1)$$

$$= 30(p^2 + 5p + 1)^9(2p + 5).$$

(b) Although the expression $at^2 + St^{-1} + S^2$ contains several letters, the notation $\dfrac{d}{dt}$ indicates that all letters except t are to be considered as constants. Hence

$$\frac{d}{dt}(at^2 + St^{-1} + S^2) = \frac{d}{dt}(at^2) + \frac{d}{dt}(St^{-1}) + \frac{d}{dt}(S^2)$$

$$= a \cdot \frac{d}{dt}(t^2) + S \cdot \frac{d}{dt}(t^{-1}) + 0$$

$$= 2at - St^{-2}.$$

$$\left[\text{The derivative } \frac{d}{dt}(S^2) \text{ is zero because } S^2 \text{ is a constant.}\right]$$

The Second Derivative   When we differentiate a function $f(x)$, we obtain a new function $f'(x)$ that is a formula for the slope of the curve $y = f(x)$. If we differentiate the function $f'(x)$, we obtain what is called the *second derivative* of $f(x)$, denoted by $f''(x)$. That is,

$$\frac{d}{dx}f'(x) = f''(x).$$

EXAMPLE 2   Find the second derivatives of the following functions.

(a) $f(x) = x^3 + (1/x)$ (b) $f(x) = 2x + 1$ (c) $f(t) = t^{1/2} + t^{-1/2}$

Solution   (a) $f(x) = x^3 + (1/x) = x^3 + x^{-1}$
$f'(x) = 3x^2 - x^{-2}$
$f''(x) = 6x + 2x^{-3}.$

(b) $f(x) = 2x + 1$
$f'(x) = 2$ (a constant function whose value is 2)
$f''(x) = 0.$ (The derivative of a constant function is zero.)

(c) $f(t) = t^{1/2} + t^{-1/2}$
$f'(t) = \frac{1}{2}t^{-1/2} - \frac{1}{2}t^{-3/2}$
$f''(t) = -\frac{1}{4}t^{-3/2} + \frac{3}{4}t^{-5/2}.$

The first derivative of a function $f(x)$ gives the slope of the graph of $f(x)$ at any point. The second derivative of $f(x)$ gives important additional information about the shape of the curve near any point. We shall examine this subject carefully in the next chapter.

Other Notation for Derivatives   Unfortunately, the process of differentiation does not have a standardized notation. Consequently, it is important to become familiar with alternative terminology.
   If y is a function of x, say $y = f(x)$, then we may denote the first and second derivatives of this function in several ways.

Prime notation	$\dfrac{d}{dx}$ notation
$f'(x)$	$\dfrac{d}{dx} f(x)$
y'	$\dfrac{dy}{dx}$
$f''(x)$	$\dfrac{d^2}{dx^2} f(x)$
y''	$\dfrac{d^2 y}{dx^2}$

The notation $\dfrac{d^2}{dx^2}$ is purely symbolic. It reminds us that the second derivative is obtained by differentiating $\dfrac{d}{dx} f(x)$; that is,

$$f'(x) = \frac{d}{dx} f(x),$$

$$f''(x) = \frac{d}{dx} \left[\frac{d}{dx} f(x) \right].$$

If we evaluate the derivative $f'(x)$ at a specific value of x, say $x = a$, we get a number $f'(a)$ that gives the slope of the curve $y = f(x)$ at the point $(a, f(a))$. Another way of writing $f'(a)$ is

$$\frac{dy}{dx}\bigg|_{x=a}.$$

If we have a second derivative $f''(x)$, then its value when $x = a$ is written

$$f''(a) \quad \text{or} \quad \frac{d^2 y}{dx^2}\bigg|_{x=a}.$$

EXAMPLE 3   If $y = x^4 - 5x^3 + 7$, find $\dfrac{d^2 y}{dx^2}\bigg|_{x=3}$.

Solution

$$\frac{dy}{dx} = \frac{d}{dx}(x^4 - 5x^3 + 7) = 4x^3 - 15x^2$$

$$\frac{d^2 y}{dx^2} = \frac{d}{dx}(4x^3 - 15x^2) = 12x^2 - 30x$$

$$\frac{d^2 y}{dx^2}\bigg|_{x=3} = 12(3)^2 - 30(3) = 108 - 90 = 18.$$

EXAMPLE 4   If $s = t^3 - 2t^2 + 3t$, find

$$\frac{ds}{dt}\bigg|_{t=-2} \quad \text{and} \quad \frac{d^2s}{dt^2}\bigg|_{t=-2}.$$

Solution

$$\frac{ds}{dt} = \frac{d}{dt}(t^3 - 2t^2 + 3t) = 3t^2 - 4t + 3$$

$$\frac{ds}{dt}\bigg|_{t=-2} = 3(-2)^2 - 4(-2) + 3 = 12 + 8 + 3 = 23.$$

To find the value of the second derivative at $t = -2$, we must first differentiate $\dfrac{ds}{dt}$.

$$\frac{d^2s}{dt^2} = \frac{d}{dt}(3t^2 - 4t + 3) = 6t - 4$$

$$\frac{d^2s}{dt^2}\bigg|_{t=-2} = 6(-2) - 4 = -12 - 4 = -16.$$

PRACTICE PROBLEMS 7

1. Let $f(t) = t + (1/t)$. Find $f''(2)$.

2. Differentiate $g(r) = 2\pi rh$.

EXERCISES 7

Find the first derivatives.

1. $f(t) = (t^2 + 1)^5$

2. $f(P) = P^4 - P^3 + 4P^2 - P$

3. $v = \sqrt{2t - 1}$

4. $g(z) = (z^3 - z + 1)^2$

5. $y = (T^3 + 5T)^{2/3}$

6. $s = \sqrt{t} + \dfrac{1}{\sqrt{t}}$

7. Find $\dfrac{d}{dP}(3P^2 - \frac{1}{2}P + 1)$.

8. Find $\dfrac{d}{dz}(\sqrt{z^2 - 1})$.

9. Find $\dfrac{d}{dt}(a^2t^2 + b^2t + c^2)$.

10. Find $\dfrac{d}{dx}(x^3 + t^3)$.

Find the first and second derivatives.

11. $f(x) = \frac{1}{2}x^2 - 7x + 2$

12. $y = \dfrac{1}{x^2} + 1$

13. $y = \sqrt{x}$

14. $f(t) = t^{100} + t + 1$

15. $f(r) = \pi hr^2 + 2\pi r$

16. $v = t^{3/2} + t$

17. $g(x) = 2 - 5x$

18. $V(r) = \frac{4}{3}\pi r^3$

19. $f(P) = (3P + 1)^5$

20. $u = \dfrac{t^6}{30} - \dfrac{t^4}{12}$

Compute the following.

21. $\dfrac{d}{dx}(2x^2 - 3)\Big|_{x=5}$

22. $\dfrac{d}{dt}(1 - 2t - 3t^2)\Big|_{t=-1}$

23. $\dfrac{d}{dz}(z^2 - 4)^3\Big|_{z=1}$

24. $\dfrac{d}{dT}\left(\dfrac{1}{3T + 1}\right)\Big|_{T=2}$

25. $\dfrac{d^2}{dx^2}(3x^3 - x^2 + 7x - 1)\Big|_{x=2}$

26. $\dfrac{d}{dt}\left(\dfrac{dv}{dt}\right)$, where $v = 2t^{-3}$

27. $\dfrac{d}{dP}\left(\dfrac{dy}{dP}\right)$, where $y = \dfrac{k}{2P - 1}$ $k(2P-1)^{-1}$

28. $\dfrac{d^2V}{dr^2}\Big|_{r=2}$, where $V = ar^3$

29. $f'(3)$ and $f''(3)$, when $f(x) = \sqrt{10 - 2x}$

30. $g'(2)$ and $g''(2)$, when $g(T) = (3T - 5)^{10}$

31. Suppose a company finds that the revenue R generated by spending x dollars on advertising is given by $R = 1000 + 80x - .02x^2$, for $0 \le x \le 2000$. Find $\dfrac{dR}{dx}\Big|_{x=1500}$.

32. A supermarket finds that its average daily volume of business V (in thousands of dollars) and the number of hours t the store is open for business each day are approximately related by the formula

$$V = 20\left(1 - \dfrac{100}{100 + t^2}\right), \qquad 0 \le t \le 24.$$

Find $\dfrac{dV}{dt}\Big|_{t=10}$.

33. The *third derivative* of a function $f(x)$ is the derivative of the second derivative $f''(x)$ and is denoted $f'''(x)$. Compute $f'''(x)$ for the following functions.

(a) $f(x) = x^5 - x^4 + 3x$ (b) $f(x) = 4x^{5/2}$

34. Compute the third derivatives of the following functions.

(a) $f(t) = t^{10}$ (b) $f(z) = \dfrac{1}{z + 5}$

SOLUTIONS TO PRACTICE PROBLEMS 7

1.
$$f(t) = t + t^{-1}$$
$$f'(t) = 1 + (-1)t^{(-1)-1} = 1 - t^{-2}$$
$$f''(t) = -(-2)t^{(-2)-1} = 2t^{-3} = \dfrac{2}{t^3}$$

Therefore, $f''(2) = \dfrac{2}{2^3} = \dfrac{1}{4}$. [*Note:* It is essential to first compute the function $f''(t)$ and *then* evaluate the function at $t = 2$.]

2. The expression $2\pi rh$ contains two numbers, 2 and π, and two letters, r and h. The notation "$g(r)$" tells us that the expression $2\pi rh$ is to be regarded as a function of r. Therefore, h and hence $2\pi h$ is to be treated like a constant, and differentiation is done with respect to

the variable r. That is,

$$g(r) = (2\pi h)r$$
$$g'(r) = 2\pi h.$$

1.8. The Derivative as a Rate of Change

An important interpretation of the derivative of a function is as a rate of change. Consider the following situation, for example. The weight of an animal changes as time passes and we may think of the weight as a function of time, say $W(t)$. Suppose that we measure the weight at some time $t = a$ and at a later time $t = a + h$. Then the change in weight over this time interval is

$$W(a + h) - W(a).$$

The average *rate* of change of weight with respect to time during this time interval is found by dividing the change in weight by the length of the time interval:

$$\left. \begin{array}{l} \text{average rate of change} \\ \text{of } W(t) \text{ with respect to} \\ t, \text{from } a \text{ to } a + h \end{array} \right\} = \frac{W(a + h) - W(a)}{h}$$

This average rate of change takes into account the change in weight as t increases from $t = a$ to $t = a + h$. In order to study the change in weight $W(t)$ only for t close to a, we must take h very small. But we know that as h becomes small, the quantity $[W(a + h) - W(a)]/h$ approaches the derivative $W'(a)$. Thus

$$\left. \begin{array}{l} \text{average rate of change} \\ \text{of } W(t) \text{ with respect to} \\ t, \text{from } a \text{ to } a + h \end{array} \right\} \approx W'(a),$$

for h very small.

In this sense, the derivative $W'(a)$ gives the rate of change of $W(t)$ with respect to t precisely at the point when $t = a$.

In geometric terms, the average rate of change of $W(t)$ from a to $a + h$ is the slope of the secant line in Fig. 1. The derivative $W'(a)$ is the slope of the tangent

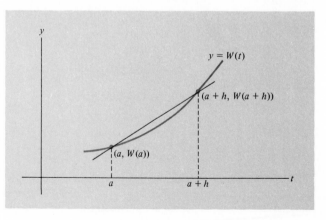

FIGURE 1

line through $(a, W(a))$ and gives the best measure of the rate of change of the weight at time a.

The derivative of any function $f(t)$ may be interpreted as a rate of change, just as we did for the weight function $W(t)$. We state it formally in the following definition.

> The derivative $f'(a)$ measures the rate of change of $f(t)$ with respect to t, at the point $t = a$.

The examples and exercises that follow show why it is often convenient to use a derivative to measure a rate of change. Other illustrations will be given later in the book.

EXAMPLE 1   A flu epidemic hits a midwestern town. Public Health officials estimate that the number of persons sick with the flu at time t (measured in days from the beginning of the epidemic) is approximated by $P(t) = 60t^2 - t^3$, provided that $0 \le t \le 40$. At what rate is the flu spreading at time $t = 30$?

Solution   The rate at which the flu spreads is given by the rate of change of $P(t)$, measured in people per day. Now

$$P'(t) = 120t - 3t^2$$

$$P'(30) = 120(30) - 3(30)^2 = 900.$$

Thus 30 days after the beginning of the epidemic the flu is spreading at the rate of 900 people per day.

EXAMPLE 2   A common clinical procedure for studying a person's calcium metabolism (the rate at which the body assimilates and uses calcium) is to inject some chemically "labeled" calcium into the bloodstream and then measure how fast this calcium is removed from the blood by the person's bodily processes. Suppose that t days after an injection of calcium, the amount A of the labeled calcium remaining in the blood is $A = t^{-3/2}$ for $t \ge .5$, where A is measured in suitable units.* How fast (in "units of calcium per day") is the body removing calcium from the blood when $t = 1$ day?

Solution   The rate of change (per day) of calcium in the blood is given by the derivative

$$\frac{dA}{dt} = -\tfrac{3}{2}t^{-5/2}.$$

When $t = 1$, this rate equals

$$\frac{dA}{dt}\bigg|_{t=1} = -\tfrac{3}{2}(1)^{-5/2} = -\tfrac{3}{2}.$$

---

* For a discussion of this mathematical model, see J. Defares, I. Sneddon, and M. Wise, *An Introduction to the Mathematics of Medicine and Biology* (Chicago: Year Book Publishers, Inc., 1973), pp. 609–619.

The amount of calcium in the blood is changing at the rate of $-\frac{3}{2}$ units per day when $t = 1$. The negative sign indicates that the amount of calcium is decreasing rather than increasing.

Approximating the Change in a Function As we have seen above, when h is near 0

$$\frac{f(a + h) - f(a)}{h} \approx f'(a),$$

where the symbol "$\approx$" means "is approximately equal to." Suppose, for example, that $a = 2$ and $f'(a) = 5$. Then for small values of h, we would have

$$\frac{f(2 + h) - f(2)}{h} \approx 5$$

or

$$f(2 + h) - f(2) \approx 5h.$$

In other words, if x is increased from 2 to $2 + h$, the change in the value of $f(x)$ [namely $f(2 + h) - f(2)$] is approximately $5h$. That is, as x varies near 2, the value of $f(x)$ varies at a rate equal to 5 times the change in x. We may generalize this observation to yield the following important fact.

If h is small, then

$$f(a + h) - f(a) \approx f'(a) \cdot h. \tag{1}$$

The Marginal Concept in Economics Suppose a company determines that the cost of producing x units of its product is $C(x)$ dollars. Recall from Section 3 that $C'(x)$ is called the marginal cost function. The value of the derivative of $C(x)$ at $x = a$—that is, $C'(a)$—is called the *marginal cost at production level a*. Since the marginal cost is just a derivative, its value gives the rate at which costs are increasing with respect to the level of production, assuming that production is at level a. The use of marginal cost is illustrated in the next example.

EXAMPLE 3 Suppose that the cost function is $C(x) = .005x^3 - 3x$ and production is proceeding at 1000 units per day.

(a) What is the marginal cost at this production level?

(b) What is the cost of increasing production from 1000 to 1001 units per day?

Solution (a) The marginal cost at production level 1000 is $C'(1000)$. Thus

$$C'(x) = .015x^2 - 3$$

$$C'(1000) = 14{,}997.$$

(b) The cost of increasing production by one unit, from 1000 to 1001, is $C(1001) - C(1000)$, which equals (rounded to the nearest dollar)

$$[.005(1001)^3 - 3(1001)] - [.005(1000)^3 - 3(1000)] = 5,012,012 - 4,997,000$$

$$= 15,012.$$

Note that the marginal cost approximately equals the cost of increasing production by one unit. This fact is true of marginal cost functions generally.

Velocity and Acceleration   An everyday illustration of rate of change is given by the velocity of a moving object. Suppose that we are driving a car along a straight road and at each time t we let $s(t)$ be our position on the road, measured from some convenient reference point. See Fig. 2, where distances are positive to

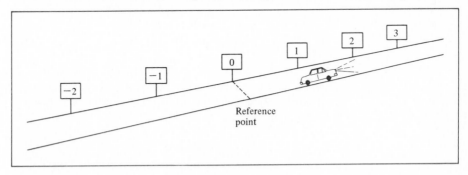

FIGURE 2

the right of the reference point and negative to the left. We call $s(t)$ the *position function* of the car. For the moment we shall assume that we are proceeding in only the positive direction along the road.

At any instant, the car's speedometer tells us how fast we are moving—that is, how fast our position $s(t)$ is changing. To show how the speedometer reading is related to our calculus concept of a derivative, let us examine what is happening at a specific time, say $t = 1$. Consider a short time interval of duration h from $t = 1$ to $t = 1 + h$. Our car will move from position $s(1)$ to position $s(1 + h)$, a distance of $s(1 + h) - s(1)$. Thus the *average velocity from* $t = 1$ *to* $t = 1 + h$ is

$$\frac{[\text{distance traveled}]}{[\text{time elapsed}]} = \frac{s(1 + h) - s(1)}{h}. \tag{2}$$

If the car is traveling at a steady speed during this time period, then the speedometer reading will equal the average velocity in equation (2).

From our discussion in Section 3 the ratio (2) approaches the derivative $s'(1)$ as h approaches zero. For this reason we call $s'(1)$ *the* (instantaneous) *velocity at* $t = 1$. This number will agree with the speedometer reading at $t = 1$ because when h is very small, the car's speed will be nearly steady over the time interval from $t = 1$ to $t = 1 + h$, and so the average velocity over this time interval will be nearly the same as the speedometer reading at $t = 1$.

The reasoning used for $t = 1$ holds for an arbitrary t as well. Thus the following definition makes sense:

> If $s(t)$ denotes the position function of an object moving in a straight line, then the velocity $v(t)$ of the object at time t is given by
>
> $$v(t) = s'(t).$$

In our discussion we assumed that the car moved in the positive direction. If the car moves in the opposite direction, the ratio (2) and the limiting value $s'(1)$ will be negative. So we interpret negative velocity as movement in the negative direction along the road.

The derivative of the velocity function $v(t)$ is called the *acceleration* function and is often written as $a(t)$:

> $$a(t) = v'(t).$$

Since $v'(t)$ measures the rate of change of the velocity $v(t)$, this use of the word acceleration agrees with our common usage in connection with automobiles. Note that since $v(t) = s'(t)$, the acceleration is actually the second derivative of the position function $s(t)$:

$$a(t) = s''(t).$$

PRACTICE PROBLEMS 8

When a ball is thrown straight up into the air, it travels along a straight line and its motion can be described in the same manner as the motion of a car. Regard "up" as the positive direction and let $s(t)$ be the height of the ball in feet after t seconds. Suppose that $s(t) = -16t^2 + 128t + 5$.

1. What will be the velocity after 2 seconds?

2. What will be the acceleration after 2 seconds?

3. After how many seconds will the ball attain its greatest height?

EXERCISES 8

1. If $f(t) = t^2 + 3t - 7$, what is the rate of change of $f(t)$ with respect to t when $t = 5$?

2. If $f(t) = 3t + 2 - \dfrac{5}{t}$, what is the rate of change of $f(t)$ with respect to t when $t = 2$?

3. An analysis of the daily output of a factory assembly line shows that about $60t + t^2 - \frac{1}{12}t^3$ units are produced after t hours of work, $0 \le t \le 8$. What is the rate of production (in units per hour) when $t = 2$?

4. Liquid is pouring into a large vat. After t hours, there are $5t - t^{1/2}$ gallons in the vat. At what rate is the liquid flowing into the vat (in gallons per hour) when $t = 4$?

5. Suppose that the weight in grams of a cancerous tumor at time t is $W(t) = .1t^2$, where t is measured in weeks. What is the rate of growth of the tumor (in grams per week) when $t = 5$?

6. After an advertising campaign, the sales of a product often increase and then decrease. Suppose that t days after the end of the advertising, the daily sales are $-3t^2 + 30t + 100$ units. At what rate (in units per day) are the sales changing when $t = 2$?

7. A sewage treatment plant accidentally discharged untreated sewage into a lake for a few days. This temporarily decreased the amount of dissolved oxygen in the lake. Let $f(t)$ be the amount of oxygen in the lake (measured in suitable units) t days after the sewage started flowing into the lake. Experimental data suggest that $f(t)$ is given approximately by

$$f(t) = 500\left[1 - \frac{10}{t + 10} + \frac{100}{(t + 10)^2}\right].$$

Find the rate of change (in units per day) of the oxygen content of the lake at $t = 5$ and at $t = 15$. Is the oxygen content increasing or decreasing when $t = 15$?

8. Suppose that t hours after being placed in a freezer, the temperature of a piece of meat is given by $T(t) = 70 - 12t + 4/(t + 1)$ degrees, where $0 \le t \le 5$. How fast is the temperature falling after 1 hour?

9. A manufacturer estimates that the hourly cost of producing x units of a product on an assembly line is $.1x^3 - 6x^2 + 136x + 200$ dollars.

   (a) Find the marginal cost when the production level is 20 units.

   (b) Show that this marginal cost is approximately the extra cost of raising the production from 20 to 21 units.

10. A market finds that if it prices a certain product so as to sell x units each week, then the revenue received will be approximately $2x - .001x^2$ dollars.

    (a) Find the marginal revenue at a sales level of 600 units.

    (b) Show that this marginal revenue is approximately the extra revenue produced by raising the sales from 600 to 601 units.

11. Suppose that the profit from producing x units of a product is given by $P(x) = .0003x^3 + .01x$.

    (a) Compute the marginal profit at a production level of 100 units.

    (b) Compute the additional profit gained from increasing sales from 100 to 101 units.

12. Suppose that the revenue from producing x units of a product is given by $R(x) = .01x^2 - 3x$.

    (a) Compute the marginal revenue at a production level of 20,000.

    (b) Compute the additional revenue gained by increasing production from 20,000 to 20,001 units.

13. An object moving in a straight line travels $s(t)$ kilometers in t hours, where $s(t) = \frac{1}{2}t^2 + 4t$.

    (a) What is the object's velocity when $t = 3$?

    (b) How far has the object traveled in 3 hours?

14. Suppose that the position of a car at time t is given by $s(t) = 50t - 7/(t + 1)$, where the position is measured in kilometers. Find the velocity and acceleration of the car at $t = 0$.

15. A toy rocket fired straight up into the air has height $s(t) = 160t - 16t^2$ feet after t seconds.

    (a) What is the rocket's initial velocity (when $t = 0$)?

    (b) What is the velocity after 2 seconds?

    (c) What is the acceleration when $t = 3$?

    (d) At what time will the rocket hit the ground?

    (e) At what velocity will the rocket be traveling just as it smashes into the ground?

16. A helicopter is rising straight up in the air. Its distance from the ground t seconds after takeoff is $s(t)$ feet, where $s(t) = t^2 + t$.

    (a) How long will it take for the helicopter to rise 20 feet?

    (b) Find the velocity and the acceleration of the helicopter when it is 20 feet above the ground.

17. During the epidemic in Example 1, is there any day when approximately 1200 people become sick with the flu?

18. Suppose that the cost of producing x units of some product is $\frac{1}{3}x^3 - 5x^2 + 30x + 10$ dollars, while the revenue received from the sale of x units is $39x - x^2$ dollars. At what value(s) of x will the marginal revenue equal the marginal cost?

19. A subway train travels from station A to station B in 2 minutes. Its distance from station A is $s(t) = t^2 - \frac{1}{3}t^3$ kilometers after t minutes. At what time(s) will the train be traveling at the rate of $\frac{1}{2}$ kilometer per minute?

20. Let $f(t)$ be the function in Exercise 7 that gives the amount of oxygen in the lake at time t. Find the time when the amount of oxygen stops decreasing and begins to increase.

21. Let $f(x) = x^3$. Use formula (1) with $f(x) = x^3$ to approximate $f(2.01) - f(2)$.

22. Suppose that the side of a cube is increased from 5 to 5.001 meters. Use the derivative to estimate the increase in volume.

23. Use the derivative to estimate $\sqrt{4.01}$. [*Hint:* Let $f(x) = \sqrt{x}$. Then $f(4 + h) - f(4) \approx f'(4) \cdot (.01)$.]

24. Use the derivative to estimate $\sqrt[3]{8.024}$.

25. The monthly payment on a \$10,000 3-year car loan is a function $f(r)$ of the interest rate, $r\%$. Now, $f(16) = 351.57$ and $f'(16) = 4.94$. What is the significance of the amount \$4.94?

SOLUTIONS TO PRACTICE PROBLEMS 8

1. Since velocity is rate of change of position, $v(t) = s'(t) = -32t + 128$. The velocity after 2 seconds is $v(2) = -32(2) + 128 = 64$. Therefore, after 2 seconds the ball will be traveling upward at a velocity of 64 feet per second.

2. The acceleration is the rate of change of velocity, so $a(t) = v'(t) = -32$. The acceleration is always -32. This constant acceleration is due to the downward (and therefore negative) force of gravity.

3. At the instant at which the ball attains its maximum height it will have zero velocity. Therefore, we can determine this time by setting $v(t) = 0$ and solving for t.

$$-32t + 128 = 0$$

$$t = 4.$$

So the ball attains its maximum height after 4 seconds.

Chapter 1: CHECKLIST

- ☐ Slope of a line
- ☐ y-intercept
- ☐ Slope Properties 1 to 5
- ☐ Slope of a curve at a point
- ☐ The derivative of a constant function is zero
- ☐ Limit definition of the derivative
- ☐ Power rule:

$$\frac{d}{dx}(x^r) = rx^{r-1}$$

- ☐ General power rule:

$$\frac{d}{dx}([g(x)]^r) = r \cdot [g(x)]^{r-1} \cdot \frac{d}{dx}[g(x)]$$

- ☐ Constant-multiple rule:

$$\frac{d}{dx}[k \cdot f(x)] = k \cdot \frac{d}{dx}[f(x)]$$

- ☐ Sum rule:

$$\frac{d}{dx}[f(x) + g(x)] = \frac{d}{dx}[f(x)] + \frac{d}{dx}[g(x)]$$

- ☐ Notation for first and second derivatives
- ☐ The derivative $f'(a)$ measures the rate of change of $f(t)$ with respect to t at $t = a$
- ☐ $f(a + h) - f(a) \approx f'(a) \cdot h$
- ☐ Marginal concept in economics
- ☐ Position, velocity, and acceleration functions
- ☐ Differentiable at $x = a$
- ☐ Continuous at $x = a$
- ☐ $\lim\limits_{x \to a} f(x)$
- ☐ $\lim\limits_{x \to \infty} f(x)$, $\lim\limits_{x \to -\infty} f(x)$

Chapter 1: SUPPLEMENTARY EXERCISES

Find the equation and sketch the graph of the following lines.

1. With slope -2, y-intercept $(0, 3)$

2. With slope $\frac{3}{4}$, y-intercept $(0, -1)$

3. Through (2, 0), with slope 5

4. Through (1, 4), with slope $-\frac{1}{3}$

5. Parallel to $y = -2x$, passing through (3, 5)

6. Parallel to $-2x + 3y = 6$, passing through (0, 1)

7. Through $(-1, 4)$, and (3, 7)

8. Through (2, 1) and (5, 1).

Differentiate.

9. $y = x^7 + x^3$

10. $y = 5x^8$

11. $y = 6\sqrt{x}$

12. $y = x^7 + 3x^5 + 1$

13. $y = 3/x$

14. $y = x^4 - 4/x$

15. $y = (3x^2 - 1)^8$

16. $y = \frac{3}{4}x^{4/3} + \frac{4}{3}x^{3/4}$

17. $y = \dfrac{1}{5x - 1}$

18. $y = (x^3 + x^2 + 1)^5$

19. $y = \sqrt{x^2 + 1}$

20. $y = \dfrac{5}{7x^2 + 1}$

21. $f(x) = 1/\sqrt[4]{x}$

22. $f(x) = (2x + 1)^3 - 5(2x + 1)^2$

23. $f(x) = 5$

24. $f(x) = \dfrac{5x}{2} - \dfrac{2}{5x}$

25. $f(x) = [x^5 - (x - 1)^5]^{10}$

26. $f(t) = t^{10} - 10t^9$

27. $g(t) = 3\sqrt{t} - \dfrac{3}{\sqrt{t}}$

28. $h(t) = 3\sqrt{2}$

29. $f(t) = \dfrac{2}{t - 3t^3}$

30. $g(P) = 4P^{.7}$

31. $h(x) = \frac{3}{2}x^{3/2} - 6x^{2/3}$

32. $f(x) = \sqrt{x + \sqrt{x}}$

33. If $f(t) = 3t^3 - 2t^2$, find $f'(2)$.

34. If $V(r) = 15\pi r^2$, find $V'(\frac{1}{3})$.

35. If $g(u) = 3u - 1$, find $g(5)$ and $g'(5)$.

36. If $h(x) = -\frac{1}{2}$, find $h(-2)$ and $h'(-2)$.

37. If $f(x) = x^{5/2}$, what is $f''(4)$?

38. If $g(t) = \frac{1}{4}(2t - 7)^4$, what is $g''(3)$?

39. Find the slope of the graph of $y = (3x - 1)^3 - 4(3x - 1)^2$ at $x = 0$.

40. Find the slope of the graph of $y = (4 - x)^5$ at $x = 5$.

Compute.

41. $\dfrac{d}{dx}(x^4 - 2x^2)$

42. $\dfrac{d}{dt}(t^{5/2} + 2t^{3/2} - t^{1/2})$

43. $\dfrac{d}{dP}(\sqrt{1 - 3P})$

44. $\dfrac{d}{dn}(n^{-5})$

45. $\dfrac{d}{dz}(z^3 - 4z^2 + z - 3)\bigg|_{z=-2}$

46. $\dfrac{d}{dx}(4x - 10)^5\bigg|_{x=3}$

47. $\dfrac{d^2}{dx^2}(5x + 1)^4$

48. $\dfrac{d^2}{dt^2}(2\sqrt{t})$

49. $\dfrac{d^2}{dt^2}(t^3 + 2t^2 - t)\bigg|_{t=-1}$

50. $\dfrac{d^2}{dP^2}(3P + 2)\bigg|_{P=4}$

51. $\dfrac{d^2 y}{dx^2}$ where $y = 4x^{3/2}$

52. $\dfrac{d}{dt}\left(\dfrac{dy}{dt}\right)$, where $y = \dfrac{1}{3t}$

53. What is the slope of the graph of $f(x) = x^3 - 4x^2 + 6$ at $x = 2$? Write the equation of the line tangent to the graph of $f(x)$ at $x = 2$.

54. What is the slope of the curve $y = 1/(3x - 5)$ at $x = 1$? Write the equation of the line tangent to this curve at $x = 1$.

55. Find the equation of the tangent line to the curve $y = x^2$ at the point $(\tfrac{3}{2}, \tfrac{9}{4})$. Sketch the graph of $y = x^2$ and sketch the tangent line at $(\tfrac{3}{2}, \tfrac{9}{4})$.

56. Find the equation of the tangent line to the curve $y = x^2$ at the point $(-2, 4)$. Sketch the graph of $y = x^2$ and sketch the tangent line at $(-2, 4)$.

57. Determine the equation of the tangent line to the curve $y = 3x^3 - 5x^2 + x + 3$ at $x = 1$.

58. Determine the equation of the tangent line to the curve $y = (2x^2 - 3x)^3$ at $x = 2$.

59. A helicopter is rising at a rate of 32 feet per second. At a height of 128 feet the pilot drops a pair of binoculars. After t seconds, the binoculars have height $s(t) = -16t^2 + 32t + 128$ feet from the ground. How fast will they be falling when they hit the ground?

60. Each day, the total output of a coal mine after t hours of operation is approximately $40t + t^2 - \frac{1}{15}t^3$ tons, $0 \le t \le 12$. What is the rate of output (in tons of coal per hour) at $t = 5$ hours?

Determine whether the following limits exist. If so, compute the limit.

61. $\displaystyle\lim_{x\to 2} \dfrac{x^2 - 4}{x - 2}$

62. $\displaystyle\lim_{x\to 3} \dfrac{1}{x^2 - 4x + 3}$

63. $\displaystyle\lim_{x\to 4} \dfrac{x - 4}{x^2 - 8x + 16}$

64. $\displaystyle\lim_{x\to 5} \dfrac{x - 5}{x^2 - 7x + 2}$

Use limits to compute the following derivatives.

65. $f'(5)$, where $f(x) = 1/(2x)$

66. $f'(3)$, where $f(x) = x^2 - 2x + 1$

67. If you deposit $100 into a savings account at the end of each month for 2 years, the balance will be a function $f(r)$ of the interest rate, $r\%$. At 7% interest (compounded monthly), $f(7) = 2568.10$ and $f'(7) = 25.06$. Approximately how much additional money would you earn if the bank paid $7\frac{1}{2}\%$ interest?

Applications of the Derivative

Chapter 2

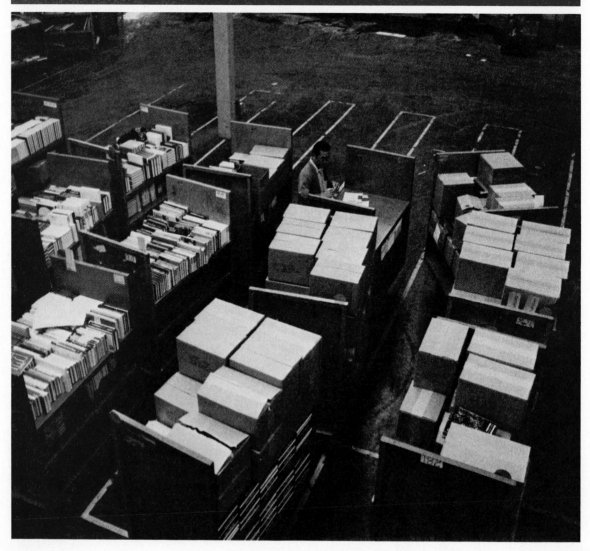

Calculus techniques can be applied to a wide variety of problems in real life. We consider many examples in this chapter. In each case we construct a function as a "mathematical model" of some problem and then analyze the function and its derivatives in order to gain information about the original problem. Our principal method for analyzing a function will be to sketch its graph. For this reason we devote the first part of the chapter to curve sketching.

2.1 Describing Graphs of Functions

The graph of the function in Fig. 1 is either rising or falling, depending on whether we look at it from left to right or from right to left. To avoid confusion, we shall always follow the accepted practice of reading a graph from left to right. We say that a function $f(x)$ is *increasing* at $x = a$ if, in some small region near the point $(a, f(a))$, the graph is rising as we move from left to right. The function is *decreasing* at $x = a$ if, in some small region near the point $(a, f(a))$, the graph is falling. (See Fig. 2.)

An *extreme point* of a function is a point at which its graph changes from increasing to decreasing, or vice versa. We distinguish the two possibilities in an obvious way. A *maximum point* is a point at which the graph changes from increasing to decreasing; a *minimum point* is a point at which the graph changes from decreasing to increasing (Fig. 3).

The *maximum value* (or *absolute maximum value*) of a function is the largest value that the function assumes on its domain. The *minimum value* (or *absolute minimum value*) of a function is the smallest value that the function assumes on

FIGURE 1

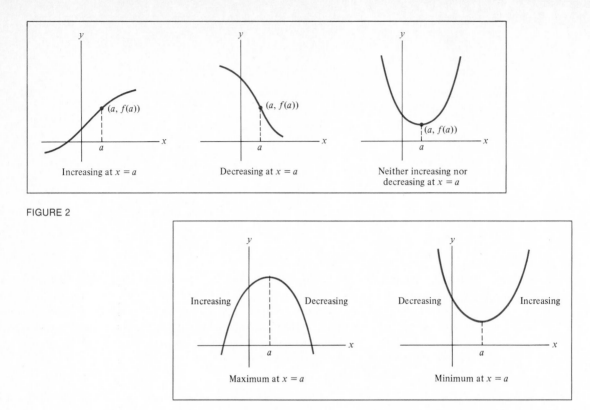

FIGURE 2

FIGURE 3

its domain. Functions may or may not have maximum and or minimum values. (See Fig. 4.) However, it can be shown that a continuous function whose domain is an interval of the form $a \le x \le b$ has both a maximum value and a minimum value.

Maximum values and minimum values of functions usually occur at maximum points and minimum points, as in Fig. 4(a) and (b). However, they can occur at endpoints of the domain, as in Fig. 4(c). If so, we say that the function has an *endpoint extreme value* (or *endpoint extremum*).

EXAMPLE 1 When a drug is injected intramuscularly (into a muscle), the concentration of the drug in the veins has the time-concentration curve shown in Fig. 5. Describe this graph, using the terms introduced above.

Solution Initially (when $t = 0$), there is no drug in the veins. When the drug is injected into the muscle, it begins to diffuse into the bloodstream. The concentration in the veins increases until it reaches its maximum value at $t = 2$. After this time the concentration begins to decrease, as the body's metabolic processes remove the drug from the blood. Eventually the drug concentration decreases to a level so small that, for all practical purposes, it is zero.

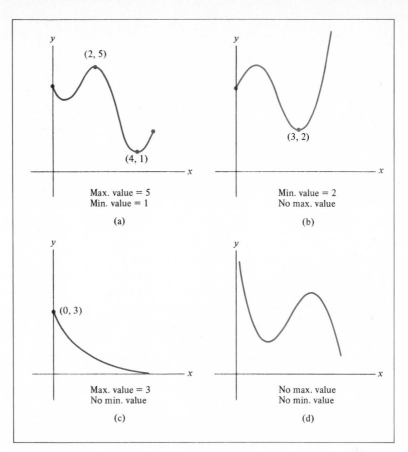

Max. value = 5
Min. value = 1

(a)

Min. value = 2
No max. value

(b)

Max. value = 3
No min. value

(c)

No max. value
No min. value

(d)

FIGURE 4

FIGURE 5

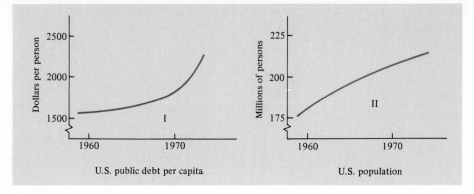

FIGURE 6

The graphs in Fig. 6 are both increasing, but there is a fundamental difference in the way they are increasing. Graph I, which describes the U.S. national debt per person, is steeper for 1970 than for 1960. That is, the *slope* of graph I is *increasing* as we move from left to right. On the other hand, the *slope* of graph II is *decreasing* as we move from left to right. Although the U.S. population is rising each year, the rate of increase was not as great in 1970 as it was in 1960.

The difference between the two graphs in Fig. 6 can also be described in geometric terms: Graph I opens up and lies above its tangent line at each point, whereas graph II opens down and lies below its tangent line at each point (Fig. 7).

FIGURE 7

We say that a function $f(x)$ is *concave up* at $x = a$ if, in some small region near the point $(a, f(a))$, the graph of $f(x)$ lies above its tangent line. Equivalently, $f(x)$ is concave up at $x = a$ if the slope of the graph increases as we move from left to right through $(a, f(a))$. Graph I is an example of a function that is concave up at each point.

Similarly, we say that a function $f(x)$ is *concave down* at $x = a$ if, in some small region near $(a, f(a))$, the graph of $f(x)$ lies below its tangent line. Equivalently, $f(x)$ is concave down at $x = a$ if the slope of the graph decreases as we move from left to right through $(a, f(a))$. Graph II is concave down at each point.

An *inflection point* of a graph is a point at which the graph changes from concave up to concave down, or vice versa. At such a point, the graph crosses its tangent line (Fig. 8).

FIGURE 8

EXAMPLE 2 Use the terms defined earlier to describe the graph shown in Fig. 9.

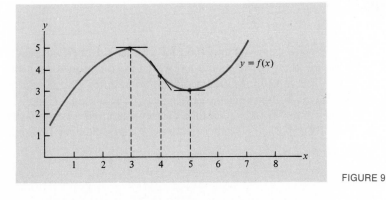

FIGURE 9

Solution (a) For $x < 3$, $f(x)$ is increasing and concave down.
(b) Maximum point at $x = 3$.
(c) For $3 < x < 4$, $f(x)$ is decreasing and concave down.
(d) Inflection point at $x = 4$.
(e) For $4 < x < 5$, $f(x)$ is decreasing and concave up.
(f) Minimum point at $x = 5$.
(g) For $x > 5$, $f(x)$ is increasing and concave up.

FIGURE 10

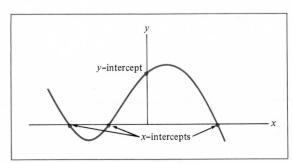

Intercepts, Undefined Points, Asymptotes A point at which a graph crosses the y-axis is called a *y-intercept*, and a point at which it crosses the x-axis is called an *x-intercept*. The x-coordinate of an x-intercept is sometimes called a "zero" of the function, since the function has the value zero there. (See Fig. 10.)

Some functions are not defined for all values of x. For instance, $f(x) = 1/x$ is not defined for $x = 0$, and $f(x) = \sqrt{x}$ is not defined for $x < 0$. (See Fig. 11.) Many functions that arise in

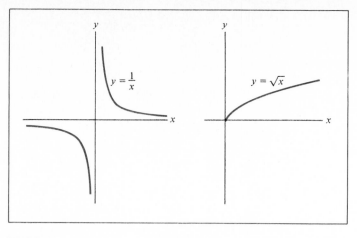

FIGURE 11

applications are defined only for $x \geq 0$. A properly drawn graph should leave no doubt as to the values of x for which the function is defined.

Graphs in applied problems sometimes straighten out and approach some straight line as x gets large (Fig. 12). Such a straight line is called an *asymptote* of the curve. The most common asymptotes are horizontal as in (a) and (b) of Fig. 12. In Example 1 the t-axis is an asymptote of the drug time-concentration curve.

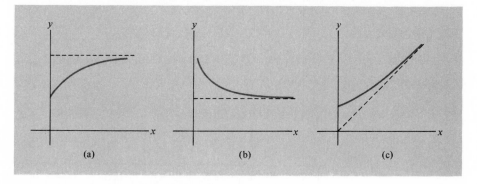

FIGURE 12   Graphs that approach asymptotes as x gets large.

The horizontal asymptotes of a graph may be determined by calculating the limits

$$\lim_{x \to \infty} f(x) \quad \text{and} \quad \lim_{x \to -\infty} f(x).$$

If either limit exists, then the value of the limit determines a horizontal asymptote.

Occasionally, a graph will approach a vertical line as x approaches some fixed value, as in Fig. 13. Such a line is a *vertical asymptote*. Most often, we expect a vertical asymptote at a value x which would result in division by zero in the definition of $f(x)$. For example, $f(x) = 1/(x - 3)$ has a vertical asymptote $x = 3$.

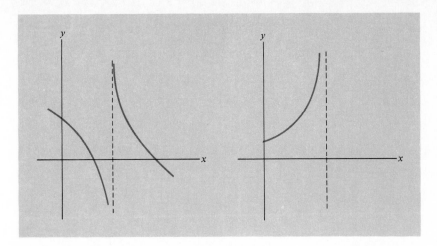

FIGURE 13   Examples of vertical asymptotes.

We now have six categories for describing the graph of a function:

1. Increasing, decreasing, maximum points, minimum points
2. Maximum value, minimum value
3. Concave up, concave down, inflection points
4. x-intercept, y-intercept
5. Undefined points
6. Asymptotes

For us, the first three categories will be the most important. However, the last three categories should not be forgotten.

PRACTICE PROBLEMS 1

1. Does the slope of the curve in Fig. 14 increase or decrease as x increases?

2. At what value of x is the slope of the curve in Fig. 15 minimized?

FIGURE 14   FIGURE 15

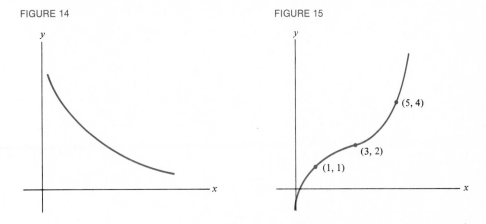

Exercises 1–4 refer to graphs (a)–(f) in Fig. 16.

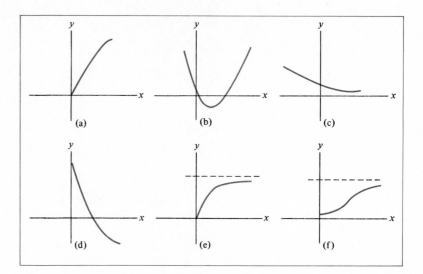

FIGURE 16

1. Which functions are increasing for all x?

2. Which functions are decreasing for all x?

3. Which functions have the property that the slope always increases as x increases?

4. Which functions have the property that the slope always decreases as x increases?

Describe each graph below. Your description should include each of the six categories mentioned above.

5.

6.

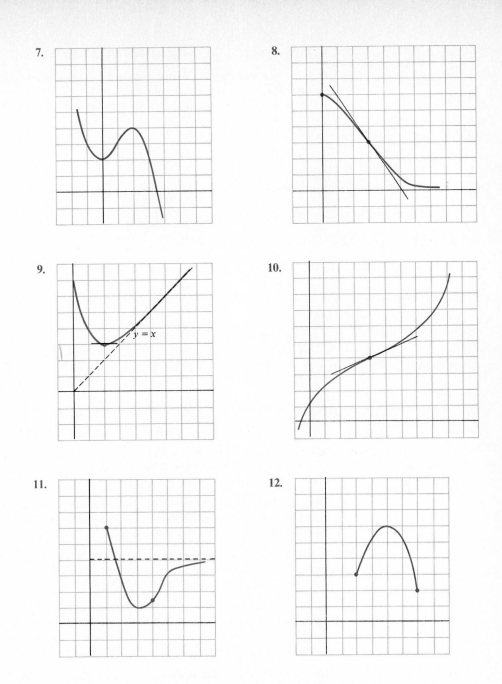

7.

8.

9.

$y = x$

10.

11.

12.

13. Describe the way the *slope* changes as you move along the graph (from left to right) in Exercise 5.

14. Describe the way the *slope* changes on the graph in Exercise 6.

15. Describe the way the *slope* changes on the graph in Exercise 8.

16. Describe the way the *slope* changes on the graph in Exercise 10.

17. Suppose that some organic waste products are dumped into a lake at time $t = 0$, and suppose that the oxygen content of the lake at time t is given by the graph in Fig. 17. Describe the graph in physical terms. Indicate the significance of the inflection point at $t = b$.

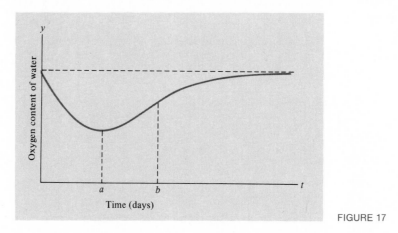

FIGURE 17

18. Let $C(x)$ denote the total cost of manufacturing x units of some product. Then $C(x)$ is an increasing function for all x. For small values of x, the rate of increase of $C(x)$ decreases. (This is because of the savings that are possible with "mass production.") Eventually, however, for large values of x, the cost $C(x)$ increases at an increasing rate. (This happens when production facilities are strained and become less efficient.) Sketch a graph that could represent $C(x)$.

19. The annual world consumption of oil rises each year. Furthermore, the amount of the annual *increase* in oil consumption is also rising each year. Sketch a graph that could represent the annual world consumption of oil.

20. In certain professions the average annual income has been rising at an increasing rate. Let $f(T)$ denote the average annual income at year T for persons in one of these professions and sketch a graph that could represent $f(T)$.

21. Let $s(t)$ be the distance (in feet) traveled by a parachutist after t seconds from the time of opening the chute and suppose that $s(t)$ has the line $y = -15t$ as an asymptote. What does this imply about the velocity of the parachutist? [*Note:* Distance traveled downward is given a negative value.]

22. Let $P(t)$ be the population of a bacteria culture after t days and suppose that $P(t)$ has the line $y = 25{,}000{,}000$ as an asymptote. What does this imply about the size of the population?

23. Consider a smooth curve with no undefined points.

   (a) If it has two maximum points, must it have a minimum point?

   (b) If it has two extreme points, must it have an inflection point?

24. Can a point of a graph be both

    (a) a minimum point and an x-intercept?

    (b) an x-intercept and a y-intercept?

    (c) a maximum point and a minimum point?

    (d) an inflection point and an x-intercept?

In Exercises 25–30, sketch the graph of a function having the given properties.

25.   (i) minimum point at $x = 0$

    (ii) inflection point at $x = 3$

    (iii) no maximum point

26.   (i) maximum points at $x = 1$ and $x = 5$

    (ii) minimum point at $x = 3$

    (iii) inflection points at $x = 2$ and $x = 4$

27.   (i) defined and increasing for all $x \geq 0$

    (ii) inflection point at $x = 5$

    (iii) asymptotic to the line $y = \frac{3}{4}x + 5$

28.   (i) inflection point at $x = 1$

    (ii) increasing for all x

29.   (i) defined for $0 \leq x \leq 10$

    (ii) maximum point at $x = 3$

    (iii) maximum value at $x = 10$

30.   (i) defined for $x \geq 0$

    (ii) minimum value at $x = 0$

    (iii) maximum point at $x = 4$

    (iv) asymptotic to the line $y = (x/2) + 1$

The difference between the pressure inside the lungs and the pressure surrounding the lungs is called the *transmural pressure* (or transmural pressure gradient). The figure below shows for three different persons how the volume of the lungs is related to the transmural pressure, based on static measurements taken while there is no air flowing through the mouth. (The functional reserve capacity mentioned on the vertical axis is the volume of air in the lungs at the end of a normal expiration.) The rate of change of lung volume with respect to transmural pressure is called the *lung compliance.*

31. If the lungs are less flexible than normal, an increase in pressure will cause a smaller change in lung volume than in a normal lung. In this case, is the compliance relatively high or low?

32. Most lung diseases cause a decrease in lung compliance. However, the compliance of a person with emphysema is higher than normal. Which curve (I or II) in the figure could correspond to a person with emphysema?

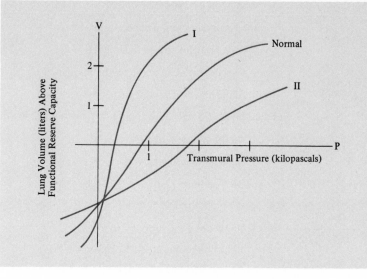

SOLUTIONS TO PRACTICE PROBLEMS 1

1. The curve is concave up, so the slope increases. Even though the curve itself is decreasing, the slope becomes less negative as we move from left to right.

2. At $x = 3$. We have drawn in tangent lines at various points. Note that as we move from left to right, the slopes decrease steadily until the point (3, 2), at which time they start to increase. This is consistent with the fact that the graph is concave down (hence, slopes are decreasing) to the left of (3, 2) and concave up (hence, slopes are increasing) to the right of (3, 2). Extreme values of slopes always occur at inflection points.

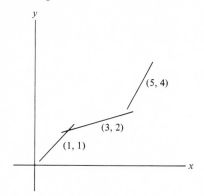

2.2 The First and Second Derivative Rules

We shall now show how properties of the graph of a function $f(x)$ are determined by properties of the derivatives, $f'(x)$ and $f''(x)$. These relationships will provide the key to the curve-sketching and optimization problems discussed in the rest of the chapter.*

---

* Throughout this chapter, we shall assume that we are dealing with functions which are not "too badly behaved." More precisely, it suffices to assume that all of our functions have continuous first and second derivatives in the interval(s) (in x) where we are considering their graphs.

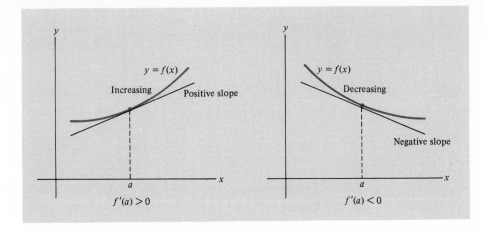

FIGURE 1

We begin with a discussion of the first derivative of a function $f(x)$. Suppose that for some value of x, say $x = a$, the derivative $f'(a)$ is positive. Then the tangent line at $(a, f(a))$ has positive slope and is a rising line (moving from left to right, of course). Since the graph of $f(x)$ near $(a, f(a))$ resembles its tangent line, the function must be increasing at $x = a$. Similarly, when $f'(a) < 0$, the function is decreasing at $x = a$. (See Fig. 1.)

Thus we have the following useful result.

> *First Derivative Rule* If $f'(a) > 0$, then $f(x)$ is increasing at $x = a$. If $f'(a) < 0$, then $f(x)$ is decreasing at $x = a$.

When $f'(a) = 0$, the first derivative rule is not decisive. In this case the function $f(x)$ might be increasing or decreasing, or have an extreme point, at $x = a$.

EXAMPLE 1 Sketch the graph of a function $f(x)$ that has all of the following properties.

(a) $f(3) = 4$

(b) $f'(x) > 0$ for $x < 3$, $f'(3) = 0$ and $f'(x) < 0$ for $x > 3$

Solution The only specific point on the graph is $(3, 4)$ [property (a)]. We plot this point and then use the fact that $f'(3) = 0$ to sketch the tangent line at $x = 3$ (Fig. 2).

From property (b) and the first derivative rule, we know that $f(x)$ must be increasing for x less than 3 and decreasing for x greater than 3. A graph with these properties might look like the curve in Fig. 3.

FIGURE 2 FIGURE 3

The second derivative of a function $f(x)$ gives useful information about the concavity of the graph of $f(x)$. Suppose that $f''(a)$ is negative. Then since $f''(x)$ is the derivative of $f'(x)$, we conclude that $f'(x)$ has a negative derivative at $x = a$. In this case, $f'(x)$ must be a decreasing function at $x = a$; that is, the slope of the graph of $f(x)$ is decreasing as we move from left to right on the graph near $(a, f(a))$. (See Fig. 4.) This means that the graph of $f(x)$ is concave down at $x = a$. A similar

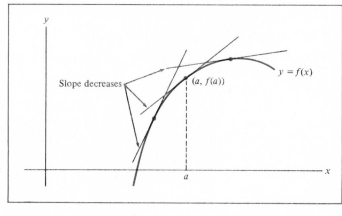

FIGURE 4

analysis shows that if $f''(a)$ is positive, then $f(x)$ is concave up at $x = a$. Thus we have the following rule.

> *Second Derivative Rule* If $f''(a) > 0$, then $f(x)$ is concave up at $x = a$. If $f''(a) < 0$, then $f(x)$ is concave down at $x = a$.

When $f''(a) = 0$, the second derivative rule gives no information. In this case, the function might be concave up, concave down, or neither, at $x = a$.

The chart below shows how a graph may combine the properties of increasing, decreasing, concave up, and concave down.

Condition on the derivatives	Description of $f(x)$ at $x = a$	Graph of $y = f(x)$ near $x = a$
1. $f'(a)$ positive $f''(a)$ positive	$f(x)$ increasing $f(x)$ concave up	
2. $f'(a)$ positive $f''(a)$ negative	$f(x)$ increasing $f(x)$ concave down	
3. $f'(a)$ negative $f''(a)$ positive	$f(x)$ decreasing $f(x)$ concave up	
4. $f'(a)$ negative $f''(a)$ negative	$f(x)$ decreasing $f(x)$ concave down	

EXAMPLE 2 Sketch the graph of a function $f(x)$ with all of the following properties.

(a) $(2, 3)$, $(4, 5)$, and $(6, 7)$ are on the graph

(b) $f'(6) = 0$ and $f'(2) = 0$

(c) $f''(x) > 0$ for $x < 4$, $f''(4) = 0$ and $f''(x) < 0$ for $x > 4$

Solution First we plot the three points from property (a) and then sketch two tangent lines, using the information from property (b). (See Fig. 5.) From condition (c) and the second derivative rule, we know that $f(x)$ is concave up for $x < 4$. In particular, $f(x)$ is concave up at $(2, 3)$. Also, $f(x)$ is concave down for $x > 4$ and, in particular, at $(6, 7)$. Note that $f(x)$ must have an inflection point at $x = 4$ because the concavity changes there. We now sketch small portions of the curve near $(2, 3)$ and $(6, 7)$. (See Fig. 6.) We can now complete the sketch (Fig. 7), taking care to make the curve concave up for $x < 4$ and concave down for $x > 4$.

FIGURE 5

FIGURE 6

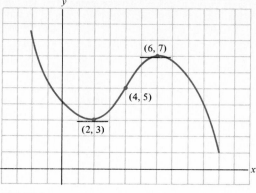

FIGURE 7

PRACTICE PROBLEMS 2

1. Make a good sketch of the function $f(x)$ near the point where $x = 2$, given that $f(2) = 5$, $f'(2) = 1$, and $f''(2) = -3$.

2. The graph of $f(x) = x^3$ is on the right.

    (a) Is the function increasing at $x = 0$?

    (b) Compute $f'(0)$.

    (c) Reconcile your answers to parts (a) and (b) with the first derivative rule.

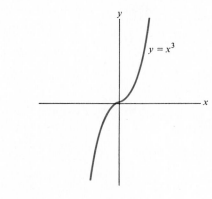

EXERCISES 2

Exercises 1–4 refer to the functions whose graphs are given in Fig. 8.

1. Which functions have a positive first derivative for all x?

2. Which functions have a negative first derivative for all x?

3. Which functions have a positive second derivative for all x?

4. Which functions have a negative second derivative for all x?

In Exercises 5–10, sketch the graph of a function $f(x)$ that has the properties described.

5. $f(2) = 1$; $f'(2) = 0$; concave up for all x.

6. $f(-1) = 0$; $f'(x) < 0$ for $x < -1$, $f'(-1) = 0$ and $f'(x) > 0$ for $x > -1$.

7. $f(3) = 5$; $f'(x) > 0$ for $x < 3$, $f'(3) = 0$ and $f'(x) > 0$ for $x > 3$.

8. $(-2, -1)$ and $(2, 5)$ are on the graph; $f'(-2) = 0$ and $f'(2) = 0$; $f''(x) > 0$ for $x < 0$; $f''(0) = 0$, $f''(x) < 0$ for $x > 0$.

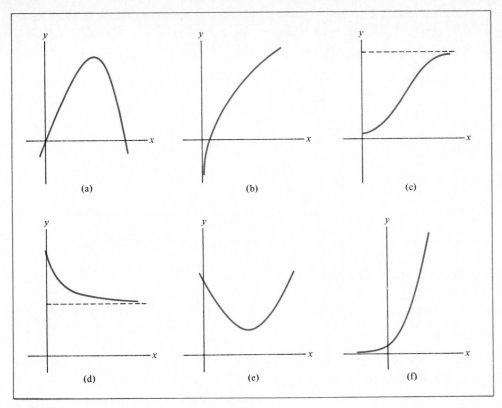

(a) (b) (c)

(d) (e) (f)

FIGURE 8

9. $(0, 6)$, $(2, 3)$, and $(4, 0)$ are on the graph; $f'(0) = 0$ and $f'(4) = 0$; $f''(x) < 0$ for $x < 2$; $f''(2) = 0$, $f''(x) > 0$ for $x > 2$.

10. $f(x)$ defined only for $x \geq 0$; $(0, 0)$ and $(5, 6)$ are on the graph; $f'(x) > 0$ for $x \geq 0$; $f''(x) < 0$ for $x < 5$; $f''(5) = 0$, $f''(x) > 0$ for $x > 5$.

11. By looking at the first derivative, decide which of the curves in Fig. 9 could *not* be the graph of $f(x) = (3x^2 + 1)^4$ for $x \geq 0$.

FIGURE 9

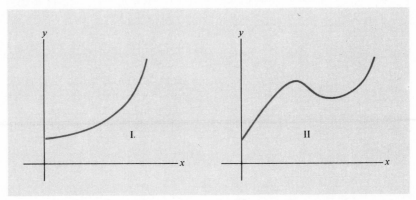

I. II

12. By looking at the first derivative, decide which of the curves in Fig. 9 could *not* be the graph of $f(x) = x^3 - 9x^2 + 24x + 1$ for $x \geq 0$. [*Hint:* Factor the formula for $f'(x)$.]

13. By looking at the second derivative, decide which of the curves in Fig. 10 could be the graph of $f(x) = \sqrt{x}$.

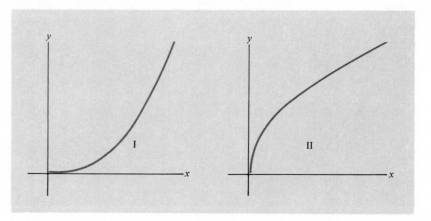

FIGURE 10

14. By looking at the second derivative, decide which of the curves in Fig. 10 could be the graph of $f(x) = x^{5/2}$.

In Exercises 15–20, use the given information to make a good sketch of the function $f(x)$ near $x = 3$.

15. $f(3) = 4, f'(3) = -\frac{1}{2}, f''(3) = 5$

16. $f(3) = -2, f'(3) = 0, f''(3) = 1$

17. $f(3) = 1, f'(3) = 0, f''(3) = 0$, inflection point at $x = 3, f'(x) > 0$ for $x > 3$

18. $f(3) = 4, f'(3) = -\frac{3}{2}, f''(3) = -2$

19. $f(3) = -2, f'(3) = 2, f''(3) = 3$

20. $f(3) = 3, f'(3) = 1, f''(3) = 0$, inflection point at $x = 3, f''(x) < 0$ for $x > 3$

SOLUTIONS TO PRACTICE PROBLEMS 2

1. Since $f(2) = 5$, the point $(2, 5)$ is on the graph [Fig. 11(a)]. Since $f'(2) = 1$, the tangent line at the point $(2, 5)$ has slope 1. Draw in the tangent line [Fig. 11(b)]. Near the point $(2, 5)$ the graph looks approximately like the tangent line. Since $f''(2) = -3$, a negative number, the graph is concave down at the point $(2, 5)$. Now we are ready to sketch the graph [Fig. 11(c)].

2. (a) Yes. The graph is steadily increasing as we pass through the point $(0, 0)$.

   (b) Since $f'(x) = 3x^2$, $f'(0) = 3 \cdot 0^2 = 0$.

FIGURE 11

(c) There is no contradiction here. The first derivative rule says that if the derivative is positive, the function is increasing. However, it does not say that this is the only condition under which a function is increasing. As we have just seen, sometimes we can have the first derivative zero and the function still increasing.

2.3 Curve Sketching (Introduction)

In this section and the next we develop our ability to sketch the graphs of functions. There are two important reasons for doing so. First, a geometric "picture" of a function is often easier to comprehend than its abstract formula. Second, the material in this section will provide a foundation for the applications in Sections 5 to 7.

A "sketch" of the graph of a function $f(x)$ should convey the general shape of the graph—it should show where $f(x)$ is increasing and decreasing and it should indicate, insofar as possible, where $f(x)$ is concave up and concave down. In addition, one or more key points should be accurately located on the graph. These points usually include extreme points, inflection points, and x- and y-intercepts. Other features of a graph may be important, too, but we shall discuss them as they arise in examples and applications.

Our general approach to curve sketching will involve four main steps:

1. Starting with $f(x)$, we compute $f'(x)$ and $f''(x)$.
2. Next, we locate all maximum and minimum points and make a partial sketch.
3. We study the concavity of $f(x)$ and locate all inflection points.
4. We consider other properties of the graph, such as the intercepts, and complete the sketch.

The first step was the main subject of the preceding chapter. We discuss the second and third steps in this section, and then present several completely worked examples that combine all four steps in the next.

Locating Extreme Points   The tangent line at a maximum or a minimum point of a function $f(x)$ has zero slope; that is, the derivative is zero there. Thus we may state the following useful rule.

> Look for possible extreme points of $f(x)$ by setting $f'(x) = 0$ and solving for x. (1)

Suppose that $f'(a) = 0$. Then $x = a$ is a candidate for an extreme point of $f(x)$. There are several ways to determine if $f(x)$ has a maximum or a minimum (or neither) at $x = a$. The method that works in most cases is described in our first two examples.

EXAMPLE 1   The graph of the quadratic function $f(x) = \frac{1}{4}x^2 - x + 2$ is a parabola and so has one extreme point. Find it and sketch the graph.

Solution   We begin by computing the first and second derivatives of $f(x)$.

$$f(x) = \tfrac{1}{4}x^2 - x + 2$$
$$f'(x) = \tfrac{1}{2}x - 1$$
$$f''(x) = \tfrac{1}{2}.$$

Setting $f'(x) = 0$, we have $\frac{1}{2}x - 1 = 0$, so that $x = 2$. Thus $f'(2) = 0$. Geometrically, this means that the graph of $f(x)$ will have a horizontal tangent line at the point where $x = 2$. To plot this point, we substitute the value 2 for x in the original expression for $f(x)$.

$$f(2) = \tfrac{1}{4}(2)^2 - (2) + 2 = 1.$$

Figure 1 shows the point $(2, 1)$ together with the horizontal tangent line. Is $(2, 1)$ an extreme point? In order to decide, we look at $f''(x)$. Since $f''(2) = \frac{1}{2}$, which is

FIGURE 1   FIGURE 2

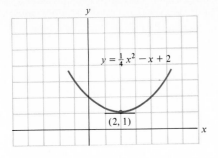

FIGURE 3

positive, the graph of $f(x)$ is concave up at $x = 2$. So a partial sketch of the graph near $(2, 1)$ should look something like Fig. 2.

We see that $(2, 1)$ is a minimum point. In fact, it is the only extreme point, for there is no other place where the tangent line is horizontal. Since the graph has no other "turning points," it must be decreasing before it gets to $(2, 1)$ and then increasing to the right of $(2, 1)$. Note that since $f''(x)$ is positive (and equal to $\frac{1}{2}$) for all x, the graph is concave upward at each point. A completed sketch is given in Fig. 3.

EXAMPLE 2 Locate all possible extreme points on the graph of the function $f(x) = x^3 - 3x^2 + 5$. Check the concavity at these points and use this information to sketch the graph of $f(x)$.

Solution We have

$$f(x) = x^3 - 3x^2 + 5$$
$$f'(x) = 3x^2 - 6x$$
$$f''(x) = 6x - 6.$$

The easiest way to find those values of x for which $f'(x)$ is zero is to factor the expression for $f'(x)$:

$$3x^2 - 6x = 3x(x - 2).$$

From this factorization it is clear that $f'(x)$ will be zero if and only if $x = 0$ or $x = 2$. In other words, the graph will have horizontal tangent lines when $x = 0$ and $x = 2$, and nowhere else.

To plot the points on the graph where $x = 0$ and $x = 2$, we substitute these values back into the original expression for $f(x)$. That is, we compute

FIGURE 4

$$f(0) = (0)^3 - 3(0)^2 + 5 = 5$$
$$f(2) = (2)^3 - 3(2)^2 + 5 = 1.$$

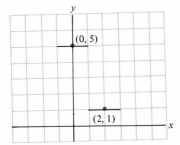

Figure 4 shows the points $(0, 5)$ and $(2, 1)$, along with the corresponding tangent lines.

Next, we check the concavity of the graph at these points by evaluating $f''(x)$ at $x = 0$ and $x = 2$:

$$f''(0) = 6(0) - 6 = -6$$
$$f''(2) = 6(2) - 6 = +6.$$

Since $f''(0)$ is negative, the graph is concave down at $x = 0$; since $f''(2)$ is positive, the graph is concave up at $x = 2$. A partial sketch of the graph is given in Fig. 5.

It is clear from Fig. 5 that $(0, 5)$ is a maximum point and $(2, 1)$ is a minimum point. Since they are the only turning points, the graph must be increasing before it gets to $(0, 5)$, decreasing from $(0, 5)$ to $(2, 1)$, and then increasing again to the right of $(2, 1)$. A sketch incorporating these properties appears in Fig. 6.

The facts that we used to sketch Fig. 6 could equally well be used to produce the graph in Fig. 7. Which graph really corresponds to $f(x) = x^3 - 3x^2 + 5$? The answer will be clear when we find the inflection points on the graph of $f(x)$.

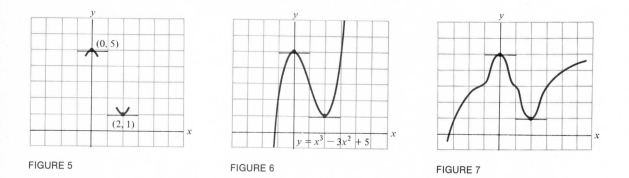

FIGURE 5

FIGURE 6

FIGURE 7

Locating Inflection Points   An inflection point of a function $f(x)$ can occur only at a value of x for which $f''(x)$ is zero, because the curve is concave up where $f''(x)$ is positive and concave down where $f''(x)$ is negative. Thus we have the following test.

> Look for possible inflection points by setting $f''(x) = 0$ and solving for x.   (2)

Once we have a value of x where the second derivative is zero, say at $x = b$, we must check the concavity of $f(x)$ at nearby points to see if the concavity really changes at $x = b$.

EXAMPLE 3   Find the inflection points of the function $f(x) = x^3 - 3x^2 + 5$ and explain why the graph in Fig. 6 has the correct shape.

Solution   From Example 2 we have $f''(x) = 6x - 6 = 6(x - 1)$. Clearly, $f''(x) = 0$ if and only if $x = 1$. We will want to plot the corresponding point on the graph, so we

$(0, 5)$

$(1, 3)$

$(2, 1)$

$y = x^3 - 3x^2 + 5$

FIGURE 8

compute

$$f(1) = (1)^3 - 3(1)^2 + 5 = 3.$$

Therefore, the only possible inflection point is $(1, 3)$.

Now look back at Fig. 5, where we indicated the concavity of the graph at the extreme points. Since $f(x)$ is concave down at $(0, 5)$ and concave up at $(2, 1)$, the concavity must reverse somewhere between these points. Hence $(1, 3)$ must be an inflection point. Furthermore, since the concavity of $f(x)$ reverses nowhere else, the concavity at all points to the left of $(1, 3)$ must be the same (i.e., concave down). Similarly, the concavity at all points to the right of $(1, 3)$ must be the same (i.e., concave up). Thus the graph in Fig. 6 has the correct shape. The graph in Fig. 7 has too many "wiggles," caused by frequent changes in concavity; that is, there are too many inflection points. A correct sketch showing the one inflection point at $(1, 3)$ is given in Fig. 8.

EXAMPLE 4   Sketch the graph of $y = -\frac{1}{3}x^3 + 3x^2 - 5x$.

Solution   Let

$$f(x) = -\tfrac{1}{3}x^3 + 3x^2 - 5x.$$

Then

$$f'(x) = -x^2 + 6x - 5$$
$$f''(x) = -2x + 6.$$

We set $f'(x) = 0$ and solve for x:

$$-(x^2 - 6x + 5) = 0$$
$$-(x - 1)(x - 5) = 0$$
$$x = 1 \quad \text{or} \quad x = 5.$$

Substituting these values of x back into $f(x)$, we find that

$$f(1) = -\tfrac{1}{3}(1)^3 + 3(1)^2 - 5(1) = -\tfrac{7}{3}$$
$$f(5) = -\tfrac{1}{3}(5)^3 + 3(5)^2 - 5(5) = \tfrac{25}{3}.$$

The information we have so far is given in Fig. 9(a). The sketch in Fig. 9(b) is obtained by computing

$$f''(1) = -2(1) + 6 = 4$$
$$f''(5) = -2(5) + 6 = -4.$$

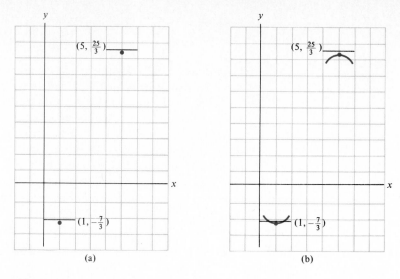

(a)   (b)

FIGURE 9

The curve is concave up at $x = 1$ because $f''(1)$ is positive, and the curve is concave down at $x = 5$ because $f''(5)$ is negative.

Since the concavity reverses somewhere between $x = 0$ and $x = 5$, there must be at least one inflection point. If we set $f''(x) = 0$, we find that

$$-2x + 6 = 0$$

$$x = 3.$$

FIGURE 10

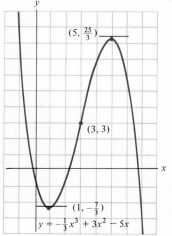

So the inflection point must occur at $x = 3$. In order to plot the inflection point, we compute

$$f(3) = -\tfrac{1}{3}(3)^3 + 3(3)^2 - 5(3) = 3.$$

The final sketch of the graph is given in Fig. 10.

The argument in Example 4 that there must be an inflection point because concavity reverses is valid whenever $f(x)$ is a polynomial. However, it does not always apply to a function whose graph has a break in it. For example, the function $f(x) = 1/x$ is concave down at $x = -1$ and concave up at $x = 1$, but there is no inflection point in between.

A summary of curve-sketching techniques appears at the end of the next section. You may find steps 1, 2, and 3 helpful when working the exercises.

PRACTICE PROBLEMS 3

1. Which of the curves below could possibly be the graph of a function of the form $f(x) = ax^2 + bx + c$, where $a \neq 0$?

(a) (b) (c) (d)

2. Which of the curves below could possibly be the graph of a function of the form $f(x) = ax^3 + bx^2 + cx + d$, where $a \neq 0$?

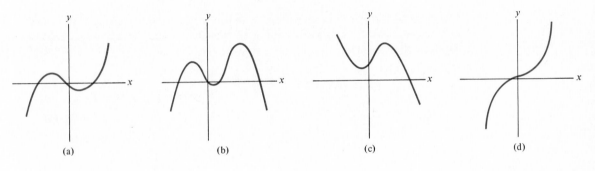

(a) (b) (c) (d)

EXERCISES 3

Each of the graphs of the functions in Exercises 1–8 has one extreme point. Plot this point and check the concavity there. Using only this information, sketch the graph. [As you work the problems, observe that if $f(x) = ax^2 + bx + c$, then $f(x)$ has a minimum point when $a > 0$ and a maximum point when $a < 0$.]

1. $f(x) = 2x^2 - 8$

2. $f(x) = 3x^2 + 6x - 5$

3. $f(x) = \frac{1}{2}x^2 + x - 4$

4. $f(x) = -\frac{1}{2}x^2 + x - 4$

5. $f(x) = 1 + 6x - x^2$

6. $f(x) = 1 + x + x^2$

7. $f(x) = -x^2 - 8x - 10$

8. $f(x) = -3x^2 + 18x - 20$

Each of the graphs of the functions in Exercises 9–16 has one maximum and one minimum point. Plot these two points and check the concavity there. Using only this information, sketch the graph.

9. $f(x) = x^3 + 6x^2 + 9x$

10. $f(x) = \frac{1}{9}x^3 - x^2$

11. $f(x) = x^3 - 12x$

12. $f(x) = -\frac{1}{3}x^3 + 9x - 2$

13. $f(x) = -\frac{1}{9}x^3 + x^2 + 9x$

14. $f(x) = 2x^3 - 15x^2 + 36x - 24$

15. $f(x) = -\frac{1}{3}x^3 + 2x^2 - 12$

16. $f(x) = \frac{1}{3}x^3 + 2x^2 - 5x + \frac{8}{3}$

Sketch the following curves, indicating all extreme points and inflection points.

17. $y = x^3 - 3x + 2$

18. $y = x^3 - 6x^2 + 9x + 3$

19. $y = 1 + 3x^2 - x^3$

20. $y = -x^3 + 12x - 4$

21. $y = \frac{1}{3}x^3 - x^2 - 3x + 5$

22. $y = x^3 + \frac{3}{2}x^2 - 6x + 4$

23. $y = 2x^3 - 3x^2 - 36x + 20$

24. $y = 11 + 9x - 3x^2 - x^3$

25. Let a, b, c be fixed numbers with $a \neq 0$ and let $f(x) = ax^2 + bx + c$. Is it possible for the graph of $f(x)$ to have an inflection point? Explain your answer.

26. Let a, b, c, d be fixed numbers with $a \neq 0$ and let $f(x) = ax^3 + bx^2 + cx + d$. Is it possible for the graph of $f(x)$ to have more than one inflection point? Explain your answer.

The graph of each function below has one extreme point. Find it (giving both x- and y-coordinates) and determine if it is a maximum or a minimum point. Do not include a sketch of the graph of the function.

27. $f(x) = \frac{1}{4}x^2 - 2x + 7$

28. $f(x) = 5 - 12x - 2x^2$

29. $g(x) = 3 + 4x - 2x^2$

30. $g(x) = x^2 + 10x + 10$

31. $f(x) = 5x^2 + x - 3$

32. $f(x) = 30x^2 - 1800x + 29{,}000$

SOLUTIONS TO PRACTICE PROBLEMS 3

1. Answer: (a) and (d). Curve (b) has the shape of a parabola, but it is not the graph of any function, since vertical lines cross it twice. Curve (c) has two extreme points, but the derivative of $f(x)$ is a linear function which could not be zero for two different values of x.

2. Answer: (a), (c), (d). Curve (b) has three extreme points, but the derivative of $f(x)$ is a quadratic function which could not be zero for three different values of x.

2.4 Curve Sketching (Conclusion)

In Section 3 we discussed the main techniques for curve sketching. Here we add a few finishing touches and examine some slightly more complicated curves.

The more points we plot on a graph, the more accurate the graph becomes. This statement is true even for the simple quadratic and cubic curves in Section 3. Of course, the most important points on a curve are the extreme points and the inflection points. In addition, the x- and y-intercepts often have some intrinsic interest in an applied problem. The y-intercept is $(0, f(0))$. To find the x-intercepts on the graph of $f(x)$, we must find those values of x for which $f(x) = 0$. Since this can be a difficult (or impossible) problem, we shall find x-intercepts only when they are easy to find or when a problem specifically requires us to find them.

When $f(x)$ is a quadratic function, as in Example 1, we can easily compute the x-intercepts (if they exist) either by factoring the expression for $f(x)$ or by using the quadratic formula.

EXAMPLE 1   Sketch the graph of $y = \frac{1}{2}x^2 - 4x + 7$.

Solution   Let

$$f(x) = \tfrac{1}{2}x^2 - 4x + 7.$$

Then

$$f'(x) = x - 4$$
$$f''(x) = 1.$$

Since $f'(x) = 0$ only when $x = 4$ and since $f''(4)$ is positive, $f(x)$ must have a minimum point at $x = 4$. The minimum point is $(4, f(4)) = (4, -1)$.

The y-intercept is $(0, f(0)) = (0, 7)$. To find the x-intercepts, we set $f(x) = 0$ and solve for x:

$$\tfrac{1}{2}x^2 - 4x + 7 = 0.$$

The expression for $f(x)$ is not easily factored, so we use the quadratic formula to solve the equation.

$$x = \frac{-(-4) \pm \sqrt{(-4)^2 - 4(\frac{1}{2})(7)}}{2(\frac{1}{2})} = 4 \pm \sqrt{2}.$$

The x-intercepts are $(4 - \sqrt{2}, 0)$ and $(4 + \sqrt{2}, 0)$. To plot these points we use the approximation: $\sqrt{2} \approx 1.4$. (See Fig. 1.)

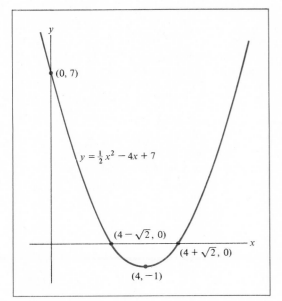

FIGURE 1

EXAMPLE 2   Sketch the graph of

$$f(x) = \tfrac{1}{6}x^3 - \tfrac{3}{2}x^2 + 5x + 1.$$

Solution
$$f(x) = \tfrac{1}{6}x^3 - \tfrac{3}{2}x^2 + 5x + 1$$
$$f'(x) = \tfrac{1}{2}x^2 - 3x + 5$$
$$f''(x) = x - 3.$$

Let us set $f'(x) = 0$ and try to solve for x:

$$\tfrac{1}{2}x^2 - 3x + 5 = 0. \tag{1}$$

If we apply the quadratic formula with $a = \tfrac{1}{2}$, $b = -3$, and $c = 5$, we see that $b^2 - 4ac$ is negative, and so there is no solution to (1). In other words, $f'(x)$ is never zero. Thus the graph cannot have extreme points. If we evaluate $f'(x)$ at some x, say $x = 0$, we see that the first derivative is positive, and so $f(x)$ is increasing there. Since the graph of $f(x)$ is a smooth curve with no extreme points and no breaks, $f(x)$ must be increasing for all x. [For if $f(x)$ is increasing at $x = a$ and decreasing at $x = b$, then $f(x)$ has an extreme point between a and b.]

Now let us check the concavity.

$y = \tfrac{1}{6}x^3 - \tfrac{3}{2}x^2 + 5x + 1$

(3, 7)

(0, 1)

	$f''(x) = x - 3$	Graph of $f(x)$
$x < 3$	Negative	Concave down
$x = 3$	Zero	Concavity reverses
$x > 3$	Positive	Concave up

The inflection point is $(3, f(3)) = (3, 7)$. The y-intercept is $(0, f(0)) = (0, 1)$. We omit the x-intercept because it is difficult to solve the cubic equation $\tfrac{1}{6}x^3 - \tfrac{3}{2}x^2 + 5x + 1 = 0$.

The quality of our sketch of the curve will be improved if we first sketch the tangent line at the inflection point. To do this, we need to know the slope of the graph at (3, 7):

$$f'(3) = \tfrac{1}{2}(3)^2 - 3(3) + 5 = \tfrac{1}{2}.$$

We draw a line through (3, 7) with slope $\tfrac{1}{2}$ and then complete the sketch as shown in Fig. 2.

FIGURE 2

The methods used so far will work for most of the functions we shall study. Occasionally, however, $f'(x)$ and $f''(x)$ are both zero at some value of x, say $x = a$, and we cannot tell from the second derivative if the function has a minimum or a maximum at $x = a$. We show how to handle this unusual situation in the next example.

EXAMPLE 3 Sketch the graph of $f(x) = (x - 2)^4 - 1$.

Solution

$$f(x) = (x - 2)^4 - 1$$
$$f'(x) = 4(x - 2)^3$$
$$f''(x) = 12(x - 2)^2.$$

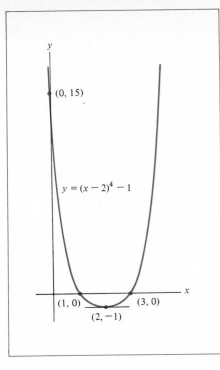

(0, 15)

$y = (x - 2)^4 - 1$

(1, 0)   (3, 0)

(2, −1)

FIGURE 3

Clearly, $f'(x) = 0$ only if $x = 2$. So the curve has a horizontal tangent at $(2, f(2)) = (2, -1)$. Since $f''(2) = 0$, the second derivative rule tells us nothing about the concavity at $x = 2$. Let us look at $f''(x)$ for x near 2. Since $f''(x) = 12(x - 2)^2 > 0$ for $x \neq 2$, the graph is concave up at all points close to $(2, -1)$. In other words, the graph lies above its tangent line at each point near $(2, -1)$. Consequently, the graph cannot drop below its tangent line at $(2, -1)$, and it is concave up at $(2, -1)$.

The y-intercept is $(0, f(0)) = (0, 15)$. To find the x-intercepts, we set $f(x) = 0$ and solve for x:

$$(x - 2)^4 - 1 = 0$$

$$(x - 2)^4 = 1$$

$$x - 2 = 1 \quad \text{or} \quad x - 2 = -1$$

$$x = 3 \quad \text{or} \quad x = 1.$$

(See Fig. 3.)

A Graph with Asymptotes   Graphs similar to the one in the next example will arise in several applications later in this chapter.

EXAMPLE 4   Sketch the graph of $f(x) = x + (1/x)$, for $x > 0$.

Solution

$$f(x) = x + \frac{1}{x}$$

$$f'(x) = 1 - \frac{1}{x^2}$$

$$f''(x) = \frac{2}{x^3}.$$

We set $f'(x) = 0$ and solve for x:

$$1 - \frac{1}{x^2} = 0$$

$$1 = \frac{1}{x^2}$$

$$x^2 = 1$$

$$x = 1.$$

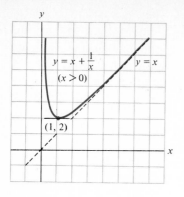

FIGURE 4

(We exclude the case $x = -1$ because we are only considering positive values of x.) The graph has a horizontal tangent at $(1, f(1)) = (1, 2)$. Now, $f''(1) = 2 > 0$, and so the graph is concave up at $x = 1$ and $(1, 2)$ is a minimum point. In fact, $f''(x) = (2/x^3) > 0$ for all positive x, and therefore the graph is concave up at all points.

Before sketching the graph, notice that as x approaches zero (a point at which $f(x)$ is not defined), the term $1/x$ in the formula for $f(x)$ becomes arbitrarily large. Thus $f(x)$ has the y-axis as an asymptote. For large values of x, $f(x) = x + (1/x)$ is only slightly larger than x; that is, the graph of $f(x)$ is slightly above the graph of $y = x$. As x increases, the graph of $f(x)$ has the line $y = x$ as an asymptote. (See Fig. 4.)

Summary of Curve-Sketching Techniques

1. Compute $f'(x)$ and $f''(x)$.
2. Find all maximum and minimum points.
   (a) Set $f'(x) = 0$ and solve for x. Suppose that $f'(a) = 0$.
     (i) If $f''(a) > 0$, the curve is concave up at $x = a$, so $f(x)$ has a minimum point at $x = a$. The minimum point is $(a, f(a))$.
     (ii) If $f''(a) < 0$, the curve is concave down at $x = a$, so $f(x)$ has a maximum point at $x = a$. The maximum point is $(a, f(a))$.
     (iii) If $f''(a) = 0$, examine $f'(x)$ to the left and right of $x = a$ in order to determine if the function changes from increasing to decreasing, or vice versa.
   (b) Make a partial sketch of the graph near each point where $f(x)$ has a horizontal tangent line.
3. Determine the concavity of $f(x)$.
   (a) Set $f''(x) = 0$ and solve for x. Suppose that $f''(b) = 0$. Next, test the concavity for x near to b. If the concavity changes at $x = b$, then $(b, f(b))$ is an inflection point. Otherwise, the concavity at $x = b$ is the same as at other nearby points.
4. Consider other properties of the function and complete the sketch.
   (a) If $f(x)$ is defined at $x = 0$, the y-intercept is $(0, f(0))$.
   (b) Does the partial sketch suggest that there are x-intercepts? If so, they are found by setting $f(x) = 0$ and solving for x. (Solve only in easy cases or when a problem essentially requires you to calculate the x-intercepts.)
   (c) Observe where $f(x)$ is defined. Sometimes the function is given only for restricted values of x. Sometimes the formula for $f(x)$ is meaningless for certain values of x.
   (d) Look for possible asymptotes.
     (i) Examine the formula for $f(x)$. If some terms become insignificant as x gets large and if the rest of the formula gives the equation of a straight line, then that straight line is an asymptote.
     (ii) Suppose that there is some point a such that $f(x)$ is defined for x near a but not at a (e.g., $1/x$ at $x = 0$). If $f(x)$ gets arbitrarily large (in the positive or negative sense) as x approaches a, then the vertical line $x = a$ is an asymptote for the graph.
   (e) Complete the sketch.

PRACTICE PROBLEMS 4

1. Determine whether each of the following functions has an asymptote as x gets large. If so, give the equation of the straight line which is the asymptote.

(a) $f(x) = (3/x) - 2x + 1$

(b) $f(x) = \sqrt{x} + x$

(c) $f(x) = 7 + 1/(2x + 6)$

2. Sketch the graph of $f(x) = 1/(2 - x) + 5$, for $x > 2$.

EXERCISES 4

Find the x-intercepts of the following curves.

1. $y = x^2 - 3x + 1$ 2. $y = x^2 + 5x + 5$

3. $y = 2x^2 + 5x + 2$ 4. $y = 4 - 2x - x^2$

5. $y = 4x - 4x^2 - 1$ 6. $y = 3x^2 + 7x + 2$

7. Show that the function $f(x) = \frac{1}{3}x^3 - 2x^2 + 5x$ has no extreme points.

8. Show that the function $f(x) = 5 - 11x + 6x^2 - \frac{4}{3}x^3$ is always decreasing.

Sketch the graphs of the following functions.

9. $f(x) = x^3 - 6x^2 + 12x - 6$ 10. $f(x) = -x^3$

11. $f(x) = x^3 + 3x + 1$ 12. $f(x) = 4 - x - x^3$

13. $f(x) = 5 - 13x + 6x^2 - x^3$ 14. $f(x) = 2x^3 + x - 2$

15. $f(x) = \frac{4}{3}x^3 - 2x^2 + x$ 16. $f(x) = 1 - 4x - 3x^2 - x^3$

17. $f(x) = 1 - 3x + 3x^2 - x^3$ 18. $f(x) = x^3 - 6x^2 + 12x - 5$

19. $f(x) = x^4 - 6x^2$ 20. $f(x) = 1 + 6x^2 - 3x^4$

21. $f(x) = (x - 3)^4$ 22. $f(x) = (x + 2)^4 - 1$

Sketch the graphs of the following functions.

23. $y = \dfrac{1}{x} + 3$ 24. $y = \dfrac{1}{x - 1}$

25. $y = \dfrac{1}{x} + \frac{1}{4}x$ 26. $y = \dfrac{1}{x} + 9x$

27. $y = \dfrac{9}{x} + x + 1 \quad (x > 0)$ 28. $y = \dfrac{12}{x} + 3x + 1 \quad (x > 0)$

29. $y = \dfrac{360}{x} + 10x \quad (x > 0)$

30. $y = \dfrac{40{,}000}{x} + 2500x + 320{,}000 \quad (x > 0)$

31. $y = \dfrac{1}{x - 2} + \dfrac{x}{4} \quad (x > 2)$

32. $y = \dfrac{4}{2x - 5} + 2x \quad (x > \tfrac{5}{2})$

33. $y = 6\sqrt{x} - x \quad (x > 0)$

34. $y = \dfrac{1}{x} + \dfrac{1}{4 - x} \quad (0 < x < 4)$

SOLUTIONS TO PRACTICE PROBLEMS 4

1. Functions with asymptotes as x gets large have the form $f(x) = g(x) + mx + b$, where $g(x)$ approaches zero as x gets large. The function $g(x)$ often looks like c/x or $c/(ax + d)$. The asymptote will be the straight line $y = mx + b$.

   (a) Here $g(x)$ is $3/x$ and the asymptote is $y = -2x + 1$.

   (b) This function has no asymptote as x gets large. Of course, it can be written as $g(x) + mx + b$, where $m = 1$ and $b = 0$. However, $g(x) = \sqrt{x}$ does not approach zero as x gets large.

   (c) Here $g(x)$ is $1/(2x + 6)$ and the asymptote is the line $y = 7$.

2.
$$f(x) = \frac{1}{2 - x} + 5 = (2 - x)^{-1} + 5, \quad x > 2$$

$$f'(x) = (2 - x)^{-2} = \frac{1}{(2 - x)^2}$$

$$f''(x) = 2(2 - x)^{-3} = \frac{2}{(2 - x)^3}$$

The first derivative, $1/(2 - x)^2$ is never zero, so there are no extreme points. The second derivative, $2/(2 - x)^3$, is also never zero, so there are no inflection points.

   Just to have a point of reference, let us find some point on the graph. For instance, let $x = 3$; then $f(3) = 4$. So the point $(3, 4)$ is on the graph. [See Fig. 5(a).] The first derivative is always positive. Thus the graph is always increasing. The second derivative

FIGURE 5

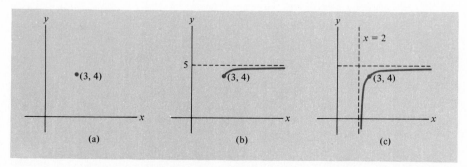

(a)   (b)   (c)

is always negative, since $x > 2$ implies that $2 - x$, and hence $(2 - x)^3$, is negative. Therefore, the graph is concave down. Now for the asymptotes. As x gets large, the term $1/(2 - x)$ approaches zero, so the graph of $f(x)$ has the straight line $y = 5$ as an asymptote [See Fig. 5(b).]

The function also has a vertical asymptote at $x = 2$. For values of x just to the right of 2, $1/(2 - x)$ is negative and very large in magnitude. [See Fig. 5(c).]

2.5 Optimization Problems

One of the most important applications of the derivative concept is to "optimization" problems, in which some quantity must be maximized or minimized. Examples of such problems abound in many areas of life. An airline must decide how many daily flights to schedule between two cities in order to maximize its profits. A doctor wants to find the minimum amount of a drug that will produce a desired response in one of her patients. A manufacturer needs to determine how often to replace certain equipment in order to minimize maintenance and replacement costs.

Our purpose in this section is to illustrate how calculus can be used to solve optimization problems. In each example we will find or construct a function that provides a "mathematical model" for the problem. Then, by sketching the graph of this function, we will be able to determine the answer to the original optimization problem by locating the maximum value or the minimum value of the function.

The first two examples are quite simple because the functions to be studied are explicitly given.

EXAMPLE 1 Find the minimum value of the function $f(x) = 2x^3 - 15x^2 + 24x + 19$ for $x \geq 0$.

Solution Using the curve-sketching techniques from Section 3, we obtain the graph in Fig. 1. The lowest point on the graph is the minimum point (4, 3). The minimum *value* of the function $f(x)$ is the y-coordinate of this point—namely, 3.

EXAMPLE 2 Suppose that a ball is thrown straight up into the air and its height after t seconds is $4 + 48t - 16t^2$ feet. Determine how long it will take for the ball to reach its maximum height and determine the maximum height.

Solution Consider the function $f(t) = 4 + 48t - 16t^2$. For each value of t, $f(t)$ is the height of the ball at time t. We want to find the value of t for which $f(t)$ is the greatest. Using the techniques of Section 3, we sketch the graph of $f(t)$. (See Fig. 2.) Note that we may neglect the portions of the graph corresponding to points for which either $t < 0$ or $f(t) < 0$. [A negative value of $f(t)$ would correspond to the ball being underneath the ground.] We see that $f(t)$ is greatest when $t = \frac{3}{2}$. At this

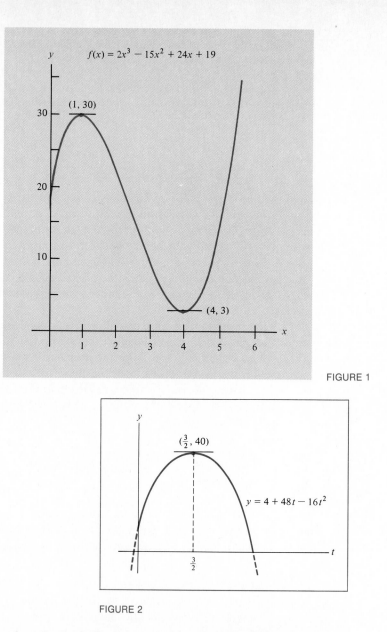

$f(x) = 2x^3 - 15x^2 + 24x + 19$

(1, 30)

30

20

10

(4, 3)

FIGURE 1

$(\frac{3}{2}, 40)$

$y = 4 + 48t - 16t^2$

$\frac{3}{2}$

FIGURE 2

value of t, the ball attains a height of 40 feet. [Note that the curve in Fig. 2 is the graph of $f(t)$, *not* a picture of the physical path of the ball.]

Answer   The ball reaches its maximum height of 40 feet in 1.5 seconds.

EXAMPLE 3   A man wants to plant a rectangular garden along one side of his house, with a picket fence on the other three sides of the garden. Find the dimensions of the largest garden that can be enclosed using 40 feet of fencing.

Solution   The first step is to make a simple diagram and assign letters to the quantities that may vary. Let us denote the dimensions of the rectangular garden by w and x (Fig. 3). The fencing on three sides must total 40 running feet; that is,

$$2x + w = 40. \tag{1}$$

House

Garden

x

w

FIGURE 3

The phrase "largest garden" above indicates that we must maximize the area, A, of the garden. In terms of the variables w and x,

$$A = wx. \tag{2}$$

We now solve equation (1) for w in terms of x:

$$w = 40 - 2x. \tag{3}$$

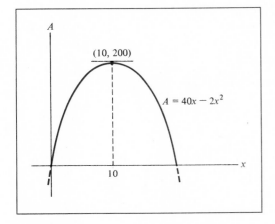

A

(10, 200)

$A = 40x - 2x^2$

10

x

FIGURE 4

Substituting this expression for w into equation (2), we have

$$A = (40 - 2x)x = 40x - 2x^2. \tag{4}$$

We now have a formula for the area A that depends on just one variable, and so we may graph A as a function of x.

Using curve-sketching techniques, we obtain the graph in Fig. 4. We see from the graph that the area is maximized when $x = 10$. (The maximum area is 200 square feet, but this fact is not needed for the problem.) From equation (3) we find that when $x = 10$,

$$w = 40 - 2(10) = 20.$$

Answer $w = 20$ feet, $x = 10$ feet.

Equation (1) in Example 3 is called a *constraint equation* because it places a limit or constraint on the way x and w may vary. Equation (2) is called the *objective equation.* It expresses the quantity to be maximized (the area of the garden) in terms of the variables w and x. The key step in the solution was to use the constraint equation to simplify the objective equation—that is, to write down the objective equation in a way that involves only one variable, as in (4).

EXAMPLE 4   The manager of a department store wants to build a 600-square-foot rectangular enclosure on the store's parking lot in order to display some equipment. Three sides of the enclosure will be built of redwood fencing, at a cost of $7 per running foot. The fourth side will be built of cement blocks, at a cost of $14 per running foot. Find the dimensions of the enclosure that will minimize the total cost of the building materials.

Solution   Let x be the length of the side built out of cement blocks and let y be the length of an adjacent side, as shown in Fig. 5. Since the area of the enclosure must be 600 square feet, the constraint equation is

FIGURE 5

$$xy = 600. \tag{5}$$

The phrase "minimizes the total cost ..." tells us that the objective equation should be a formula giving the total cost of the building materials.

[cost of redwood] = [length of redwood fencing]
$$\times \text{[cost per foot]}$$
$$= (x + 2y) \cdot 7 = 7x + 14y.$$

[cost of cement blocks] = [length of cement wall] · [cost per foot]
$$= x \cdot 14.$$

If C denotes the total cost of the materials, then

$$C = (7x + 14y) + 14x$$
$$C = 21x + 14y \qquad \text{(objective equation).} \tag{6}$$

We simplify this objective equation by solving (5) for one of the variables, say y, and substituting into (6):

$$C = 21x + 14\left(\frac{600}{x}\right) = 21x + \frac{8400}{x}.$$

We may now graph C as a function of the one variable x, where $x > 0$ (Fig. 6). (A similar curve was sketched in Example 4 of Section 4.) The minimum cost of $840 occurs where $x = 20$. From equation (5) we find that the corresponding value of y is $\frac{600}{20} = 30$.

Answer $x = 20$ feet, $y = 30$ feet.

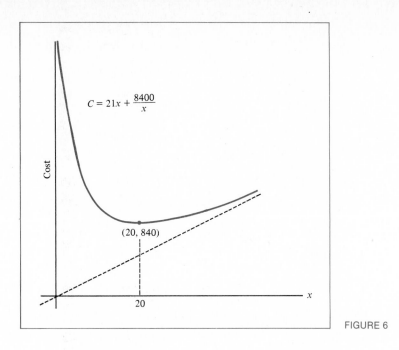

$$C = 21x + \frac{8400}{x}$$

Cost

(20, 840)

20

x

FIGURE 6

EXAMPLE 5 U.S. parcel post regulations state that packages must have length plus girth of no more than 84 inches. Find the dimensions of the cylindrical package of greatest volume that is mailable by parcel post.

Solution Let l be the length of the package and let r be the radius of the circular end. (See Fig. 7.) The girth equals the circumference of the end—that is, $2\pi r$. Since we want the package to be as large as possible, we must use the entire 84 inches allowable:

$$\text{length} + \text{girth} = 84$$

$$l + 2\pi r = 84 \qquad \text{(constraint equation)}. \tag{7}$$

FIGURE 7

The phrase "greatest volume . . ." tells us that the objective equation should express the volume of the package in terms of the dimensions l and r. Let V denote the volume. Then

$$V = [\text{area of base}] \cdot [\text{length}]$$

$$V = \pi r^2 l \qquad \text{(objective equation)}. \tag{8}$$

Girth

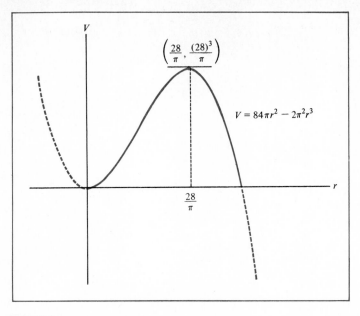

FIGURE 8

We now solve equation (7) for one of the variables, say $l = 84 - 2\pi r$. Substituting this expression into (8), we obtain

$$V = \pi r^2(84 - 2\pi r) = 84\pi r^2 - 2\pi^2 r^3. \tag{9}$$

Let $f(r) = 84\pi r^2 - 2\pi^2 r^3$. Then, for each value of r, $f(r)$ is the volume of the parcel with end radius r that meets the postal regulations. We want to find that value of r for which $f(r)$ is as large as possible.

Using curve-sketching techniques, we obtain the graph in Fig. 8. [The graph is shown with a dashed curve where r is negative and where $f(r)$ is negative, for these portions of the graph have no physical significance.] We see that the volume is greatest when $r = 28/\pi$. From (7) we find that the corresponding value of l is

$$l = 84 - 2\pi r = 84 - 2\pi\left(\frac{28}{\pi}\right)$$

$$= 84 - 56 = 28.$$

The girth when $r = 28/\pi$ is

$$2\pi r = 2\pi\left(\frac{28}{\pi}\right) = 56.$$

Answer $l = 28$ inches, $r = 28/\pi$ inches, girth $= 56$ inches.

Suggestions for Solving an Optimization Problem

1. Draw a picture, if possible
2. Decide what quantity Q is to be maximized or minimized.
3. Assign letters to other quantities that may vary.
4. Find the "constraint equation" that relates the variables to each other and to any constants that are given in the problem.
5. Determine the "objective equation" that expresses Q as a function of the variables assigned in step 3.
6. Use the constraint equation to simplify the objective equation in such a way that Q becomes a function of only one variable.
7. Sketch the graph of the function obtained in step 6 and use this graph to solve the optimization problem.

Note Optimization problems often involve geometric formulas. The most common formulas are:

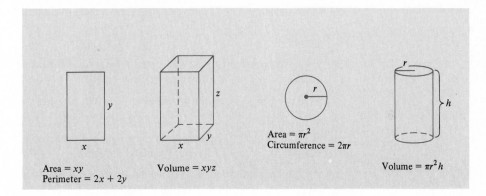

Area = xy
Perimeter = $2x + 2y$

Volume = xyz

Area = πr^2
Circumference = $2\pi r$

Volume = $\pi r^2 h$

PRACTICE PROBLEMS 5

1. A canvas wind shelter for the beach has a back, two square sides, and a top (Fig. 9). Suppose that 96 square feet of canvas are to be used. Find the dimensions of the shelter for which the space inside the shelter (i.e., the volume) will be maximized.

FIGURE 9

2. In Practice Problem 1, what are the objective equation and the constraint equation?

EXERCISES 5

1. Find the maximum value of the function $f(x) = 12x - x^2$.

2. Find the minimum value of the function $g(x) = \frac{1}{2}x^2 - 8x + 35$.

3. Find the minimum value of the function $f(t) = t^3 - 6t^2 + 40$, $t \geq 0$, and give the value of t where this minimum occurs.

4. Find the maximum value of the function $f(s) = 9s^2 - s^3$, $s \geq 0$, and give the value of s where this maximum occurs.

5. For what x does the function $g(x) = 10 + 40x - x^2$ have its maximum value?

6. For what t does the function $f(t) = t^2 - 24t$ have its minimum value?

7. A ball thrown straight up into the air has a height of $-16t^2 + 96t + 6$ feet after t seconds. How high will the ball go?

8. The height of a bullet t seconds after it is fired from a rifle is $-16t^2 + 50t$ meters. How soon will the bullet reach its maximum height?

9. A manufacturer estimates that his profit from the daily production and sale of x power saws will be approximately $-.1x^2 + 50x - 1000$ dollars. How many saws should be produced each day in order to maximize the profit?

10. A manufacturer of costume jewelry estimates that the profit from producing x hundred necklaces per week is $300x - x^3 - 500$ dollars. What is the largest possible weekly profit?

11. Find two positive numbers whose sum is 100 and whose product is as large as possible. [*Note:* Denote the two numbers by x and h; then $x + h = 100$ and the product is $x \cdot h$.]

12. Find two positive numbers whose product is 100 and whose sum is as small as possible.

13. Figure 10(a) shows an open rectangular box with a square base. Find x and h such that the volume is 32 cubic inches and the amount of material needed to construct the box is minimal. [*Note:* The amount of material needed is $x^2 + 4xh$ square inches.]

14. A large soup can is to be designed so that the can will hold 16π cubic inches (about 28 fluid ounces) of soup. [See Fig. 10(b).] Find the values of x and h such that the amount

FIGURE 10

(a)

Side unrolled

(b)

of metal needed is as small as possible. [*Note:* The side requires $2\pi x h$ square inches of metal and each end requires πx^2 square inches.]

15. Figure 11(a) shows a "Norman window," which consists of a rectangle capped by a semicircular region. Find the value of x such that the perimeter of the window will be 14 feet and the area of the window will be as large as possible. [*Note:* The perimeter of the window is $2x + 2h + \pi x$.]

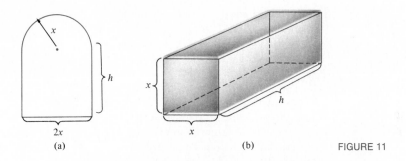

(a)   (b)   FIGURE 11

16. Postal requirements specify that parcels must have length plus girth of at most 84 inches. Find the dimensions of the square-ended rectangular package [Fig. 11(b)] of greatest volume that is mailable. [*Note:* The length plus girth is $h + 4x$.]

17. A rectangular garden of area 75 square feet is to be surrounded on three sides by a brick wall costing \$10 per foot and on one side by a fence costing \$5 per foot. Find the dimensions of the garden such that the cost of the materials is minimized. [*Note:* Letting x be the length of the fence and h be the length of the other dimension, the cost of the materials is $5x + 10(2h + x)$.]

18. A closed rectangular box with square base and a volume of 12 cubic feet is to be constructed using two different types of materials. The top is made of a metal costing \$2 per square foot and the remainder of wood costing \$1 per square foot. Find the dimensions of the box for which the cost of materials is minimized.

19. Find the dimensions of the closed rectangular box with square base and volume 8000 cubic centimeters that can be constructed using the least amount of material.

20. Find the dimensions of the rectangular garden of greatest area that can be fenced off (all four sides) with 300 meters of fencing.

21. A storage shed is to be built in the shape of a box with a square base. It is to have a volume of 150 cubic feet. The concrete for the base costs \$4 per square foot, the material for the roof costs \$2 per square foot, and the material for the sides costs \$2.50 per square foot. Find the dimensions of the most economical shed.

22. A canvas wind shelter for the beach has a back, two square sides, and a top. Find the dimensions for which the volume will be 250 cubic feet and which requires the least possible amount of canvas.

23. In Example 3 one can solve the constraint equation (1) for x instead of w to get $x = 20 - \frac{1}{2}w$. Substituting this for x in (2), one has $A = xw = (20 - \frac{1}{2}w)w$. Sketch the

graph of the equation $A = 20w - \frac{1}{2}w^2$ and show that the maximum area occurs when $w = 20$ and $x = 10$.

24. In Example 4 one can solve the constraint equation (5) for x instead of y to get $x = 600/y$. Substituting this for x in (6), one has $C = 21x + 14y = (12,600/y) + 14y$. Sketch the graph of C as a function of y and show that the minimum cost occurs when $y = 30$ feet.

SOLUTIONS TO PRACTICE PROBLEMS 5

1. Since the sides of the wind shelter are square, we may let x represent the length of each side of the square. The remaining dimension of the wind shelter can be denoted by the letter h. (See Fig. 12.) The volume of the shelter is x^2h, and this is to be maximized. Since we have learned to maximize only functions of a single variable, we must express h in terms of x. We must use the information that 96 feet of canvas are used—that is, $2x^2 + 2xh = 96$. [*Note:* The roof and the back each have area xh, and each end has area x^2.] We now solve this equation for h.

FIGURE 12

$$2x^2 + 2xh = 96$$

$$2xh = 96 - 2x^2$$

$$h = \frac{96}{2x} - \frac{2x^2}{2x} = \frac{48}{x} - x.$$

The volume, V, is

$$x^2h = x^2\left(\frac{48}{x} - x\right) = 48x - x^3.$$

By sketching the graph of $V = 48x - x^3$, we see that V has a maximum value when $x = 4$. Then, $h = \frac{48}{4} - 4 = 12 - 4 = 8$. So each end of the shelter should be a 4-foot-by-4-foot square and the top should be 8 feet long.

2. The objective equation is: $V = x^2h$, since it expresses the volume (the quantity to be maximized) in terms of the variables. The constraint equation is $2x^2 + 2xh = 96$, for it relates the variables to each other; that is, it can be used to express one of the variables in terms of the other.

2.6 Further Optimization Problems

In this section we apply the optimization techniques developed in the preceding section to some practical situations.

EXAMPLE 1 Suppose that, on a certain route, an airline carries 8000 passengers per month, each paying $50. The airline wants to increase the fare. However, the market research department estimates that for each $1 increase in fare, the airline will lose 100 passengers. Determine the price which maximizes the airline's revenue.

Solution Since the problem calls for setting an optimum price, let x be the price per ticket. The other variable is the number of passengers, which we can denote by n. The goal is to maximize revenue.

$$[\text{revenue}] = [\text{number of passengers}][\text{price per ticket}]$$
$$= n \cdot x.$$

If R denotes revenue, then the objective equation is

$$R = nx.$$

Now the constraint equation is given by

$$[\text{number of passengers}] = [\text{original number}] - [\text{loss due to fare increase}]$$
$$n \quad = \quad 8000 \quad\quad (x - 50) \cdot 100$$
$$= 13,000 - 100x.$$

Therefore,

$$R = nx = (13,000 - 100x)x = 13,000x - 100x^2.$$

Figure 1 shows that revenue is maximized when the price per ticket is $65.

$R = 13000x - 100x^2$

FIGURE 1

EXAMPLE 2 A supermarket manager wants to establish an optimal inventory policy for frozen orange juice. It is estimated that a total of 1200 cases will be sold at a steady rate during the next quarter-year. Since the orange juice requires costly freezer space for storage, she does not want to order all 1200 cases at once. On the other hand, since each delivery incurs additional handling costs, she cannot place too many separate orders. She decides to place several orders of the same size, say each consisting of x cases, equally spaced throughout the quarter-year. Use the data below to determine the optimal reorder quantity x that minimizes total purchasing and inventory costs.

1. The handling cost for each delivery is $75.
2. Inventory cost per quarter is $4x$ dollars. [Here is how such a model arises. Suppose that it costs $8 to store one case for one quarter-year. Since the stock begins at each delivery with x cases and is depleted steadily, there will be an average of $x/2$ cases in stock at any time. Therefore, inventory cost for the quarter will be $(x/2) \cdot 8$ or $4x$ dollars.]

Solution   Let r be the number of orders placed during the quarter. Since each order calls for x cases, the constraint equation is

$$r \cdot x = 1200.$$

Now

$$[\text{total cost}] = [\text{handling cost}] + [\text{inventory cost}]$$
$$= 75r + 4x.$$

If C denotes the total cost, then the objective equation is

$$C = 75r + 4x.$$

The constraint equation says that $r = 1200/x$. Substitution into the objective equation yields

$$C = \frac{90{,}000}{x} + 4x.$$

Figure 2 is the graph of C as a function of x, for $x > 0$. The total cost is at a minimum when $x = 150$. Therefore, the optimum inventory policy is to order 150 cases at a time and to place $1200/150 = 8$ orders during the quarter-year.

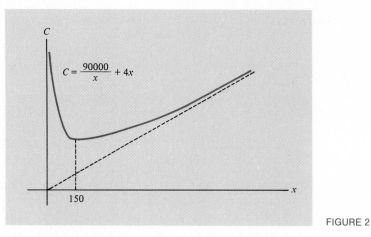

FIGURE 2

EXAMPLE 3   What should the inventory policy of Example 2 be if sales of frozen orange juice increase fourfold (i.e., 4800 cases are sold each quarter), but all other conditions are the same?

Solution   The only change in our previous solution is in the constraint equation, which now becomes

$$r \cdot x = 4800.$$

The objective equation is, as before,

$$C = 75r + 4x.$$

Since $r = 4800/x$,

$$C = 75 \cdot \frac{4800}{x} + 4x$$

$$= \frac{360,000}{x} + 4x.$$

Now

$$C' = -\frac{360,000}{x^2} + 4.$$

Setting $C' = 0$ yields

$$\frac{360,000}{x^2} = 4$$

$$90,000 = x^2$$

$$x = 300.$$

Therefore, the optimal reorder quantity is 300 cases.

Notice that although the sales increased by a factor of 4, the optimal reorder quantity only increased by a factor of 2 $(=\sqrt{4})$. In general, a store's inventory of an item should be proportional to the square root of the expected sales. (See Exercise 16 for a derivation of this result.) Many stores tend to keep their average inventories at a fixed percentage of sales. For example, each order may contain enough goods to last for 4 or 5 weeks. This policy is likely to create excessive inventories of high-volume items and uncomfortably low inventories of slower-moving items.

EXAMPLE 4   When a person coughs, the trachea (windpipe) contracts. (See Fig. 3.) Let

r_0 = normal radius of the trachea,

r = radius during a cough,

P = increase in air pressure in the trachea during cough,

v = velocity of air through trachea during cough.

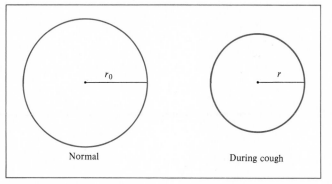

Normal   During cough

FIGURE 3

Use the principles of fluid flow given below to determine how much the trachea should contract in order to create the greatest air velocity—that is, the most effective condition for clearing the lungs and the trachea.

1. $r_0 - r = aP$, for some positive constant a. (Experiment has shown that, during coughing, the decrease in the radius of the trachea is nearly proportional to the increase in the air pressure.)

2. $v = b \cdot P \cdot \pi r^2$, for some positive constant b. (The theory of fluid flow requires that the velocity of the air forced through the trachea is proportional to the product of the increase in the air pressure and the area of a cross section of the trachea.)

Solution In this problem the constraint equation (1) and the objective equation (2) are given directly. Solving equation (1) for P and substituting this result into equation (2), we have

$$v = b\left(\frac{r_0 - r}{a}\right)\pi r^2 = k(r_0 - r)r^2,$$

where $k = b\pi/a$. To find the radius at which the velocity v is a maximum, we first compute the derivatives:

$$v = k(r_0 r^2 - r^3)$$

$$\frac{dv}{dr} = k(2r_0 r - 3r^2) = kr(2r_0 - 3r)$$

$$\frac{d^2v}{dr^2} = k(2r_0 - 6r).$$

We see that $\dfrac{dv}{dr} = 0$ when $r = 0$ or when $2r_0 - 3r = 0$; that is, when $r = \frac{2}{3}r_0$. It is easy to see that $\dfrac{d^2v}{dr^2}$ is positive at $r = 0$ and is negative at $r = \frac{2}{3}r_0$. The graph of v as a function of r is drawn in Fig. 4. The air velocity is maximized at $r = \frac{2}{3}r_0$.

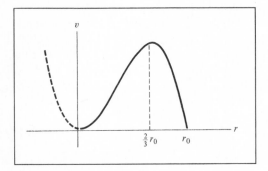

FIGURE 4

When solving optimization problems, we have always made a rough sketch of the graph of the function being optimized before drawing any conclusions. This procedure is necessary, as the next example shows.

EXAMPLE 5 A rancher has 204 meters of fencing from which to build two corrals; one square and the other rectangular with length that is twice the width. Find the dimensions which result in the greatest combined area.

Solution Let x be the width of the rectangular corral and h the length of each side of the square corral. (See Fig. 5.) Then

$$204 = [\text{perimeter of square}] + [\text{perimeter of rectangle}] = 4h + 6x.$$

h

h

2x

x

FIGURE 5

Let A be the combined area. Then

$$A = [\text{area of square}] + [\text{area of rectangle}]$$
$$= h^2 + 2x^2.$$

Since the perimeter of the rectangle cannot exceed 204, we must have $0 \leq 6x \leq 204$ or $0 \leq x \leq 34$. Solving the constraint equation for h and substituting into the objective equation leads to the function graphed in Fig. 6. The graph reveals that the area is minimized when $x = 18$. However, the problem asks for the *maximum* possible area. From Fig. 6 we see that this occurs at the endpoint where $x = 0$. Therefore, the rancher should build only the square corral, with $h = 204/4 = 51$ meters. In this example, the objective function has an endpoint extremum. Namely, the maximum value occurs at the endpoint $x = 0$.

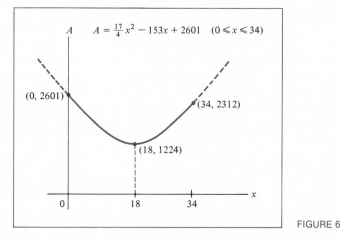

A $A = \frac{17}{4}x^2 - 153x + 2601$ $(0 \leq x \leq 34)$

$(0, 2601)$

$(34, 2312)$

$(18, 1224)$

0 18 34 x

FIGURE 6

PRACTICE PROBLEMS 6

1. An apple orchard produces a profit of $40 a tree when planted with 1000 trees. Because of overcrowding, the profit per tree (for each tree in the orchard) is reduced by 2 cents for each additional tree planted. How many trees should be planted in order to maximize the total profit from the orchard?

2. In the inventory problem of Example 2, suppose that the sales of frozen orange juice increase ninefold; that is, 10,800 cases are sold each quarter. What is now the optimal reorder quantity?

EXERCISES 6

1. Find the dimensions of the rectangular garden of area 100 square meters for which the amount of fencing needed to surround the garden is as small as possible.

2. A certain airline requires that rectangular packages carried on an airplane by passengers be such that the sum of the three dimensions is at most 120 centimeters. Find the dimensions of the square-ended rectangular package of greatest volume that meets this requirement.

(a)   (b)

FIGURE 7

3. An athletic field [Fig. 7(a)] consists of a rectangular region with a semicircular region at each end. The perimeter will be used for a 440-yard track. Find the value of x for which the area of the rectangular region is as large as possible.

4. An open rectangular box is to be constructed by cutting square corners out of a 16 × 16-inch piece of cardboard and folding up the flaps. [See Fig. 7(b).] Find the value for x for which the volume of the box will be as large as possible.

5. A long rectangular sheet of metal 30 inches wide is to be made into a gutter by turning up strips vertically along the two sides. How many inches should be turned up on each side so as to maximize the amount of water that the gutter can carry?

6. Design an open rectangular box with square ends, having volume 36 cubic inches, that minimizes the amount of material required for construction.

7. A supermarket is to be designed in the form of a rectangle whose area is 12,000 square feet. The front of the building will be mostly glass and will cost $70 per running foot for materials. The other three sides will be constructed of brick and cement block, at a cost of $50 per running foot. Ignore all other costs (labor, cost of foundation and roof, etc.) and find the dimensions of the building that will minimize the cost of the materials for the four sides of the building.

8. A manufacturer can produce x units of a certain commodity at a cost of $C(x) = .01x^3 - .04x^2 + 2x + 100$ dollars and receives a revenue of $R(x) = 50x - .04x^2$ dollars. How many units should be produced in order to maximize the profit? [*Note:* The profit from producing x units is $P(x) = R(x) - C(x)$.]

9.

FIGURE 8

A rectangular corral of 54 square meters is to be fenced off and then divided by a fence into two sections. (See Fig. 8.) Find the dimensions of the corral so that the amount of fencing required is minimized.

10. Referring to Exercise 9, suppose that the cost of the fencing for the boundary is $5 per meter and the dividing fence costs $2 per meter. Find the dimensions of the corral that minimize the cost of the fencing.

11. A swimming club offers memberships at the rate of $200, provided that a minimum of 100 people join. For each member in excess of 100, the membership fee will be reduced by $1 per person (for every member). At most 160 memberships will be sold. How many memberships should the club try to sell in order to maximize its revenue?

12. In the planning of a sidewalk cafe, it is estimated that if there are 12 tables, the daily profit will be $10 per table. Because of overcrowding, for each additional table the profit per table (for every table in the cafe) will be reduced by $.50. How many tables should be provided in order to maximize the profit from the cafe?

13. A certain toll road averages 36,000 cars per day when charging $1 per car. A survey concludes that increasing the toll will result in 300 fewer cars for each cent of increase. What toll should be charged in order to maximize the revenue?

14. A small orchard yields 25 bushels of fruit per tree when planted with 40 trees. Because of overcrowding, the yield per tree (for each tree in the orchard) is reduced by $\frac{1}{2}$ bushel for each additional tree that is planted. How many trees should be planted in order to maximize the total yield of the orchard?

15. A publishing company sells 400,000 copies of a certain book each year. Ordering the entire amount printed at the beginning of the year ties up valuable storage space. However, running off the copies in several partial runs throughout the year results in added costs for setting up each printing run. Using the data below, find the production run size, x, that minimizes the total setting up and storage costs.

(i) Setting up each production run costs $1000.

(ii) The storage cost is $x/4$, where x is the size of each production run. (There will be an average of $x/2$ books in storage during the year and storage costs are 50 cents per book per year.)

16. A store manager wants to establish an optimal inventory policy for an item. Sales are expected to be at a steady rate and should total Q items sold during the year. Each time an order is placed, a handling cost of h dollars is incurred. If the size of each order is x, then storage costs for the year will be $(x/2) \cdot s$ dollars. (That is, there will be an average of $x/2$ items in storage during the year, and the cost of storage is s dollars per item per year.) Show that total costs for handling and storage are minimized when each order calls for $\sqrt{2hQ/s}$ items.

17. Foggy Optics, Inc., makes laboratory microscopes and similar items. Setting up each production run costs $2500. The expense of holding one microscope in inventory for 1 year is $50. Suppose that the company expects to sell 1600 microscopes at a fairly uniform rate throughout the year. Determine the production run size that will minimize the company's expenses.

18. The Great American Tire Co. expects to sell 600,000 tires of a particular size and grade during the next year. Sales tend to be roughly the same from month to month. Setting up each production run costs the company an additional $15,000. The inventory expense for one tire is $5 per year. What size production runs should be scheduled for the coming year in order to minimize the overall cost of producing these tires?

19. Let $f(t)$ be the amount of oxygen in a lake t days after sewage is dumped into the lake, and suppose that $f(t)$ is given approximately by

$$f(t) = 500\left[1 - \frac{10}{t + 10} + \frac{100}{(t + 10)^2}\right].$$

At what time is the oxygen content increasing the fastest?

20. The daily output of a coal mine after t hours of operation is approximately $40t + t^2 - \frac{1}{15}t^3$ tons, $0 \leq t \leq 12$. Find the maximum rate of output (in tons of coal per hour).

21. Consider a parabolic arch whose shape may be represented by the graph of $y = 9 - x^2$, where the base of the arch lies on the x-axis from $x = -3$ to $x = 3$. Find the dimensions of the rectangular window of maximum area that can be constructed inside the arch.

22. Advertising for a certain product is terminated, and t weeks later the weekly sales are $f(t)$ cases, where $f(t) = 1000(t + 8)^{-1} - 4000(t + 8)^{-2}$. At what time is the weekly sales amount falling the fastest?

SOLUTIONS TO PRACTICE PROBLEMS 6

1. Since the question asks for the optimum "number of trees," let x be the number of trees to be planted. The other quantity that varies is "profit per tree." So let p be the profit per tree. The objective is to maximize total profit, call it T. Then

$$[\text{total profit}] = [\text{profit per tree}] \cdot [\text{number of trees}]$$

$$T = p \cdot x.$$

Now,

$$[\text{profit per tree}] = [\text{original profit per tree}] - [\text{loss per tree due to increase}]$$

$$p = 40 - (x - 1000)(.02)$$

$$p = 60 - .02x.$$

Therefore,

$$T = p \cdot x = (60 - .02x)x = 60x - .02x^2.$$

By computing first and second derivatives and sketching the graph, we easily find that total profit is maximum when $x = 1500$. Therefore, 1500 trees should be planted.

2. This problem can be solved in the same manner that Example 3 was solved. However, the comment made at the end of Example 3 indicates that the optimal reorder quantity should increase by a factor of 3, since $3 = \sqrt{9}$. Therefore, the optimal reorder quantity is $3 \cdot 150 = 450$ cases.

2.7 Applications of Calculus to Business and Economics

In recent years economic decision making has become more and more mathematically oriented. Faced with huge masses of statistical data, depending on hundreds or even thousands of different variables, business analysts and economists have increasingly turned to mathematical methods to help them describe what is happening, predict the effects of various policy alternatives, and choose reasonable courses of action from the myriad possibilities. Among the mathematical methods employed is calculus. In this section we illustrate just a few of the many applications of calculus to business and economics. All our applications will center around what economists call *the theory of the firm*. In other words, we study the activity of a business (or possibly a whole industry) and restrict our analysis to a time period during which background conditions (such as supplies of raw materials, wage rates, taxes) are fairly constant. We then show how calculus can help the management of such a firm make vital production decisions.

Management, whether or not it knows calculus, utilizes many functions of the sort we have been considering. Examples of such functions are

$C(x)$ = cost of producing x units of the product,

$R(x)$ = revenue generated by producing x units of the product,

$P(x) = R(x) - C(x)$ = the profit (or loss) generated by producing x units of the product.

Note that the functions $C(x)$, $R(x)$, $P(x)$ often are defined only for nonnegative integers—that is, for $x = 0, 1, 2, 3, \ldots$. The reason is that it does not make sense to speak about the cost of producing -1 cars or the revenue generated by selling 3.62 refrigerators. Thus each of these functions may give rise to a set of discrete points on a graph, as in Fig. 1(a). In studying these functions, however, economists usually draw a smooth curve through the points and assume that $C(x)$ is actually defined for all positive x. Of course, we must often interpret answers to problems in light of the fact that x is, in most cases, a nonnegative integer.

FIGURE 1

Cost Functions   If we assume that a cost function $C(x)$ has a smooth graph as in Fig. 1(b), we can use the tools of calculus to study it. A typical cost function is analyzed in Example 1.

EXAMPLE 1   Suppose that the cost function for a manufacturer is given by $C(x) = (10^{-6})x^3 - .003x^2 + 5x + 1000$ dollars.

(a) Describe the behavior of the marginal cost.

(b) Sketch the graph of $C(x)$.

Solution   The first two derivatives of $C(x)$ are given by

$$C'(x) = (3 \cdot 10^{-6})x^2 - .006x + 5$$

$$C''(x) = (6 \cdot 10^{-6})x - .006.$$

Let us sketch the marginal cost $C'(x)$ first. From the behavior of $C'(x)$, we will be able to graph $C(x)$. The marginal cost function $y = (3 \cdot 10^{-6})x^2 - .006x + 5$ has as its graph a parabola that opens upward. Since $y' = C''(x) = .000006(x - 1000)$, we see that the parabola has a horizontal tangent at $x = 1000$. So the minimum value of y occurs at $x = 1000$. The corresponding y-coordinate is

$$(3 \cdot 10^{-6})(1000)^2 - .006 \cdot (1000) + 5 = 3 - 6 + 5 = 2.$$

FIGURE 2

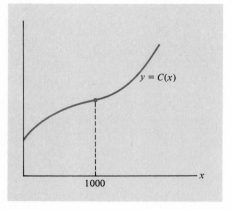

FIGURE 3

The graph of $y = C'(x)$ is shown in Fig. 2. Consequently, at first the marginal cost decreases. It reaches a minimum of 2 at production level 1000 and increases thereafter. This answers part (a). Let us now graph $C(x)$. Since the graph shown in Fig. 2 is the graph of the derivative of $C(x)$, we see that $C'(x)$ is never zero, so that there are no extreme points. Since $C'(x)$ is always positive, $C(x)$ is always increasing (as any cost curve should be). Moreover, since $C'(x)$ decreases for x less than 1000 and increases for x greater than 1000, we see that $C(x)$ is concave down for x less than 1000, concave up for x greater than 1000, and has an inflection point at $x = 1000$. The graph of $C(x)$ is drawn in Fig. 3. Note that the inflection point of $C(x)$ occurs at the value of x for which marginal cost is a minimum.

Actually, most marginal cost functions have the same general shape as the marginal cost curve of Example 1. For when x is small, production of additional units is subject to economies of production, which lower unit costs. Thus, for x small, marginal cost decreases. However, increased production eventually leads to overtime, use of less efficient, older plants, and competition for scarce raw materials. As a result, the cost of additional units will increase for very large x. So we see that $C'(x)$ initially decreases and then increases.

Revenue Functions   In general, a business is concerned not only with its costs, but also with its revenues. Recall that if $R(x)$ is the revenue received from the sale of x units of some commodity, then the derivative $R'(x)$ is called the *marginal revenue*. Economists use this to measure the rate of increase in revenue per unit increase in sales.

If x units of a product are sold at a price p per unit, then the total revenue $R(x)$ is given by

$$R(x) = x \cdot p.$$

If a firm is small and is in competition with many other companies, its sales have little effect on the market price. Then since the price is constant as far as the one firm is concerned, the marginal revenue $R'(x)$ equals the price p [that is, $R'(x)$ is the amount that the firm receives from the sale of one additional unit]. In this case, the revenue function will have a graph as in Fig. 4.

FIGURE 4

An interesting problem arises when a single firm is the only supplier of a certain product or service—that is, when the firm has a "monopoly." Consumers will buy large amounts of the commodity if the price per unit is low and less if the price is raised. For each quantity x, let $f(x)$ be the highest price per unit that can be set in order to sell all x units to consumers. Since selling greater quantities requires a lowering of the price, $f(x)$ will be a decreasing function. Figure 5 shows a typical "demand curve" that relates the quantity demanded, x, to the price $p = f(x)$.

The *demand equation* $p = f(x)$ determines the total revenue function. If the firm wants to sell x units, the highest price it can set is $f(x)$ dollars per unit, and so the total revenue from the sale of x units is

$$R(x) = x \cdot p = x \cdot f(x).$$

The concept of a demand curve applies to an entire industry (with many producers) as well as to a single monopolistic firm. In this case, many producers offer the same product for sale. If x denotes the total output of the industry, then $f(x)$ is the market price per unit of output and $x \cdot f(x)$ is the total revenue earned from the sale of the x units.

p = f(x)

Price

Quantity

FIGURE 5

EXAMPLE 2 The demand equation for a certain product is $p = 6 - \frac{1}{2}x$. Find the level of production that results in maximum revenue.

Solution The revenue function $R(x)$, in this case, is

$$R(x) = x \cdot p = x(6 - \tfrac{1}{2}x) = 6x - \tfrac{1}{2}x^2.$$

The marginal revenue is given by

$$R'(x) = 6 - x.$$

The graph of $R(x)$ is a parabola that opens downward (Fig. 6). It has a horizontal tangent precisely at those x for which $R'(x) = 0$—that is, for those x at which marginal revenue is 0. The only such x is $x = 6$. The corresponding value of revenue is

$$R(6) = 6 \cdot 6 - \tfrac{1}{2}(6)^2 = 18.$$

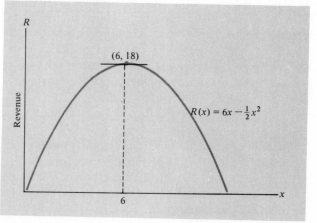

(6, 18)

Revenue

$R(x) = 6x - \frac{1}{2}x^2$

6

FIGURE 6

Thus the rate of production resulting in maximum revenue is $x = 6$, which results in total revenue of 18.

EXAMPLE 3   The WMA Bus Lines offers sightseeing tours of Washington, D.C. One of the tours, priced at $7 per person, had an average demand of about 1000 customers per week. When the price was lowered to $6, the weekly demand jumped to about 1200 customers. Assuming that the demand equation is linear, find the tour price that should be charged per person in order to maximize the total revenue each week.

Solution   First we must find the demand equation. Let x be the number of customers per week and let p be the price of a tour ticket. Then $(x, p) = (1000, 7)$ and $(x, p) = (1200, 6)$ are on the demand curve (Fig. 7). Using the point-slope formula for the line through these two points, we have

$$p - 7 = \frac{7 - 6}{1000 - 1200} \cdot (x - 1000)$$

$$= -\frac{1}{200}(x - 1000)$$

$$= -\frac{1}{200}x + 5,$$

so

$$p = 12 - \frac{1}{200}x. \tag{1}$$

From equation (1) we obtain the revenue function

$$R(x) = x(12 - \tfrac{1}{200}x) = 12x - \tfrac{1}{200}x^2.$$

The marginal revenue is

$$R'(x) = 12 - \tfrac{1}{100}x = -\tfrac{1}{100}(x - 1200).$$

FIGURE 7

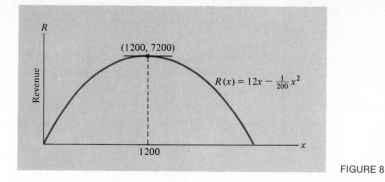

FIGURE 8

Using $R(x)$ and $R'(x)$, we can sketch the graph of $R(x)$ (Fig. 8). The maximum revenue occurs when the marginal revenue is zero—that is, when $x = 1200$. The price corresponding to this number of customers is found from the demand equation (1),

$$p = 12 - \tfrac{1}{200}(1200) = 6.$$

Thus the price of $6 is most likely to bring in the greatest revenue per week.

Profit Functions   Once we know the cost function $C(x)$ and the revenue function $R(x)$, we can compute the profit function $P(x)$ from

$$P(x) = R(x) - C(x).$$

EXAMPLE 4   Suppose that the demand equation for a monopolist is $p = 100 - .01x$ and the cost function is $C(x) = 50x + 10,000$. Find the value of x that maximizes the profit and determine the corresponding price and total profit for this level of production. (See Fig. 9.)

Solution   The total revenue function is

$$R(x) = x \cdot p = x(100 - .01x) = 100x - .01x^2.$$

FIGURE 9

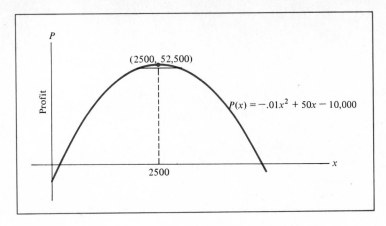

P

(2500, 52,500)

Profit

$P(x) = -.01x^2 + 50x - 10,000$

2500

x

FIGURE 10

Hence the profit function is

$$P(x) = R(x) - C(x)$$
$$= 100x - .01x^2 - (50x + 10,000)$$
$$= -.01x^2 + 50x - 10,000.$$

The graph of this function is a parabola that opens downward. (See Fig. 10.) Its highest point will be where the curve has zero slope—that is, where the marginal profit $P'(x)$ is zero. Now

$$P'(x) = -.02x + 50 = -.02(x - 2500).$$

So $P'(x) = 0$ when $x = 2500$. The profit for this level of production is

$$P(2500) = -.01(2500)^2 + 50(2500) - 10,000$$
$$= 52,500.$$

Finally, we return to the demand equation to find the highest price that can be charged per unit in order to sell all 2500 units:

$$p = 100 - .01(2500)$$
$$= 100 - 25 = 75.$$

Answer Produce 2500 units and sell them at $75 per unit. The profit will be $52,500.

EXAMPLE 5 Rework Example 4 under the condition that the government imposes an excise tax of $10 per unit.

Solution For each unit sold, the manufacturer will have to pay $10 to the government. In other words, $10x$ dollars is added to the cost of producing and selling x units. The cost function now will be

$$C(x) = (50x + 10,000) + 10x = 60x + 10,000.$$

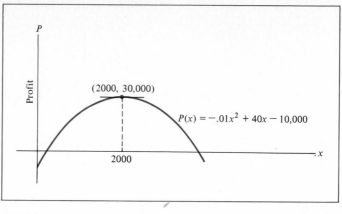

$P(x) = -.01x^2 + 40x - 10{,}000$

(2000, 30,000)

2000

FIGURE 11

The demand equation is unchanged by this tax, so the revenue function is still

$$R(x) = 100x - .01x^2.$$

Proceeding as before, we have

$$P(x) = R(x) - C(x)$$
$$= 100x - .01x^2 - (60x + 10{,}000)$$
$$= -.01x^2 + 40x - 10{,}000.$$
$$P'(x) = -.02x + 40 = -.02(x - 2000).$$

The graph of $P(x)$ is still a parabola that opens downward, and the highest point is where $P'(x) = 0$—that is, where $x = 2000$. (See Fig. 11.) The corresponding profit is

$$P(2000) = -.01(2000)^2 + 40(2000) - 10{,}000$$
$$= 30{,}000.$$

From the demand equation, $p = 100 - .01x$, we find the price that corresponds to $x = 2000$:

$$p = 100 - .01(2000) = 80.$$

Answer Produce 2000 units and sell them at $80 per unit. The profit will be $30,000.

Notice in Example 5 that the optimal price is raised from $75 to $80. If the monopolist wishes to maximize profits, he should pass only half the $10 tax on to the consumer. The monopolist cannot avoid the fact that profits will be substantially lowered by the imposition of the tax. This is one reason why industries lobby against taxation.

Setting Production Levels   Suppose that a firm has cost function $C(x)$ and revenue function $R(x)$. In a free enterprise economy the firm will set production x

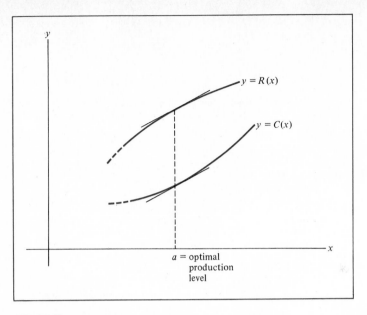

FIGURE 12

in such a way as to maximize its profit function

$$P(x) = R(x) - C(x).$$

We have seen that if $P(x)$ has a maximum at $x = a$, then $P'(a) = 0$. In other words, since $P'(x) = R'(x) - C'(x)$, we see that

$$R'(a) - C'(a) = 0$$

$$R'(a) = C'(a).$$

Thus profit is maximized at a production level for which marginal revenue equals marginal cost. (See Fig. 12.)

PRACTICE PROBLEMS 7

1. Rework Example 4 by finding the production level at which marginal revenue equals marginal cost.

2. Rework Example 4 under the condition that the fixed cost is increased from $10,000 to $15,000.

EXERCISES 7

1. Given the cost function $C(x) = x^3 - 6x^2 + 13x + 15$, find the minimum marginal cost.

2. Suppose that a total cost function is $C(x) = .0001x^3 - .06x^2 + 12x + 100$. Is the marginal cost increasing, decreasing, or not changing at $x = 100$? Find the minimum marginal cost.

3. The revenue function for a one-product firm is

$$R(x) = 200 - \frac{1600}{x + 8} - x.$$

Find the value of x that results in maximum revenue.

4. The revenue function for a particular product is $R(x) = x(4 - .0001x)$. Find the largest possible revenue.

5. A one-product firm estimates that its daily total cost function (in suitable units) is $C(x) = x^3 - 6x^2 + 13x + 15$ and its total revenue function is $R(x) = 28x$. Find the value of x that maximizes the daily profit.

6. A small tie shop sells men's ties for \$3.50 each. The daily cost function is estimated to be $C(x)$ dollars, where x is the number of ties sold on a typical day and $C(x) = .0006x^3 - .03x^2 + 2x + 80$. Find the value of x that will maximize the store's daily profit.

7. The demand equation for a certain commodity is $p = \frac{1}{12}x^2 - 10x + 300, 0 \leq x \leq 60$. Find the value of x and the corresponding price p that maximize the revenue.

8. The demand equation for a product is $p = 2 - .001x$. Find the value of x and the corresponding price p that maximize the revenue.

9. Some years ago it was estimated that the demand for steel approximately satisfied the equation $p = 256 - 50x$, and the total cost of producing x units of steel was $C(x) = 182 + 56x$. (The quantity x was measured in millions of tons and the price and total cost were measured in millions of dollars.) Determine the level of production and the corresponding price that maximize the profits.

10. Consider a rectangle in the xy-plane, with corners at $(0, 0), (a, 0), (0, b)$, and (a, b). Suppose that (a, b) lies on the graph of the equation $y = 30 - x$. Find a and b such that the area of the rectangle is maximized. What economic interpretation can be given to your answer if the equation $y = 30 - x$ represents a demand curve and y is the price corresponding to the demand x?

11. Until recently hamburgers at the city sports arena cost \$1 each. The food concessionaire sold an average of 10,000 hamburgers on a game night. When the price was raised to \$1.20, hamburger sales dropped off to an average of 8000 per night.

   (a) Assuming a linear demand curve, find the price of a hamburger that will maximize the nightly hamburger revenue.

   (b) Suppose that the concessionaire has fixed costs of \$1000 per night and the variable cost is \$.30 per hamburger. Find the price of a hamburger that will maximize the nightly hamburger profit.

12. The average ticket price for a concert at the opera house was \$32. The 4000-seat auditorium was filled for nearly every performance. When the ticket price was raised to \$34 attendance declined to an average of 3800 persons per performance. Salaries, electricity, and maintenance expenses total \$60,000 per performance. Costs that vary with the size of the audience are negligible. What should the average ticket price be in order to maximize the profit for the opera house? (Assume a linear demand curve.)

13. The monthly demand equation for an electric utility company is estimated to be

$$p = 60 - (10^{-5})x,$$

where p is measured in dollars and x is measured in thousands of kilowatt-hours. The utility has fixed costs of $7,000,000 per month and variable costs of $30 per thousand kilowatt-hours of electricity generated, so that the cost function is

$$C(x) = 7 \cdot 10^6 + 30x.$$

(a) Find the value of x and the corresponding price for a thousand kilowatt-hours that maximize the utility's profit.

(b) Suppose that rising fuel costs increase the utility's variable costs from $30 to $40 so that its new cost function is

$$C_1(x) = 7 \cdot 10^6 + 40x.$$

Should the utility pass all this increase of $10 per thousand kilowatt-hours on to consumers? Explain your answer.

14. The demand equation for a monopolist is $p = 200 - 3x$, and the cost function is $C(x) = 75 + 80x - x^2, 0 \le x \le 40$.

(a) Determine the value of x and the corresponding price that maximizes the profit.

(b) Suppose that the government imposes a tax on the monopolist of $4 per unit quantity produced. Determine the new price that maximizes the profit.

(c) Suppose that the government imposes a tax of T dollars per unit quantity produced, so that the new cost function is

$$C(x) = 75 + (80 + T)x - x^2, \qquad 0 \le x \le 40.$$

Determine the new value of x that maximizes the monopolist's profit as a function of T. Assuming that the monopolist cuts back production to this level, express the tax revenues received by the government as a function of T. Finally, determine the value of T that will maximize the tax revenue received by the government.

SOLUTIONS TO PRACTICE PROBLEMS 7

1. The revenue function is $R(x) = 100x - .01x^2$, so the marginal revenue function is $R'(x) = 100 - .02x$. The cost function is $C(x) = 50x + 10,000$, so the marginal cost function is $C'(x) = 50$. Let us now equate the two marginal functions and solve for x.

$$R'(x) = C'(x)$$

$$100 - .02x = 50$$

$$-.02x = -50$$

$$x = \frac{-50}{-.02} = \frac{5000}{2} = 2500.$$

Of course, we obtain the same level of production as before.

2. If the fixed cost is increased from $10,000 to $15,000, the new cost function will be $C(x) = 50x + 15,000$ but the marginal cost function will still be $C'(x) = 50$. Therefore, the solution will be the same: 2500 units should be produced and sold at $75 per unit. [Increases in fixed costs should not necessarily be passed on to the consumer if the objective is to maximize the profit.]

Chapter 2: CHECKLIST

- ☐ Increasing, decreasing
- ☐ Extreme point, maximum point, minimum point
- ☐ Maximum value, minimum value
- ☐ Concave up, concave down
- ☐ Inflection point
- ☐ x-intercept, y-intercept
- ☐ Asymptote
- ☐ First derivative rule
- ☐ Second derivative rule
- ☐ How to look for possible extreme points
- ☐ How to decide whether an extreme point is a maximum or minimum point
- ☐ How to look for possible inflection points
- ☐ Summary of curve-sketching techniques
- ☐ Objective equation
- ☐ Constraint equation
- ☐ Suggestions for solving an optimization problem

Chapter 2: SUPPLEMENTARY EXERCISES

Properties of various functions are described below. In each case draw some conclusion about the graph of the function.

1. $f(1) = 2, f'(1) > 0$

2. $g(1) = 5, g'(1) = -1$

3. $h'(3) = 4, h''(3) = 1$

4. $F'(2) = -1, F''(2) < 0$

5. $G(10) = 2, G'(10) = 0, G''(10) > 0$

6. $f(4) = -2, f'(4) > 0, f''(4) = -1$

7. $g(5) = -1, g'(5) = -2, g''(5) = 0$

8. $H(0) = 0, H'(0) = 0, H''(0) = 1$

9. $F(-2) = 0, F'(-2) = 0, F''(-2) = -1$

10. $h(-3) = 4, h'(-3) = 1, h''(-3) = 0$

Sketch the following parabolas. Include their x- and y-intercepts.

11. $y = 3 - x^2$

12. $y = 7 + 6x - x^2$

13. $y = x^2 + 3x - 10$

14. $y = 4 + 3x - x^2$

15. $y = -2x^2 + 10x - 10$

16. $y = x^2 - 9x + 19$

17. $y = x^2 + 3x + 2$

18. $y = -x^2 + 8x - 13$

19. $y = -x^2 + 20x - 90$

20. $y = 2x^2 + x - 1$

Sketch the following curves.

21. $y = 2x^3 + 3x^2 + 1$

22. $y = x^3 - \frac{3}{2}x^2 - 6x$

23. $y = x^3 - 3x^2 + 3x - 2$

24. $y = 100 + 36x - 6x^2 - x^3$

25. $y = \frac{11}{3} + 3x - x^2 - \frac{1}{3}x^3$

26. $y = x^3 - 3x^2 - 9x + 7$

27. $y = -\frac{1}{3}x^3 - 2x^2 - 5x$

28. $y = x^3 - 6x^2 - 15x + 50$

29. $y = x^4 - 2x^2$

30. $y = x^4 - 4x^3$

31. Let $f(x) = (x^2 + 2)^{3/2}$. Show that the graph of $f(x)$ has a possible extreme point at $x = 0$.

32. Show that the function $f(x) = (2x^2 + 3)^{3/2}$ is decreasing for $x < 0$ and increasing for $x > 0$.

33. Let $f(x)$ be a function whose *derivative* is

$$f'(x) = \frac{1}{1 + x^2}.$$

   Note that $f'(x)$ is always positive. Show that the graph of $f(x)$ definitely has an inflection point at $x = 0$.

34. Let $f(x)$ be a function whose *derivative* is

$$f'(x) = \sqrt{5x^2 + 1}.$$

   Show that the graph of $f(x)$ definitely has an inflection point at $x = 0$.

35. A closed rectangular box is to be constructed such that the base has a length twice as long as its width. Suppose that the volume must be 9 cubic feet. Find the dimensions of the box that will minimize the total surface area and show that this minimum surface area is 27 square feet.

36. A closed rectangular box is to be constructed such that the base has a length twice as long as its width. Suppose that the total surface area must be 27 square feet. Find the dimensions of the box that will maximize the volume and show that this maximum volume is 9 cubic feet.

37. The strength S of a rectangular beam is proportional to its width x and the square of its depth y, so that $S = kxy^2$, where k is a positive constant (Fig. 13). Find the dimensions of the strongest beam that can be cut from a circular log of diameter 75 centimeters.

FIGURE 13

75 cm

38. A fabric store purchases scissors for $5 a pair and resells them for $9. It costs $10 to place an order with the manufacturer, and the yearly inventory cost per pair of scissors is $.80. If the store sells about 900 pairs of scissors each year at a relatively constant rate, how many scissors should be ordered at one time?

39. A California distributor of sporting equipment expects to receive orders during the coming year for 100,000 cans of tennis balls. Yearly inventory costs per can are about $.50, and the cost of placing an order with the manufacturer is $10. Assuming a fairly constant demand for the tennis balls, determine the optimal reorder quantity for the distributor.

40. Suppose that the demand equation for a monopolist is $p = 150 - .02x$ and the cost function is $C(x) = 10x + 300$. Find the value of x that maximizes the profit.

Techniques
of Differentiation

Chapter 3

We have seen that the derivative is useful in many applications. However, our ability to differentiate functions is somewhat limited. For example, we cannot yet readily differentiate

$$(x^{3/2} + 1)\sqrt{1 - x^2}, \qquad \frac{x}{x^2 + 1}, \qquad \frac{x^3}{3x^2 - 3x + 1}.$$

In this chapter we develop differentiation techniques that apply to functions like those given above. Two new rules are the *product rule* and the *quotient rule*. In Section 2 we extend the general power rule into a powerful formula called the *chain rule*.

3.1 The Product and Quotient Rules

We observed in our discussion of the sum rule for derivatives that the derivative of the sum of two differentiable functions is the sum of the derivatives. Unfortunately, however, the derivative of the product $f(x)g(x)$ is *not* the product of the derivatives. Rather, the derivative of a product is determined from the following rule.

Product Rule

$$\frac{d}{dx}\left[f(x)g(x)\right] = f(x)g'(x) + g(x)f'(x).$$

The derivative of the product of two functions is the first function times the derivative of the second plus the second function times the derivative of the first. At the end of the section we show why this statement is true.

EXAMPLE 1 Show that the product rule works for the case $f(x) = x^2$, $g(x) = x^3$.

Solution Since $x^2 \cdot x^3 = x^5$, we know that

$$\frac{d}{dx}[x^2 \cdot x^3] = \frac{d}{dx}[x^5] = 5x^4.$$

On the other hand, using the product rule,

$$\frac{d}{dx}(x^2 \cdot x^3) = x^2 \frac{d}{dx}(x^3) + x^3 \frac{d}{dx}(x^2)$$

$$= x^2(3x^2) + x^3(2x)$$

$$= 3x^4 + 2x^4 = 5x^4.$$

Thus the product rule gives the correct answer.

EXAMPLE 2 Differentiate the product $(2x^3 - 5x)(3x + 1)$.

Solution Let $f(x) = 2x^3 - 5x$ and $g(x) = 3x + 1$. Then

$$\frac{d}{dx}[(2x^3 - 5x)(3x + 1)] = (2x^3 - 5x) \cdot \frac{d}{dx}(3x + 1) + (3x + 1) \cdot \frac{d}{dx}(2x^3 - 5x)$$

$$= (2x^3 - 5x)(3) + (3x + 1)(6x^2 - 5)$$

$$= 6x^3 - 15x + 18x^3 - 15x + 6x^2 - 5$$

$$= 24x^3 + 6x^2 - 30x - 5.$$

EXAMPLE 3 Apply the product rule to $y = g(x) \cdot g(x)$.

Solution

$$\frac{d}{dx}[g(x) \cdot g(x)] = g(x) \cdot g'(x) + g(x) \cdot g'(x)$$

$$= 2g(x)g'(x).$$

This answer is the same as that given by the general power rule:

$$\frac{d}{dx}[g(x) \cdot g(x)] = \frac{d}{dx}[g(x)]^2 = 2g(x)g'(x).$$

EXAMPLE 4 Differentiate the function $(x^3 - 1)(x^2 + 1)^5$.

Solution Let $f(x) = x^3 - 1$ and $g(x) = (x^2 + 1)^5$. When we compute $g'(x)$ we will need the general power rule. Using the product rule first, we find that

$$\frac{d}{dx}[(x^3 - 1)(x^2 + 1)^5] = (x^3 - 1) \frac{d}{dx}(x^2 + 1)^5 + (x^2 + 1)^5 \frac{d}{dx}(x^3 - 1)$$

$$= (x^3 - 1) \cdot 5(x^2 + 1)^4(2x) + (x^2 + 1)^5 \cdot (3x^2)$$

$$= 10x(x^3 - 1)(x^2 + 1)^4 + 3x^2(x^2 + 1)^5.$$

The Quotient Rule   Another useful formula for differentiating functions is the quotient rule.

> **Quotient Rule**
>
> $$\frac{d}{dx}\left[\frac{f(x)}{g(x)}\right] = \frac{g(x)f'(x) - f(x)g'(x)}{[g(x)]^2}.$$

One must be careful to remember the order of the terms in this formula because of the minus sign in the numerator.

EXAMPLE 5   Differentiate $\dfrac{x^2 + 1}{x^3 + 5}$.

Solution   Let $f(x) = x^2 + 1$ and $g(x) = x^3 + 5$.

$$\frac{d}{dx}\left(\frac{x^2 + 1}{x^3 + 5}\right) = \frac{(x^3 + 5)\dfrac{d}{dx}(x^2 + 1) - (x^2 + 1)\dfrac{d}{dx}(x^3 + 5)}{(x^3 + 5)^2}$$

$$= \frac{(x^3 + 5)(2x) - (x^2 + 1)(3x^2)}{(x^3 + 5)^2}$$

$$= \frac{-x^4 - 3x^2 + 10x}{(x^3 + 5)^2}.$$

EXAMPLE 6   Suppose that the total cost of manufacturing x units of a certain product is given by the function $C(x)$. Then the *average cost per unit, AC,* is defined by

$$AC = \frac{C(x)}{x}.$$

Recall that the *marginal cost, MC,* is defined by

$$MC = C'(x).$$

Show that at the level of production where the average cost is at a minimum, the average cost equals the marginal cost.

Solution   In practice, the marginal cost and average cost curves will have the general shapes shown in Fig. 1. So the minimum point on the average cost curve will occur when $\dfrac{d}{dx}(AC) = 0$. To compute the derivative, we need the quotient rule,

$$\frac{d}{dx}(AC) = \frac{d}{dx}\left[\frac{C(x)}{x}\right] = \frac{x \cdot C'(x) - C(x)}{x^2}.$$

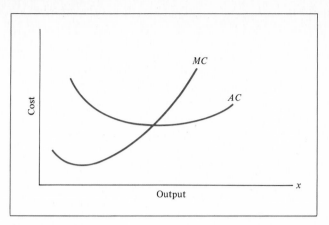

FIGURE 1   Marginal cost and average cost functions.

Setting the derivative equal to zero and multiplying by x^2, we obtain

$$0 = x \cdot C'(x) - C(x)$$

$$C(x) = x \cdot C'(x)$$

$$\frac{C(x)}{x} = C'(x)$$

$$AC = MC.$$

Thus when the output x is chosen so that the average cost is minimized, the average cost equals the marginal cost.

Verification of the Product and Quotient Rules

Verification of the Product Rule   From our discussion of limits we compute the derivative of $f(x)g(x)$ at $x = a$ as the limit

$$\frac{d}{dx}\left[f(x)g(x) \right]\Big|_{x=a} = \lim_{h \to 0} \frac{f(a + h)g(a + h) - f(a)g(a)}{h}.$$

Let us add and subtract the quantity $f(a)g(a + h)$ in the numerator. After factoring and applying Limit Theorem III, we obtain

$$\lim_{h \to 0} \frac{\left[f(a + h)g(a + h) - f(a)g(a + h) \right] + \left[f(a)g(a + h) - f(a)g(a) \right]}{h}$$

$$= \lim_{h \to 0} g(a + h) \cdot \frac{f(a + h) - f(a)}{h} + \lim_{h \to 0} f(a) \cdot \frac{g(a + h) - g(a)}{h}.$$

And this expression may be rewritten by Limit Theorem V as

$$\lim_{h \to 0} g(a + h) \cdot \lim_{h \to 0} \frac{f(a + h) - f(a)}{h} + \lim_{h \to 0} f(a) \cdot \lim_{h \to 0} \frac{g(a + h) - g(a)}{h}.$$

Note, however, that since $g(x)$ is differentiable at $x = a$, it is continuous there, so that $\lim_{h \to 0} g(a + h) = g(a)$. Therefore, the expression above equals

$$g(a)f'(a) + f(a)g'(a).$$

That is, we have proved that

$$\frac{d}{dx}[f(x)g(x)]\bigg|_{x=a} = g(a)f'(a) + f(a)g'(a),$$

which is the product rule. An alternative verification of the product rule not involving limit arguments is outlined in Exercise 45.

Verification of the Quotient Rule   From the general power rule, we know that

$$\frac{d}{dx}\left[\frac{1}{g(x)}\right] = \frac{d}{dx}[g(x)]^{-1} = (-1)[g(x)]^{-2} \cdot g'(x).$$

We can now derive the quotient rule from the product rule.

$$\frac{d}{dx}\left[\frac{f(x)}{g(x)}\right] = \frac{d}{dx}\left[\frac{1}{g(x)} \cdot f(x)\right]$$

$$= \frac{1}{g(x)} \cdot f'(x) + f(x) \cdot \frac{d}{dx}\left[\frac{1}{g(x)}\right]$$

$$= \frac{g(x)f'(x)}{[g(x)]^2} + f(x) \cdot (-1)[g(x)]^{-2} \cdot g'(x)$$

$$= \frac{g(x)f'(x) - f(x)g'(x)}{[g(x)]^2}.$$

PRACTICE PROBLEMS 1

1. Consider the function $y = (\sqrt{x} + 1)x$.

    (a) Differentiate y by the product rule.

    (b) Multiply out first and then differentiate.

2. Differentiate $y = x/(x^4 - x^3 + 1)$.

EXERCISES 1

Differentiate the following functions.

1. $(x + 1)(x^3 + 5x + 2)$

2. $(2x - 1)(x^2 - 3)$

3. $(3x^2 - x + 2)(2x^2 - 1)$

4. $(x^3 + x^2 + x)(5x - 4)$

5. $(x^4 + 1)(3x + 5)$

6. $(10x^2 - 2x + 1)(x^2 + 7x - 8)$

7. $(x^2 - 3x + 1)\left(x + \frac{2}{x}\right)$

8. $\left(4 - \frac{1}{x^2}\right)(3x^3 - 5x + 8)$

9. $(2x - 7)(x - 1)^5$

10. $(x^2 + 4)(x + 3)^6$

11. $(x^2 + 3)(x^2 - 3)^{10}$

12. $(x + 1)^2(x - 3)^3$

13. $(2x + 9)^{-2}(5 - 3x)$

14. $(1 - x^2)(x^2 - 4x - 1)^{-4}$

15. $\sqrt{x}(3x^2 - 1)^3$

16. $(5x - 1)^4\left(\sqrt{x} + \dfrac{1}{\sqrt{x}}\right)$

17. $(1 - 4x)\sqrt{x^2 - 1}$

18. $(x^{3/2} + 2x^{1/2})(4x^{5/2} - x^{3/2})$

19. $\dfrac{x^2 - 1}{x^2 + 1}$

20. $\dfrac{x^2 + x + 1}{5 - x}$

21. $\dfrac{1}{5x^2 + 2x + 5}$

22. $\dfrac{2x - 1}{x}$

23. $\dfrac{\sqrt{x}}{x + 1}$

24. $\dfrac{3x^2 + 5x + 2}{x^2 - 3}$

25. $\dfrac{x}{(x^2 + 1)^2}$

26. $\dfrac{4 + x - x^2}{1 + \sqrt{x}}$

27. $\dfrac{x - 2}{x + 2}$

28. $\dfrac{x - 1}{(2x + 1)^2}$

29. $\dfrac{3 - x}{\sqrt{4x + x^2}}$

30. $\dfrac{1 + \sqrt{2x}}{1 - \sqrt{2x}}$

31. $\dfrac{(x^5 + 1)^3}{(3x^2 + 1)^2}$

32. $\left(\dfrac{4x - 1}{3x + 1}\right)^3$

33. Find the equation of the tangent line to the curve $y = (x - 2)^5(x + 1)^2$ at the point $(3, 16)$.

34. Find the equation of the tangent line to the curve $y = (x + 1)/(x - 1)$ at the point $(2, 3)$.

35. A sugar refinery can produce x tons of sugar per week at a weekly cost of $\frac{1}{10}x^2 + 5x + 2250$ dollars. Find the level of production for which the average cost is at a minimum and show that the average cost equals the marginal cost at that level of production.

36. A cigar manufacturer produces x cases of cigars per day at a daily cost of $50x(x + 200)/(x + 100)$ dollars. Show that his cost increases and his average cost decreases as output increases.

37. An open rectangular box is 3 feet long and has a surface area of 16 square feet. Find the dimensions of the box for which the volume is as large as possible.

38. A closed rectangular box is to be constructed with one side 1 meter long. The material for the top costs $20 per square meter, and the material for the sides and bottom costs $10 per square meter. Find the dimensions of the box with largest possible volume that can be built at a cost of $240 for materials.

39. (*A Voting Model*) Let x be the proportion of the total popular vote a Democratic candidate for president receives in a U.S. national election. Political scientists have observed that a good estimate of the proportion of seats in the House of Representatives going to Democratic candidates is given by

$$f(x) = \frac{x^3}{3x^2 - 3x + 1}, \qquad 0 \le x \le 1.$$

This formula is referred to as the "cube law" by some political scientists, primarily because $f(x)$ may also be written as

$$f(x) = \frac{x^3}{x^3 + (1 - x)^3}.$$

(Note that $1 - x$ is approximately the proportion of votes received by the Republicans.)

(a) Are $f(0)$, $f(\frac{1}{2})$, and $f(1)$ what you would expect them to be?

(b) Show that $f'(\frac{1}{2}) = 3$.

(c) It can be shown (by the quotient rule) that

$$f''(x) = \frac{6x(x - 1)(2x - 1)}{(3x^2 - 3x + 1)^3}.$$

What does this tell you about $f(x)$?

(d) Sketch the graph of $y = f(x)$ for $0 \le x \le 1$.

40. Let $f(x) = 1/x$ and $g(x) = x^3$.

(a) Show that the product rule yields the correct derivative of $(1/x) \cdot x^3 = x^2$.

(b) Compute the product $f'(x)g'(x)$ and note that it is *not* the derivative of $f(x)g(x)$.

41. The derivative of $(x^3 - 4x)/x$ is obviously $2x$ for $x \ne 0$, because $(x^3 - 4x)/x = x^2 - 4$ for $x \ne 0$. Verify that the quotient rule gives the same derivative.

42. Let $f(x)$, $g(x)$, $h(x)$ be differentiable functions. Find a formula for the derivative of $f(x)g(x)h(x)$.

43. Suppose that $f(x)$ is a function whose derivative is $f'(x) = 1/x$. Find the derivative of $xf(x) - x$.

44. Suppose that $f(x)$ is a function whose derivative is $f'(x) = \dfrac{1}{1 + x^2}$. Find the derivative of $\dfrac{f(x)}{1 + x^2}$.

45. (*Alternative Verification of the Product Rule*) Apply the special case of the general power rule $\dfrac{d}{dx}[h(x)]^2 = 2h(x)h'(x)$ and the identity $fg = \frac{1}{4}[(f + g)^2 - (f - g)^2]$ to prove the product rule.

SOLUTIONS TO PRACTICE PROBLEMS 1

1. (a) Apply the product rule to $y = (\sqrt{x} + 1)x$ with

$$f(x) = \sqrt{x} + 1 = x^{1/2} + 1$$

$$g(x) = x$$

$$\frac{dy}{dx} = (x^{1/2} + 1) \cdot 1 + x \cdot \tfrac{1}{2}x^{-1/2}$$

$$= x^{1/2} + 1 + \tfrac{1}{2}x^{1/2}$$

$$= \tfrac{3}{2}\sqrt{x} + 1.$$

(b) $y = (\sqrt{x} + 1)x = (x^{1/2} + 1)x = x^{3/2} + x.$

$$\frac{dy}{dx} = \tfrac{3}{2}x^{1/2} + \tfrac{3}{2}\sqrt{x} + 1.$$

Comparing parts (a) and (b), we note that sometimes it is helpful to simplify the function before differentiating.

2. Apply the quotient rule to $y = x/(x^4 - x^3 + 1)$.

$$\frac{dy}{dx} = \frac{(x^4 - x^3 + 1) \cdot 1 - x(4x^3 - 3x^2)}{(x^4 - x^3 + 1)^2}$$

$$= \frac{x^4 - x^3 + 1 - 4x^4 + 3x^3}{(x^4 - x^3 + 1)^2}$$

$$= \frac{-3x^4 + 2x^3 + 1}{(x^4 - x^3 + 1)^2}.$$

3.2 The Chain Rule and the General Power Rule

A useful way of combining functions $f(x)$ and $g(x)$ is to replace every occurrence of the variable x in $f(x)$ by the function $g(x)$. The resulting function is called the *composition* (or *composite*) of $f(x)$ and $g(x)$ and is denoted $f(g(x))$.

EXAMPLE 1   Let $f(x) = x^2$, $g(x) = x^3 - 3x + 1$. What is $f(g(x))$?

Solution   Replace each occurrence of x in $f(x)$ by $g(x)$ to obtain

$$f(g(x)) = [g(x)]^2 = (x^3 - 3x + 1)^2.$$

EXAMPLE 2   Let $f(x) = \dfrac{x - 1}{x + 1}$, $g(x) = x^3$. What is $f(g(x))$?

Solution   Replace each occurrence of x in $f(x)$ by $g(x)$ to obtain

$$f(g(x)) = \frac{g(x) - 1}{g(x) + 1}$$

$$= \frac{x^3 - 1}{x^3 + 1}.$$

EXAMPLE 3   Let $f(x) = \dfrac{1}{x}$, $g(x) = x^2 + 3x + 1$. What are $f(g(x))$ and $g(f(x))$?

Solution

$$f(g(x)) = \frac{1}{g(x)} = \frac{1}{x^2 + 3x + 1}.$$

$$g(f(x)) = [f(x)]^2 + 3[f(x)] + 1$$

$$= \left(\frac{1}{x}\right)^2 + 3 \cdot \left(\frac{1}{x}\right) + 1$$

$$= \frac{1}{x^2} + \frac{3}{x} + 1.$$

Not only is it important to be able to compute $f(g(x))$ when $f(x)$ and $g(x)$ are given, but it is also important (see below) to be able to recognize a given function as the composite of two simpler functions.

EXAMPLE 4 Let $h(x) = (x^5 + 9x + 7)^{18}$, $k(x) = \sqrt{3x^2 + 1}$. Write $h(x)$ and $k(x)$ as composites of simpler functions.

Solution (a) Let $f(x) = x^{18}$, $g(x) = x^5 + 9x + 7$. Then $f(g(x)) = (x^5 + 9x + 7)^{18} = h(x)$.
(b) Let $f(x) = \sqrt{x}$, $g(x) = 3x^2 + 1$. Then $f(g(x)) = \sqrt{3x^2 + 1} = k(x)$.

A function of the form $[g(x)]^r$ is a composite $f(g(x))$, where $f(x) = x^r$. Moreover, we have already given a rule for differentiating this function—namely

$$\frac{d}{dx}[g(x)]^r = r[g(x)]^{r-1} \cdot g'(x). \tag{1}$$

This rule is a special case of the *chain rule*, which describes how to differentiate a composite $f(g(x))$. This rule enables us to calculate the derivative of $f(g(x))$ in terms of the derivatives of $f(x)$ and $g(x)$.

The Chain Rule To differentiate $f(g(x))$, first differentiate $f(x)$ and substitute $g(x)$ for x in the result. Then multiply by the derivative of $g(x)$. Symbolically,

$$\frac{d}{dx}[f(g(x))] = f'(g(x)) \cdot g'(x).$$

EXAMPLE 5 Use the chain rule to compute the derivative of $h(x) = (x^5 + 9x + 7)^{18}$.

Solution We saw in Example 4(a) that $h(x)$ can be represented as the composite $f(g(x))$, where $f(x) = x^{18}$, $g(x) = x^5 + 9x + 7$. Therefore, we apply the chain rule to compute $h'(x)$.

$$f'(x) = 18x^{17}, \qquad g'(x) = 5x^4 + 9$$

$$f'(g(x)) = 18(x^5 + 9x + 7)^{17}$$

$$f'(g(x)) \cdot g'(x) = 18(x^5 + 9x + 7)^{17} \cdot (5x^4 + 9).$$

Therefore,

$$h'(x) = 18(x^5 + 9x + 7)^{17} \cdot (5x^4 + 9).$$

Note that this is the same answer as derived using the general power rule.

EXAMPLE 6 Use the chain rule to compute the derivative of

$$h(x) = \frac{(x^3 + 1)^5 - 1}{(x^3 + 1)^5 + 1}.$$

Solution Note that $h(x) = f(g(x))$, where

$$f(x) = \frac{x^5 - 1}{x^5 + 1}, \qquad g(x) = x^3 + 1.$$

By the quotient rule,

$$f'(x) = \frac{(x^5 + 1) \cdot 5x^4 - (x^5 - 1) \cdot 5x^4}{(x^5 + 1)^2} = \frac{10x^4}{(x^5 + 1)^2}.$$

Therefore,

$$f'(g(x)) = \frac{10(x^3 + 1)^4}{[(x^3 + 1)^5 + 1]^2}.$$

$$g'(x) = 3x^2.$$

$$h'(x) = f'(g(x)) \cdot g'(x)$$

$$= \frac{10(x^3 + 1)^4}{[(x^3 + 1)^5 + 1]^2} \cdot 3x^2$$

$$= \frac{30x^2(x^3 + 1)^4}{[(x^3 + 1)^5 + 1]^2}.$$

Note that this answer can also be derived using the quotient rule together with the general power rule.

Here is a good mnemonic device for remembering the chain rule. Suppose that we wish to differentiate a function of the form $y = f(g(x))$. If we set $u = g(x)$, we can view y as both a function of x and a function of u. Then the chain rule can be stated:

$$\frac{dy}{dx} = \frac{dy}{du} \cdot \frac{du}{dx}.$$

That is, to compute the derivative of y with respect to x, we multiply the derivative of y with respect to u, by the derivative of u with respect to x. [Of course, we express $\frac{dy}{du}$ as a function of x by setting $u = g(x)$.] You may remember this formulation of the chain rule by thinking of the derivative as a fraction and the equation above arising from "canceling" the du's.

At this stage, the chain rule does not enable us to differentiate any functions which we could not already differentiate via the general power rule. However, as we shall see in our discussion of the exponential and logarithmic functions, the chain rule does indeed expand the repertoire of functions we can differentiate.

When the power rule and the general power rule were introduced in Chapter 1, we promised to justify these rules later in the text. The general power rule can be deduced from the chain rule and the ordinary power rule. The chain rule is justified below. We will verify the ordinary power rule in Chapter 4.

Verification of the Chain Rule Suppose that $f(x)$ and $g(x)$ are differentiable, and let $x = a$ be a number in the domain of $f(g(x))$. Since every differentiable function is continuous, we have

$$\lim_{h \to 0} g(a + h) = g(a),$$

which implies that

$$\lim_{h \to 0} \left[g(a + h) - g(a) \right] = 0. \tag{2}$$

Now $g(a)$ is a number in the domain of f, and the limit definition of the derivative gives us

$$f'(g(a)) = \lim_{k \to 0} \frac{f(g(a) + k) - f(g(a))}{k}. \tag{3}$$

Let $k = g(a + h) - g(a)$. By equation (2), k approaches zero as h approaches zero. Also, $g(a + h) = g(a) + k$. Therefore, (3) may be rewritten in the form

$$f'(g(a)) = \lim_{h \to 0} \frac{f(g(a + h)) - f(g(a))}{g(a + h) - g(a)}. \tag{4}$$

[Strictly speaking, we must assume that the denominator in (4) is never zero. This assumption may be avoided by a somewhat different and more technical argument which we omit.] Finally, we show that the function $f(g(x))$ has a derivative at $x = a$. We use the limit definition of the derivative, Limit Theorem V, and (4) above.

$$\frac{d}{dx} \left[f(g(x)) \right] \bigg|_{x=a} = \lim_{h \to 0} \frac{f(g(a + h)) - f(g(a))}{h}$$

$$= \lim_{h \to 0} \left[\frac{f(g(a + h)) - f(g(a))}{g(a + h) - g(a)} \cdot \frac{g(a + h) - g(a)}{h} \right]$$

$$= \lim_{h \to 0} \frac{f(g(a + h)) - f(g(a))}{g(a + h) - g(a)} \cdot \lim_{h \to 0} \frac{g(a + h) - g(a)}{h}$$

$$= f'(g(a)) \cdot g'(a).$$

PRACTICE PROBLEMS 2

Consider the function $h(x) = (2x^3 - 5)^5 + (2x^3 - 5)^4$.

1. Write $h(x)$ as a composite function, $f(g(x))$.

2. Compute $f'(x)$ and $f'(g(x))$.

3. Use the chain rule to differentiate $h(x)$.

EXERCISES 2

Compute $f(g(x))$, where $f(x)$ and $g(x)$ are the following.

1. $f(x) = \dfrac{1}{x + 1}, g(x) = x^3$

2. $f(x) = \dfrac{1}{x^2 - 1}, g(x) = \dfrac{1}{x}$

3. $f(x) = \sqrt{x + 1}, g(x) = \dfrac{1}{x^2 - 1}$

4. $f(x) = \sqrt{x}, g(x) = \dfrac{x - 1}{x + 1}$

5. $f(x) = x^2, g(x) = x^2$

6. $f(x) = \sqrt{x}, g(x) = \sqrt[3]{x}$

Write the following functions as composites of two functions.

7. $\sqrt{x^2 + 3x - 1}$

8. $(2x - 3)^3 + (2x - 3)^2 + \dfrac{1}{2x - 3}$

9. $\sqrt{\dfrac{x - 1}{x + 1}}$

10. $\dfrac{\sqrt{x} - 1}{\sqrt{x} + 1}$

11. $(x^3 + 9x - 2)^5$

12. $(9x^2 + 2x - 5)^7$

13. $\dfrac{x^2 - 1}{x^2 + 1}$

14. $\dfrac{x^3 + 3x^6}{1 + x^9}$

15. $\sqrt{x^{1/3} + 1}$

16. $(\sqrt{x})^{1/3}$

17. $(x^3 + 1)^2 + 1$

18. $(x^2 + 1)^3 + 1$

19. $x^2(x^4 + 1)^2$

20. $(x^2 + 1)(x^4 + 1)^3$

Differentiate the following functions using the product rule, quotient rule, and chain rule.

21. $(x^2 + 1)^{15}$

22. $(x^4 + x^2 + 1)^{10}$

23. $\sqrt{1 + x^4}$

24. $(2x^3 + 3x - 1)^{-7}$

25. $\left[\dfrac{2}{x} + 5(3x - 1)^4\right]^3$

26. $[(2x^2 + 1)^3 + 5]^4$

27. $(x^3 - 1)(x^2 + 1)^4$

28. $(2x^2 - 4)^5(5x + 1)^3$

29. $(4 - x)^3(4 + x)^3$

30. $(2x - 1)^{1/4}(2x + 1)^{3/4}$

31. $\sqrt{2 + \sqrt{1 + x^2}}$

32. $\left(\dfrac{x - 1}{x + 1}\right)^3$

33. $\dfrac{x}{\sqrt{1 + x^2}}$

34. $x^2\sqrt{x^2 - 9}$

35. $\dfrac{1 + \sqrt{2x}}{1 - \sqrt{2x}}$

36. $\dfrac{(x^5 + 1)^3}{(3x^2 + 1)^4}$

37. Sketch the graph of $y = 4x/(x + 1)^2$, $(x > -1)$.

38. Sketch the graph of $y = 2/(1 + x^2)$.

39. Suppose that $f(x)$ and $g(x)$ are differentiable functions. Find $g(x)$ if you know that

$$\frac{d}{dx} f(g(x)) = 3x^2 \cdot f'(x^3 + 1).$$

40. Suppose that $f(x)$ and $g(x)$ are differentiable functions. Find $g(x)$ if you know that $f'(x) = 1/x$ and

$$\frac{d}{dx} f(g(x)) = \frac{2x + 5}{x^2 + 5x - 4}.$$

SOLUTIONS TO PRACTICE PROBLEMS 2

1. Let $f(x) = x^5 + x^4$ and $g(x) = 2x^3 - 5$.

2. $f'(x) = 5x^4 + 4x^3$, $f'(g(x)) = 5(2x^3 - 5)^4 + 4(2x^3 - 5)^3$.

3. We have $g'(x) = 6x^2$. Then, from the chain rule and the result of Problem 2, we have

$$h'(x) = f'(g(x))g'(x) = [5(2x^3 - 5)^4 + 4(2x^3 - 5)^3](6x^2).$$

3.3 Implicit Differentiation and Related Rates

In this section we apply the chain rule to implicit differentiation and to related rates problems.

Implicit Differentiation If a function is defined directly by a formula, such as

$$y = (x^2 + 1)^3,$$

then we say that the function is defined *explicitly*. However, in many applications, functions are not defined explicitly but are specified only by an equation involving x and y. An example of such an equation is

$$x^3 + y^8 - y = 9. \tag{1}$$

We say that this equation determines y *implicitly* as a function of x. Roughly speaking, to obtain y explicitly as a function of x, it would be necessary to solve for y in terms of x. However, as the equation above illustrates, this may not be a simple task.

Using the technique of implicit differentiation, it is possible to calculate derivatives of implicitly defined functions without ever determining the functions explicitly! The basic idea is to differentiate all terms in an equation with respect to x, treating y as a function of x, say $y = f(x)$. Let us illustrate this technique for equation (1) above. The first term x^3 has derivative $3x^2$, as usual. The term y^8, which we think of as $(f(x))^8$, must be differentiated using the general power rule.

$$\frac{d}{dx}(f(x))^8 = 8(f(x))^7 f'(x)$$

$$= 8y^7 y'.$$

Finally, the derivatives of the terms y and 9 are, respectively, y' and 0. Thus differentiation of the original equation gives us the relationship

$$3x^2 + 8y^7 y' - y' = 0.$$

Let us solve this equation for y'. We place all terms involving y' on one side of the equation, factor out the y', and then solve for y'.

$$8y^7 y' - y' = -3x^2$$

$$(8y^7 - 1)y' = -3x^2$$

$$y' = \frac{-3x^2}{8y^7 - 1}.$$

The last equation gives us a formula for the derivative y'. Note, however, that this formula involves not only x, but y as well. So to evaluate y', it is necessary to have values for both x and y. This is the price we pay for our original function being given implicitly. However, it is a rather small price. The original equation cannot be solved for y in terms of x in any elementary fashion. Nevertheless, we may calculate the derivative of y.

EXAMPLE 1   (a) Use implicit differentiation to determine y' if y is defined implicitly by the equation $x^2 y + xy - 3x = 5$.

(b) Use the formula of part (a) to evaluate y' when $x = 1$, $y = 4$.

Solution   (a) We differentiate the equation term by term, taking care to differentiate $x^2 y$ and xy using the product rule and to treat y as a function of x.

$$x^2 y' + 2xy + xy' + y - 3 = 0.$$

We now solve for y' in terms of x and y.

$$x^2 y' + xy' = 3 - y - 2xy$$

$$(x^2 + x)y' = 3 - y - 2xy$$

$$y' = \frac{3 - y - 2xy}{x^2 + x}.$$

(b) When $x = 1$, $y = 4$, we have

$$y' = \frac{3 - 4 - 2(1)(4)}{1^2 + 1} = -\frac{9}{2}.$$

EXAMPLE 2 Use implicit differentiation to calculate $\dfrac{dy}{dx}$ when y and x are related by the equation

$$x^5 + y^5 = 2x^2.$$

Solution
$$5x^4 + 5y^4 \frac{dy}{dx} = 4x$$

$$5y^4 \frac{dy}{dx} = 4x - 5x^4$$

$$\frac{dy}{dx} = \frac{4x - 5x^4}{5y^4}.$$

It so happens that the equation in Example 2 can be solved explicitly for y in terms of x. Namely, we have

$$y^5 = 2x^2 - x^5$$

$$y = (2x^2 - x^5)^{1/5}.$$

We may differentiate this formula by the general power rule, obtaining

$$\frac{dy}{dx} = \frac{1}{5}(2x^2 - x^5)^{-4/5}(4x - 5x^4).$$

In some respects, this formula for $\dfrac{dy}{dx}$ is more cumbersome than the formula obtained in Example 2.

Related Rates In implicit differentiation, we differentiate an equation involving x and y, where y is viewed as a function of x. However, in many applications, x and y are related by an equation, but both are functions of a third variable t (which often represents time). Then by differentiating the equation connecting x and y we may derive a relationship between the derivatives $\dfrac{dx}{dt}$ and $\dfrac{dy}{dt}$.

EXAMPLE 3 Suppose that x and y are both functions of t and are related by the equation

$$x^2 + 5xy = 1.$$

(a) Compute $\dfrac{dy}{dt}$ in terms of $\dfrac{dx}{dt}$, x and y.

(b) Suppose that at a certain time $x = 4$, $y = -\frac{3}{4}$, and $\dfrac{dx}{dt} = 80$. Calculate $\dfrac{dy}{dt}$.

Solution (a) We differentiate the given equation with respect to t. The derivative of x^2 is then $2x \dfrac{dx}{dt}$ (by the chain rule, with x regarded as a function of t), and the derivative of $5xy$ is $5x \dfrac{dy}{dt} + 5y \dfrac{dx}{dt}$ (by the product rule). Thus we have

$$2x \frac{dx}{dt} + 5x \frac{dy}{dx} + 5y \frac{dx}{dt} = 0$$

$$5x \frac{dy}{dt} = -2x \frac{dx}{dt} - 5y \frac{dx}{dt}$$

$$= -(2x + 5y) \frac{dx}{dt}.$$

Finally,

$$\frac{dy}{dt} = -\frac{2x + 5y}{5x} \cdot \frac{dx}{dt}.$$

(b) If $x = 4$, $y = -\frac{3}{4}$, and $\dfrac{dx}{dt} = 80$, then

$$\frac{dy}{dt} = -\frac{2(4) + 5(-\frac{3}{4})}{5(4)} \cdot 80$$

$$= -17.$$

EXAMPLE 4 A corporation manufactures manual typewriters. Its research department estimates that the demand for its product t years from now will be x hundred thousand typewriters per year, where

$$x = \frac{30}{t + 2} + 1.$$

The profit P (in hundreds of thousands of dollars per year) from the sale of manual typewriters is related to the quantity sold by the equation

$$2P - 3x^2 = -5.$$

At what rate will profits be changing 3 years from now?

Solution Both P and x are functions of t. We are asked to calculate $\dfrac{dP}{dt}$ when $t = 3$. From the equation relating P and x we see that

$$2 \frac{dP}{dt} - 3 \cdot 2x \cdot \frac{dx}{dt} = 0$$

$$\frac{dP}{dt} = 3x \frac{dx}{dt}.$$

However,

$$\frac{dx}{dt} = -\frac{30}{(t + 2)^2},$$

so that

$$\frac{dP}{dt} = 3x\left[-\frac{30}{(t+2)^2}\right]$$

$$= -\frac{90x}{(t+2)^2}.$$

In particular, when $t = 3$, we have $x = 30/(t+2) + 1 = 30/5 + 1 = 7$, and

$$\frac{dP}{dt} = -\frac{90(7)}{5^2}$$

$$= -25.2.$$

In other words, 3 years from now, profits will be declining at the rate of 25.2 hundred thousand dollars per year, or 2.52 million dollars per year.

EXAMPLE 5 A corporate planning department estimates that the demand q (in thousands of units per year) for a certain type of computer terminal is related to their price by the demand equation $q = 100 - .08p$. Suppose that technological advances are causing the price to fall at a rate of \$200 per year. The current price of a terminal is \$1000. At what rate are revenues changing?

Solution The revenue R is related to the quantity sold by the equation

$$[\text{revenue}] = [\text{quantity}] \cdot [\text{price}]$$

$$R = q \cdot p$$

$$= (100 - .08p)p$$

$$= 100p - .08p^2.$$

Therefore, the rate at which revenue is changing equals

$$\frac{dR}{dt} = 100\frac{dp}{dt} - .08 \cdot \left(2p\frac{dp}{dt}\right)$$

$$= 100\frac{dp}{dt} - .16p\frac{dp}{dt}.$$

We are given that $p = 1000$ and $\frac{dp}{dt} = -200$. Therefore,

$$\frac{dR}{dt} = 100(-200) - .16(1000)(-200)$$

$$= -20{,}000 + 32{,}000$$

$$= 12{,}000.$$

Therefore, revenue is increasing at the rate of \$12,000 per year. (Can you explain why revenue is increasing even though the price is falling?)

EXAMPLE 6   An epidemiologist finds that a certain tropical disease is carried by mosquitoes. He relates the number of reported cases N of the disease to the number of mosquitoes M observed in a certain marsh, where the mosquito population is estimated using a standardized experiment. He finds that N and M are related by the equation

$$N = .0048M^{2/3}.$$

A program is initiated to destroy the breeding grounds of the mosquito. At one point the mosquito population is 8,000,000, and the program is estimated to be destroying 1,000,000 mosquitoes per day. At what rate is the number of reported cases of the disease changing?

Solution   We are asked to calculate $\dfrac{dN}{dt}$. From the given equation, we have

$$\frac{dN}{dt} = .0032M^{-1/3}\frac{dM}{dt}.$$

We are given that $M = 8{,}000{,}000$ and $\dfrac{dM}{dt} = -1{,}000{,}000$. Thus

$$\frac{dN}{dt} = .0032(8{,}000{,}000)^{-1/3}(-1{,}000{,}000)$$

$$= .0032\left(\frac{1}{200}\right)(-1{,}000{,}000)$$

$$= -16.$$

That is, the number of reported cases of the disease is decreasing at the rate of 16 cases per day.

PRACTICE PROBLEMS 3

Suppose that x and y are related by the equation

$$x^2y^2 + 4x = 3y^4.$$

1. Use implicit differentiation to find a formula for $\dfrac{dy}{dx}$.

2. Find $\dfrac{dy}{dx}$ for $x = 3$, $y = -2$.

EXERCISES 3

In Exercises 1–12, suppose that x and y are related by the given equation and use implicit differentiation to determine $\dfrac{dy}{dx}$.

1. $x^2 + y^2 = 1$ 2. $x^3 + y^3 - 6 = 0$ 3. $y^5 - 3x^2 = x$

4. $x^4 + (y + 3)^4 = x^2$ 5. $xy = 5$ 6. $xy^3 = 2$

7. $xy^{-2} = 1$ 8. $x(y + 2)^5 = 8$ 9. $(x + 1)^2(y - 1)^2 = 1$

10. $x^3y^2 - 4x^2 = 1$ 11. $y^2 - y = x^2$ 12. $x + y - xy = 0$

In Exercises 13–18, suppose that x and y are related by the given equation. Use implicit differentiation to calculate the value of $\dfrac{dy}{dx}$ for the given values of x and y.

13. $4y^3 - x^2 = -5; x = 3, y = 1$ 14. $xy = 9; x = -3, y = -3$

15. $xy + y^3 = 14; x = 3, y = 2$ 16. $y^2 - 3xy = -5; x = 2, y = 1$

17. $x^2y + y^2x = 6; x = 1, y = -3$ 18. $x^3 + y^3 - 2xy = 0; x = 1, y = 1$

In Exercises 19–24, suppose that x and y are both functions of t and are related by the given equation. Use implicit differentiation to determine $\dfrac{dy}{dt}$ in terms of x, y, and $\dfrac{dx}{dt}$.

19. $x^4 + y^4 = 1$ 20. $y^4 - x^2 = 1$ 21. $3xy - 3x^2 = 4$

22. $x^4y^4 = 10$ 23. $x^2 + 2xy = y^3$ 24. $x^2y^2 = 2y^3 + 1$

25. A manufacturer of microcomputers estimates that t months from now it will sell x thousand units of its main line of microcomputers per month, where $x = .05t^2 + 2t + 5$. Because of economies of scale, the profit P from manufacturing and selling x thousand units is estimated to be $P = .001x^2 + .1x - .25$ million dollars. Calculate the rate at which the profit will be increasing 5 months from now.

26. Ecologists estimate that the average level of carbon monoxide in the air above a certain city will be $1 + .4x + .0001x^2$ ppm (parts per million) when the population is x thousand persons. The population of the city is estimated to be $x = 750 + 25t + .1t^2$ thousand people t years from the present.

(a) Find the rate of change of carbon monoxide with respect to the population of the city.

(b) How fast (with respect to time) is the carbon monoxide level changing at time $t = 2$?

27. A town library estimates that when the town's population is x thousand persons, approximately y thousand books will be checked out of the library during 1 year, where x and y are related by the equation $y^3 - 8000x^2 = 0$. Currently, the population is 27 thousand persons, and the library's book circulation is 180 thousand books per year. At what rate will the library's book circulation rise if the population is increasing at the rate of 1.2 thousand persons per year?

28. Suppose that in a certain city the wholesale price p of oranges (in dollars per crate) and the daily supply x (in thousands of crates) are related by the equation $px + 7x + 8p = 328$. If there are 4000 crates available today at a price of \$25 per crate, and if the supply is decreasing at the rate of 300 crates (i.e., .3 thousand) crates per day, at what rate is the price changing?

29. The volume V of a sphere of radius r is $V = \frac{4}{3}\pi r^3$. When helium is pumped into a spherical balloon, both the radius and volume change with respect to time.

(a) Express $\dfrac{dr}{dt}$ in terms of r and $\dfrac{dV}{dt}$.

(b) Suppose that the volume of the balloon is increasing at the rate of 500 cm^3 per minute. At what rate is the radius of the balloon changing when the radius is 10 cm?

30. A rectangular aquarium with a 50-cm by 80-cm base is being filled with water at the rate of 5000 cm^3 per minute. At what rate is the height of the water changing when the height is 15 cm?

SOLUTIONS TO PRACTICE PROBLEMS 3

1. We differentiate term by term, regarding y as a function of x.

$$\frac{d}{dx}(x^2 y^2) + \frac{d}{dx}(4x) = \frac{d}{dx}(3y^4)$$

$$x^2 \cdot 2y\frac{dy}{dx} + y^2 \cdot 2x + 4 = 12y^3\frac{dy}{dx}.$$

Next, we move the terms involving $\dfrac{dy}{dx}$ to the left side of the equation and move the other terms to the right side. Then we can factor out $\dfrac{dy}{dx}$ and solve for $\dfrac{dy}{dx}$.

$$2x^2 y\frac{dy}{dx} - 12y^3\frac{dy}{dx} = -4 - 2xy^2$$

$$(2x^2 y - 12y^3)\frac{dy}{dx} = -4 - 2xy^2$$

$$\frac{dy}{dx} = \frac{-4 - 2xy^2}{2x^2 y - 12y^3} = \frac{-2 - xy^2}{x^2 y - 6y^3}.$$

2. Substituting $x = 3$ and $y = -2$, we have

$$\frac{dy}{dx} = \frac{-2 - (3)(-2)^2}{3^2(-2) - 6(-2)^3} = \frac{-14}{30} = -\frac{7}{15}.$$

Chapter 3: CHECKLIST

☐ $\dfrac{d}{dx}[f(x) \cdot g(x)] = f(x)g'(x) + g(x)f'(x)$

☐ $\dfrac{d}{dx}\left[\dfrac{f(x)}{g(x)}\right] = \dfrac{g(x)f'(x) - f(x)g'(x)}{[g(x)]^2}$

☐ $\dfrac{d}{dx}[f(g(x))] = f'(g(x)) \cdot g'(x)$

☐ Implicit differentiation
☐ Related rates

Chapter 3: SUPPLEMENTARY EXERCISES

Differentiate the following functions.

1. $\sqrt{x}(x^3 + 1)$

2. $\sqrt{x + 2}(x^2 + 5x + 1)$

3. $\dfrac{\sqrt{x}}{\sqrt{x + 1}}$

4. $(x^2 - 1)^5(x^3 + 5)^4$

5. $(x^2 + 5x + 9)(x^3 - 2)^{10}$

6. $(\sqrt{x} - 1)^3(\sqrt[3]{x} - 2)^2$

7. $\dfrac{(x + 2)^2 - (x - 2)^2}{x^2 + 2}$

8. $\dfrac{x - \sqrt{x}}{x + \sqrt{x}}$

9. $(x + 2 + (x + 2)^2)^2$

10. $(\sqrt{x} + \sqrt{x + 1})^2$

In Exercises 11–16, find formulas for $f(g(x))$ and $f'(g(x))$, where $g(x)$ represents an arbitrary differentiable function.

11. $f(x) = x^2 + x + 3$

12. $f(x) = x^3 - 8x$

13. $f(x) = \sqrt{x}(x + 1)$

14. $f(x) = \dfrac{x}{x + 4}$

15. $f(x) = (3x - 1)^5$

16. $f(x) = (5 - \frac{1}{2}x)^6$

In Exercises 17–22, find a formula for $\dfrac{d}{dx}[f(g(x))]$, where $g(x)$ represents an arbitrary differentiable function.

17. $f(x) = \dfrac{5}{x^3}$

18. $f(x) = \sqrt{x}$

19. $f(x) = (4 - x)^{3/4}$

20. $f(x) = (2x + 3)^{5/3}$

21. $f(x) = \dfrac{2}{\sqrt{x + 1}}$

22. $f(x) = \dfrac{8}{(x - 1)^2}$

In Exercises 23–25, find a formula for $\dfrac{d}{dx}[f(g(x))]$, where $f(x)$ is a function such that $f'(x) = 1/(x^2 + 1)$.

23. $g(x) = x^3$

24. $g(x) = \dfrac{1}{x}$

25. $g(x) = x^2 + 1$

In Exercises 26–28, find a formula for $\dfrac{d}{dx}[f(g(x))]$, where $f(x)$ is a function such that $f'(x) = x\sqrt{1 - x^2}$.

26. $g(x) = x^2$

27. $g(x) = \sqrt{x}$

28. $g(x) = x^{3/2}$

29. Suppose that $y = f(x)$ is a function with the property that the slope of the tangent line at each point (x, y) on its graph is $1/y$. In addition, suppose that $(1, 2)$ is a point on the graph. What is the slope of the tangent line to the curve $y = f(x^3)$ at $(1, 2)$?

30. Rework Exercise 29 with $1/y$ replaced by $1/x$.

31. Animal physiologists have determined experimentally that the weight W (in kilograms) and the surface area S (in square meters) of a typical horse are related by the empirical equation $S = 0.1W^{2/3}$. How fast is the surface area of a horse increasing at a time when the horse weighs 350 kg and is gaining weight at the rate of 200 kg per year? [*Hint:* Use the chain rule.]

32. When oxygen is compressed adiabatically (i.e., in such a way that no heat enters or leaves the gas), the pressure P (in torr) and the volume V (in liters) are related by the equation $P = 7.6/V^{1.4}$. Suppose that at some moment the volume is 2 liters, the pressure is 2.88 torr, and the volume is decreasing at the rate of .6 liter per minute. How fast is the pressure increasing?

In Exercises 33–38, x and y are related by the given equation. Use implicit differentiation to calculate the value of $\dfrac{dy}{dx}$ for the given values of x and y.

33. $x^2 y^2 = 9$; $x = 1$, $y = 3$

34. $xy^4 = 48$; $x = 3$, $y = 2$

35. $x^2 - xy^3 = 20$; $x = 5$, $y = 1$

36. $xy^2 - x^3 = 10$; $x = 2$, $y = 3$

37. $x^3 y^3 - x^2 y^2 = 0$; $x = 5$, $y = \frac{1}{5}$

38. $(x + y)^3 = x^3 + y^3$; $x = 1$, $y = -1$

39. Suppose that a kitchen appliance company's monthly sales and advertising expenses are approximately related by the equation $xy - 6x + 20y = 0$, where x is thousands of dollars spent on advertising and y is thousands of dishwashers sold. Currently, the company is spending 10 thousand dollars on advertising and is selling 2 thousand dishwashers each month. If the company plans to increase monthly advertising expenditures at the rate of $1.5 thousand per month, how fast will sales rise? Use implicit differentiation to answer the question.

The Exponential and Natural Logarithm Functions

Chapter 4

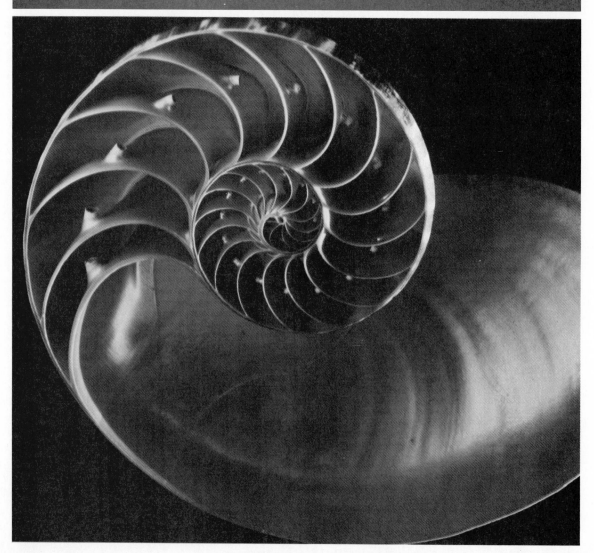

When an investment grows steadily at 15% per year, the rate of growth of the investment at any time is proportional to the value of the investment at that time. When a bacteria culture grows in a laboratory dish, the rate of growth of the culture at any moment is proportional to the total number of bacteria in the dish at that moment. These situations are examples of what is called *exponential growth*. A pile of radioactive uranium U^{235} decays at a rate that at each moment is proportional to the amount of U^{235} present. This decay of uranium (and of radioactive elements in general) is called *exponential decay*. Both exponential growth and exponential decay can be described and studied in terms of exponential functions and the natural logarithm function. The properties of these functions are investigated in this chapter. Subsequently, we shall explore a wide range of applications, in fields such as business, biology, archeology, public health, and pyschology.

4.1 Exponential Functions

Throughout this section b will denote a positive number. The function

$$f(x) = b^x$$

is called an *exponential function*, because the variable x is in the exponent. The number b is called the *base* of the exponential function. In Section 0.5 we reviewed the definition of b^x for various values of b and x (although we used the letter r there instead of x). For instance, if $f(x)$ is the exponential function with base 2,

$$f(x) = 2^x,$$

then

$$f(0) = 2^0 = 1, \quad f(1) = 2^1 = 2, \quad f(4) = 2^4 = 2 \cdot 2 \cdot 2 \cdot 2 = 16,$$

and

$$f(-1) = 2^{-1} = \tfrac{1}{2}, \qquad f(\tfrac{1}{2}) = 2^{1/2} = \sqrt{2}, \qquad f(\tfrac{3}{5}) = (2^{1/5})^3 = (\sqrt[5]{2})^3.$$

Actually, in Section 0.5, we only defined b^x for rational (i.e., integer or fractional) values of x. For other values of x (such as $\sqrt{3}$ or π), it is possible to define b^x by first approximating x with rational numbers and then applying a limiting process. We shall omit the details and simply assume henceforth that b^x can be defined for all numbers x in such a way that the usual laws of exponents remain valid.

Let us state the laws of exponents for reference.

(i)	$b^x \cdot b^y = b^{x+y}$	(iv)	$(b^y)^x = b^{xy}$
(ii)	$b^{-x} = \dfrac{1}{b^x}$	(v)	$a^x b^x = (ab)^x$
(iii)	$\dfrac{b^x}{b^y} = b^x \cdot b^{-y} = b^{x-y}$	(vi)	$\dfrac{a^x}{b^x} = \left(\dfrac{a}{b}\right)^x$

Property (iv) may be used to change the appearance of an exponential function. For instance, the function $f(x) = 8^x$ may also be written as $f(x) = (2^3)^x = 2^{3x}$, and $g(x) = (\tfrac{1}{9})^x$ may be written as $g(x) = (1/3^2)^x = (3^{-2})^x = 3^{-2x}$.

EXAMPLE 1   Use properties of exponents to write the following functions in the form 2^{kx} for a suitable constant k.

(a) $4^{5x/2}$ (b) $(2^{4x} \cdot 2^{-x})^{1/2}$ (c) $8^{x/3} \cdot 16^{3x/4}$ (d) $\dfrac{10^x}{5^x}$

Solution   (a) First express the base 4 as a power of 2, and then use Property (iv):

$$4^{5x/2} = (2^2)^{5x/2} = 2^{2(5x/2)} = 2^{5x}.$$

(b) Use Property (i) first to simplify the quantity inside the parentheses, and then use Property (iv):

$$(2^{4x} \cdot 2^{-x})^{1/2} = (2^{4x-x})^{1/2} = (2^{3x})^{1/2} = 2^{(3/2)x}.$$

(c) First express the bases 8 and 16 as powers of 2, and then use (iv) and (i):

$$8^{x/3} \cdot 16^{3x/4} = (2^3)^{x/3} \cdot (2^4)^{3x/4} = 2^x \cdot 2^{3x} = 2^{4x}.$$

(d) Use (v) to change the numerator 10^x, and then cancel the common term 5^x:

$$\frac{10^x}{5^x} = \frac{(2 \cdot 5)^x}{5^x} = \frac{2^x \cdot 5^x}{5^x} = 2^x.$$

An alternative method is to use Property (vi):

$$\frac{10^x}{5^x} = \left(\frac{10}{5}\right)^x = 2^x.$$

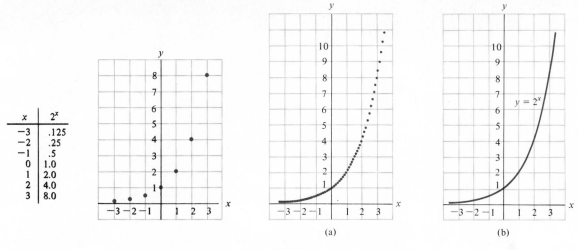

x	2^x
−3	.125
−2	.25
−1	.5
0	1.0
1	2.0
2	4.0
3	8.0

(a)

(b)

FIGURE 1

FIGURE 2

Let us now study the graph of the exponential function $y = b^x$ for various values of b. We begin with the special case $b = 2$.

We have tabulated the values of 2^x for $x = 0, \pm 1, \pm 2, \pm 3$ and plotted these values in Fig. 1. Other intermediate values of 2^x for $x = \pm .1, \pm .2, \pm .3, \ldots$, may be obtained from tables or from a calculator with a y^x key. [See Fig. 2(a).] By passing a smooth curve through these points, we obtain the graph of $y = 2^x$, in Fig. 2(b).

In the same manner, we have sketched the graph of $y = 3^x$ (Fig. 3). The graphs of $y = 2^x$ and $y = 3^x$ have the same basic shape. Also note that they both pass through the point (0, 1) (because $2^0 = 1$, $3^0 = 1$).

In Fig. 4 we have sketched the graphs of several more exponential functions. Notice that the graph of $y = 5^x$ has a large slope at $x = 0$, since the graph at

FIGURE 3

FIGURE 4

$x = 0$ is quite steep; however, the graph of $y = (1.1)^x$ is nearly horizontal at $x = 0$, and hence the slope is close to zero.

EXAMPLE 2 Let $f(x) = 3^{5x}$. Determine all x for which $f(x) = 27$.

Solution Since $27 = 3^3$, we must determine all x for which

$$3^{5x} = 3^3.$$

Equating exponents, we have

$$5x = 3$$

$$x = \tfrac{3}{5}.$$

In general, for $b > 1$, the equation $b^r = b^s$ implies that $r = s$. This is because the graph of $y = b^x$ has the same basic shape as $y = 2^x$ and $y = 3^x$. Similarly, when $0 < b < 1$, the equation $b^r = b^s$ implies that $r = s$, because the graph of $y = b^x$ resembles the graph of $y = (\tfrac{1}{3})^x$ and is always decreasing.

There is no need at this point to become familiar with the graphs of the functions b^x. We have shown a few graphs merely to make the reader more comfortable with the concept of an exponential function. The main purpose of this section has been to review properties of exponents in a context that is appropriate for our future work.

PRACTICE PROBLEMS 1

1. Can a function such as $f(x) = 5^{3x}$ be written in the form $f(x) = b^x$? If so, what is b?

2. Solve the equation $7 \cdot 2^{6-3x} = 28$.

EXERCISES 1

Write each function in Exercises 1–14 in the form 2^{kx} or 3^{kx}, for a suitable constant k.

1. $4^x, (\sqrt{3})^x, (\tfrac{1}{9})^x$

2. $27^x, (\sqrt[3]{2})^x, (\tfrac{1}{8})^x$

3. $8^{2x/3}, 9^{3x/2}, 16^{-3x/4}$

4. $9^{-x/2}, 8^{4x/3}, 27^{-2x/3}$

5. $(\tfrac{1}{4})^{2x}, (\tfrac{1}{8})^{-3x}, (\tfrac{1}{81})^{x/2}$

6. $(\tfrac{1}{9})^{2x}, (\tfrac{1}{27})^{x/3}, (\tfrac{1}{16})^{-x/2}$

7. $2^{3x} \cdot 2^{-5x/2}, 3^{2x} \cdot (\tfrac{1}{3})^{2x/3}$

8. $2^{5x/4} \cdot (\tfrac{1}{2})^x, 3^{-2x} \cdot 3^{5x/2}$

9. $(2^{-3x} \cdot 2^{-2x})^{2/5}, (9^{1/2} \cdot 9^4)^{x/9}$

10. $(3^{-x} \cdot 3^{x/5})^5, (16^{1/4} \cdot 16^{-3/4})^{3x}$

11. $\dfrac{3^{4x}}{3^{2x}}, \dfrac{2^{5x+1}}{2 \cdot 2^{-x}}, \dfrac{9^{-x}}{27^{-x/3}}$

12. $\dfrac{2^x}{6^x}, \dfrac{3^{-5x}}{3^{-2x}}, \dfrac{16^x}{8^{-x}}$

13. $6^x \cdot 3^{-x}, \dfrac{15^x}{5^x}, \dfrac{12^x}{2^{2x}}$

14. $7^{-x} \cdot 14^x, \dfrac{2^x}{6^x}, \dfrac{3^{2x}}{18^x}$

From a table we have $2^{1/2} \approx 1.414$, $2^{1/10} \approx 1.072$, $(2.7)^{1/2} \approx 1.643$, and $(2.7)^{1/10} \approx 1.104$. (These figures are accurate to three decimal places.) Compute the following numbers, rounding off your answer to two decimal places, if necessary.

15. (a) 2^x for $x = 3$ (b) 2^x for $x = -3$

(c) 2^x for $x = \frac{5}{2}$ $[Hint: \frac{5}{2} = 2 + \frac{1}{2}.]$ (d) 2^x for $x = 4.1$

(e) 2^x for $x = .2$ (f) 2^x for $x = .9$ $[Hint: .9 = 1 - .1.]$

(g) 2^x for $x = -2.5$ $[Hint: -2.5 = -3 + .5.]$

(h) 2^x for $x = -3.9$

16. (a) $(2.7)^x$ for $x = .2$ (b) $(2.7)^x$ for $x = 1.5$ (c) $(2.7)^x$ for $x = 0$

(d) $(2.7)^x$ for $x = -1$ (e) $(2.7)^x$ for $x = 1.1$ (f) $(2.7)^x$ for $x = .6$

Solve the following equations for x.

17. $5^{2x} = 5^2$

18. $10^{-x} = 10^2$

19. $(2.5)^{2x+1} = (2.5)^5$

20. $(3.2)^{x-3} = (3.2)^5$

21. $10^{1-x} = 100$

22. $2^{4-x} = 8$

23. $3(2.7)^{5x} = 8.1$

24. $4(2.7)^{2x-1} = 10.8$

25. $(2^{x+1} \cdot 2^{-3})^2 = 2$

26. $(3^{2x} \cdot 3^2)^4 = 3$

27. $2^{3x} = 4 \cdot 2^{5x}$

28. $3^{5x} \cdot 3^x - 3 = 0$

29. $(1+x)2^{-x} - 5 \cdot 2^{-x} = 0$

30. $(2 - 3x)5^x + 4 \cdot 5^x = 0$

The expressions in Exercises 31–38 may be factored as shown. Find the missing factors.

31. $2^{3+h} = 2^3 (\quad)$

32. $5^{2+h} = 25 (\quad)$

33. $2^{x+h} - 2^x = 2^x (\quad)$

34. $5^{x+h} + 5^x = 5^x (\quad)$

35. $3^{x/2} + 3^{-x/2} = 3^{-x/2} (\quad)$

36. $5^{7x/2} - 5^{x/2} = \sqrt{5^x} (\quad)$

37. $3^{10x} - 1 = (3^{5x} - 1) (\quad)$

38. $2^{6x} - 2^x = (2^{3x} - 2^{x/2}) (\quad)$

SOLUTIONS TO PRACTICE PROBLEMS 1

1. If $5^{3x} = b^x$, then when $x = 1$, $5^{3(1)} = b^1$, which says that $b = 125$. This value of b certainly works, because

$$5^{3x} = (5^3)^x = 125^x.$$

2. Divide both sides of the equation by 7. We then obtain

$$2^{6-3x} = 4.$$

Now 4 can be written as 2^2. So we have

$$2^{6-3x} = 2^2.$$

Equate exponents.

$$6 - 3x = 2$$

$$4 = 3x$$

$$x = \tfrac{4}{3}.$$

4.2 The Exponential Function e^x

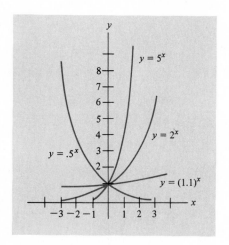

FIGURE 1

Let us begin by examining the graphs of the exponential functions shown in Fig. 1. They all pass through $(0, 1)$, but with different slopes there. Notice that the graph of 5^x is quite steep at $x = 0$, while the graph of $(1.1)^x$ is nearly horizontal at $x = 0$. It turns out that at $x = 0$, the graph of 2^x has a slope of approximately .69, while the graph of 3^x has a slope of approximately 1.1.

Evidently, there is a particular value of the base b, between 2 and 3, where the graph of b^x has slope *exactly* 1 at $x = 0$. We denote this special value of b by the letter e, and we call

$$f(x) = e^x$$

the exponential function. The number e is an important constant of nature that has been calculated to thousands of decimal places. To 10 significant digits, we have $e = 2.718281828$. For our purposes, it is usually sufficient to think of e as "approximately 2.7."

Our goal in this section is to find a formula for the derivative of e^x. It turns out that the calculations for e^x and 2^x are very similar. Since many people are more comfortable working with 2^x rather than e^x, we shall first analyze the graph of 2^x. Then we shall draw the appropriate conclusions about the graph of e^x.

Before computing the slope of $y = 2^x$ at an arbitrary x, let us consider the special case $x = 0$. Denote the slope at $x = 0$ by m. We shall use the secant-line approximation of the derivative to approximate m. We proceed by constructing a secant line in Fig. 2. The slope of the secant line through $(0, 1)$ and $(h, 2^h)$ is $\dfrac{2^h - 1}{h}$. As h approaches zero, the slope of the secant line approaches the slope of $y = 2^x$ at $x = 0$. That is,

$$m = \lim_{h \to 0} \frac{2^h - 1}{h}. \tag{1}$$

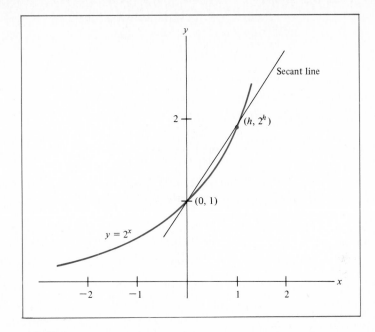

FIGURE 2

We can estimate the value of m by taking h smaller and smaller. When $h = .1$, we have $2^h \approx 1.072$ (from a table of values of 2^x), and

$$\frac{2^h - 1}{h} \approx \frac{.072}{.1} = .72.$$

When $h = .01$, we have $2^h \approx 1.00696$, and

$$\frac{2^h - 1}{h} \approx \frac{.00696}{.01} \approx .696.$$

When $h = .001$, $2^h \approx 1.0006934$, so that

$$\frac{2^h - 1}{h} \approx \frac{.0006934}{.001} \approx .693.$$

Thus it is reasonable to conclude that $m \approx .69$.* Since m equals the slope of $y = 2^x$ at $x = 0$, we have

$$m = \frac{d}{dx}(2^x)\bigg|_{x=0} \approx .69. \qquad (2)$$

* To 10 decimal places, m is .6931471806.

Now that we have estimated the slope of $y = 2^x$ at $x = 0$, let us compute the slope for an arbitrary value of x. We construct a secant line through $(x, 2^x)$ and a nearby point $(x + h, 2^{x+h})$ on the graph. The slope of the secant line is

$$\frac{2^{x+h} - 2^x}{h}. \tag{3}$$

By law of exponents (i), we have $2^{x+h} - 2^x = 2^x(2^h - 1)$ so that, by (1), we see that

$$\lim_{h \to 0} \frac{2^{x+h} - 2^x}{h} = \lim_{h \to 0} 2^x \frac{2^h - 1}{h} = 2^x \lim_{h \to 0} \frac{2^h - 1}{h} = m2^x. \tag{4}$$

However, the slope of the secant (3) approaches the derivative of 2^x as h approaches zero. Consequently, we have

$$\frac{d}{dx}(2^x) = m2^x, \qquad \text{where } m = \frac{d}{dx}(2^x)\Big|_{x=0}. \tag{5}$$

EXAMPLE 1 Calculate (a) $\dfrac{d}{dx}(2^x)\Big|_{x=3}$ and (b) $\dfrac{d}{dx}(2^x)\Big|_{x=-1}$

Solution (a) $\dfrac{d}{dx}(2^x)\Big|_{x=3} = m \cdot 2^3 = 8m \approx 8(.69) = 5.52.$

(b) $\dfrac{d}{dx}(2^x)\Big|_{x=-1} = m \cdot 2^{-1} = .5m \approx .5(.69) = .345.$

The calculations just carried out for $y = 2^x$ can be carried out for $y = b^x$, where b is any positive number. Equation (5) will read exactly the same except that 2 will be replaced by b. Thus we have the following formula for the derivative of the function $f(x) = b^x$:

$$\frac{d}{dx}(b^x) = mb^x, \qquad \text{where } m = \frac{d}{dx}(b^x)\Big|_{x=0}. \tag{6}$$

Our calculations showed that if $b = 2$, then $m \approx .69$. If $b = 3$, then it turns out that $m \approx 1.1$. (See Exercise 1.) Obviously, the derivative formula in (6) is simplest when $m = 1$, that is, when the graph of b^x has slope 1 at $x = 0$. As we said earlier, this special value of b is denoted by the letter e. Thus the number e has the property that

$$\frac{d}{dx}(e^x)\Big|_{x=0} = 1 \tag{7}$$

and

$$\frac{d}{dx}(e^x) = 1 \cdot e^x = e^x. \tag{8}$$

The graphical interpretation of (7) is that the curve $y = e^x$ has slope 1 at $x = 0$. The graphical interpretation of (8) is that the slope of the curve $y = e^x$ at an arbitrary value of x is exactly equal to the value of the function e^x at that point. (See Fig. 3.)

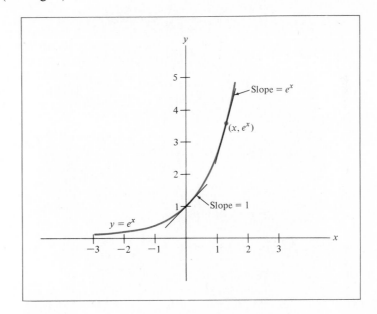

FIGURE 3

The function e^x is the same type of function as 2^x and 3^x except that taking derivatives of e^x is much easier. For this reason, functions based on e^x are used in almost all applications that require an exponential-type function to describe a physical phenomenon. Extensive tables are available from which to determine the value of e^x for a wide range of values of x. Also, scientific calculators provide for calculating e^x at the push of a button.

PRACTICE PROBLEMS 2

In the following problems, use the number 20 as the (approximate) value of e^3.

1. Find the equation of the tangent line to the graph of $y = e^x$ at $x = 3$.

2. Solve the following equation for x:
$$4e^{6x} = 80.$$

EXERCISES 2

1. Use the table below to show that

$$\frac{d}{dx}(3^x)\bigg|_{x=0} \approx 1.1.$$

That is, calculate the slope

$$\frac{3^h - 1}{h}$$

of the secant line passing through the point $(0, 1)$ and $(h, 3^h)$. Take $h = .1, .01,$ and $.001.$

x	3^x
0	1.00000
.001	1.00110
.010	1.01105
.100	1.11612

2. Use the table below to show that

$$\frac{d}{dx}(2.7)^x\bigg|_{x=0} \approx .99.$$

That is, calculate

$$\frac{(2.7)^h - 1}{h}$$

for $h = .1, .01,$ and $.001.$

x	$(2.7)^x$
0	1.00000
.001	1.00099
.010	1.00998
.100	1.10443

3. Consider the secant line on the graph of e^x passing through $(0, 1)$ and (h, e^h). Its slope is

$$\frac{e^h - 1}{h}.$$

Compute this quantity for $h = .01, .005,$ and $.001.$

x	e^x
0	1.00000
.001	1.00100
.005	1.00501
.010	1.01005

4. Use (8) and a familiar rule for differentiation to find

$$\frac{d}{dx}(5e^x).$$

5. Use (8) and a familiar rule for differentiation to find

$$\frac{d}{dx}(e^x)^{10}.$$

6. Use the fact that $e^{2+x} = e^2 \cdot e^x$ to find

$$\frac{d}{dx}(e^{2+x}).$$

[Remember that e^2 is just a constant—approximately $(2.7)^2$.]

7. Use the fact that $e^{4x} = (e^x)^4$ to find

$$\frac{d}{dx}(e^{4x}).$$

8. Find $\dfrac{d}{dx}(e^x + x^2)$.

Simplify.

9. $e^{2x}(1 + e^{3x})$ **10.** $(e^x)^2$ **11.** $e^{1-x} \cdot e^{2x}$

12. $\dfrac{5e^{3x}}{e^x}$ **13.** $\dfrac{1}{e^{-2x}}$ **14.** $e^3 \cdot e^{x+1}$

Use Table 1 of the Appendix to determine the following numbers:

15. e^2 **16.** $e^{-1.30}$ **17.** $e^{-.5}$ **18.** $e^{3/2}$

Solve the following equations for x.

19. $e^{5x} = e^{20}$ **20.** $e^{1-x} = e^2$

21. $e^{x^2-2x} = e^8$ **22.** $e^{-x} = 1$

Differentiate the following functions.

23. xe^x **24.** $\dfrac{e^x}{x}$ **25.** $\dfrac{e^x}{1 + e^x}$

26. $(1 + x^2)e^x$ **27.** $(1 + 5e^x)^4$ **28.** $(xe^x - 1)^{-3}$

SOLUTIONS TO PRACTICE PROBLEMS 2

1. When $x = 3$, $y = e^3 = 20$. So the point $(3, 20)$ is on the tangent line. Since $\dfrac{d}{dx}(e^x) = e^x$, the slope of the tangent line is e^3 or 20. Therefore, the equation of the tangent line in point-slope form is $y - 20 = 20(x - 3)$.

2. This problem is similar to the Practice Problem 2 of Section 1. First divide both sides of the equation by 4.

$$e^{6x} = 20.$$

The idea is to express 20 as a power of e and then equate exponents.

$$e^{6x} = e^3$$

$$6x = 3$$

$$x = \tfrac{1}{2}.$$

4.3 Differentiation of Exponential Functions

We have shown that $\dfrac{d}{dx}(e^x) = e^x$. Using this fact and the chain rule, we may differentiate functions of the form $e^{g(x)}$, where $g(x)$ is any differentiable function. This is because $e^{g(x)}$ is the composite of two functions. Indeed, if $f(x) = e^x$, then

$$e^{g(x)} = f(g(x)).$$

Thus, by the chain rule, we have

$$\frac{d}{dx}(e^{g(x)}) = f'(g(x))g'(x)$$

$$= f(g(x))g'(x) \qquad [\text{since } f'(x) = f(x)]$$

$$= e^{g(x)}g'(x).$$

So we have the following result:

Chain Rule for Exponential Functions   Let $g(x)$ be any differentiable function. Then

$$\frac{d}{dx}(e^{g(x)}) = e^{g(x)}g'(x).$$

(1)

EXAMPLE 1   Differentiate e^{x^2+1}.

Solution   Here $g(x) = x^2 + 1$, $g'(x) = 2x$, so

$$\frac{d}{dx}(e^{x^2+1}) = e^{x^2+1} \cdot 2x$$

$$= 2xe^{x^2+1}.$$

EXAMPLE 2   Differentiate $e^{3x^2-(1/x)}$.

Solution

$$\frac{d}{dx}(e^{3x^2-(1/x)}) = e^{3x^2-(1/x)} \cdot \frac{d}{dx}\left(3x^2 - \frac{1}{x}\right)$$

$$= e^{3x^2-(1/x)}\left(6x + \frac{1}{x^2}\right).$$

EXAMPLE 3   Differentiate e^{5x}.

Solution

$$\frac{d}{dx}(e^{5x}) = e^{5x} \cdot \frac{d}{dx}(5x) = e^{5x} \cdot 5 = 5e^{5x}.$$

Using a computation similar to that used in Example 3, we may differentiate e^{kx} for any constant k. (In Example 3 we have $k = 5$.) The result is the following useful formula.

$$\frac{d}{dx}(e^{kx}) = ke^{kx}. \tag{2}$$

Many applications involve exponential functions of the form $y = Ce^{kx}$, where C and k are constants. In the next example we differentiate such functions.

EXAMPLE 4   Differentiate the following exponential functions.

(a) $3e^{5x}$

(b) $3e^{kx}$, where k is a constant

(c) Ce^{kx}, where C and k are constants

Solution (a) $\dfrac{d}{dx}(3e^{5x}) = 3\dfrac{d}{dx}(e^{5x}) = 3 \cdot 5e^{5x} = 15e^{5x}.$

(b) $\dfrac{d}{dx}(3e^{kx}) = 3\dfrac{d}{dx}(e^{kx})$

$$= 3 \cdot ke^{kx} \qquad [\text{by (2)}]$$

$$= 3ke^{kx}.$$

(c) $\dfrac{d}{dx}(Ce^{kx}) = C\dfrac{d}{dx}(e^{kx}) = Cke^{kx}.$

The result of part (c) may be summarized in an extremely useful fashion as follows: Suppose that we let $y = Ce^{kx}$. By part (c) we have

$$y' = Cke^{kx}$$

$$= k \cdot (Ce^{kx})$$

$$= ky.$$

In other words, the derivative of the function Ce^{kx} is k times the function itself. Let us record this fact.

> Let C, k be any constants and let $y = Ce^{kx}$. Then y satisfies the equation
> $$y' = ky.$$

The equation $y' = ky$ expresses a relationship between the function y and its derivative y'. Any equation expressing a relationship between a function y and one or more of its derivatives is called a *differential equation*.

Very often an applied problem will involve a function $y = f(x)$ which satisfies the differential equation $y' = ky$. It can be shown that y must then necessarily be an exponential function of the form Ce^{kx}. That is, we have the following result.

> Suppose that $y = f(x)$ satisfies the differential equation
> $$y' = ky.$$
> Then y is an exponential function of the form
> $$y = Ce^{kx}, \qquad C \text{ a constant.}$$

(3)

We shall verify this result in Exercise 47.

EXAMPLE 5 Determine all functions $y = f(x)$ such that $y' = -.2y$.

Solution The equation $y' = -.2y$ has the form $y' = ky$ with $k = -.2$. Therefore, any solution of the equation has the form

$$y = Ce^{-.2x},$$

where C is a constant.

EXAMPLE 6 Determine all functions $y = f(x)$ such that $y' = y/2$ and $f(0) = 4$.

Solution The equation $y' = y/2$ has the form $y' = ky$ with $k = \frac{1}{2}$. Therefore,

$$f(x) = Ce^{(1/2)x}$$

for some constant C. We also require that $f(0) = 4$. That is,

$$4 = f(0) = Ce^{(1/2) \cdot 0} = Ce^0 = C.$$

So $C = 4$ and

$$f(x) = 4e^{(1/2)x}.$$

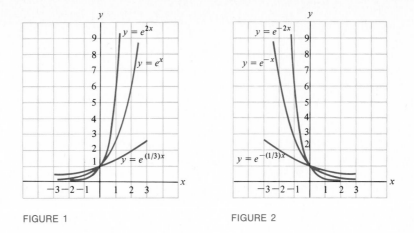

FIGURE 1   FIGURE 2

The Functions e^{kx} Exponential functions of the form e^{kx} occur in many applications. Figure 1 shows the graphs of several functions of this type when k is a positive number. These curves $y = e^{kx}$, k positive, have several properties in common:

1. $(0, 1)$ is on the graph.
2. The graph lies strictly above the x-axis (e^{kx} is never zero).
3. The x-axis is an asymptote as x becomes large negatively.
4. The graph is always increasing and concave up.

When k is negative, the graph of $y = e^{kx}$ is decreasing. (See Fig. 2.) Note the following properties of the curves $y = e^{kx}$, k negative:

1. $(0, 1)$ is on the graph.
2. The graph lies strictly above the x-axis.
3. The x-axis is an asymptote as x becomes large positively.
4. The graph is always decreasing and concave up.

The Functions b^x If b is a positive number, then the function b^x may be written in the form e^{kx} for some k. For example, take $b = 2$. From Fig. 3 of the preceding section it is clear that there is some value of x such that $e^x = 2$. Call this value k, so that $e^k = 2$. Then

$$2^x = (e^k)^x = e^{kx}$$

for all x. In general, if b is any positive number, there is a value of x, say $x = k$, such that $e^k = b$. In this case, $b^x = (e^k)^x = e^{kx}$. Thus all the curves $y = b^x$ discussed in Section 1 can be written in the form $y = e^{kx}$. This is one reason why we have focused on exponential functions with base e instead of studying 2^x, 3^x, and so on.

PRACTICE PROBLEMS 3

1. Differentiate $[e^{-3x}(1 + e^{6x})]^{12}$.

2. Determine all functions $y = f(x)$ such that $y' = -y/20$, $f(0) = 2$.

EXERCISES 3

Differentiate the following.

1. $y = e^{-x}$

2. $f(x) = e^{10x}$

3. $f(x) = 5e^x$

4. $y = \dfrac{e^x + e^{-x}}{2}$

5. $f(t) = e^{t^2}$

6. $f(t) = e^{-2t}$

7. $f(x) = \dfrac{e^x - e^{-x}}{2}$

8. $f(x) = 2e^{1-x}$

9. $y = e^{-2x} - 2x$

10. $f(x) = \frac{1}{10}e^{-x^2/2}$

11. $g(x) = (e^x + e^{-x})^3$

12. $y = (e^{-x})^2$

13. $y = \frac{1}{3}e^{3-2x}$

14. $g(x) = e^{1/x}$

15. $f(t) = e^t(e^{2t} - e^{-t})$

16. $f(t) = \dfrac{e^t + e^{-t}}{e^t}$

[*Hint*: In Exercises 15 and 16, simplify $f(t)$ before differentiating.]

17. $y = e^{x^3 + x - (1/x)}$

18. $y = (e^{x^2} + x^2)^5$

19. $f(x) = (2x + 1 - e^{2x+1})^4$

20. $f(x) = e^{1/(3x-7)}$

21. $x^3 e^{x^2}$

22. $x^2 e^{-3x}$

23. e^{x^3}/x

24. $e^{(x-1)/x}$

25. $(x + 1)e^{-x+2}$

26. $x\sqrt{2 + e^x}$

27. $\left(\dfrac{1}{x} + 3\right)e^x$

28. $\dfrac{e^{-3x}}{1 - 3x}$

29. $\dfrac{e^x - 1}{e^x + 1}$

30. $\dfrac{xe^x - 3}{x + 1}$

In Exercises 31–40, find all values of x such that the function has a possible maximum or minimum point. Use the second derivative test to determine the nature of the function at these points. (Recall that e^x is positive for all x.)

31. $f(x) = (1 + x)e^{-x/2}$

32. $f(x) = (1 - x)e^{-x/2}$

33. $f(x) = \dfrac{3 - 2x}{e^{x/4}}$

34. $f(x) = \dfrac{4x - 3}{e^{x/2}}$

35. $f(x) = (8 - 2x)e^{x+5}$

36. $f(x) = (4x - 1)e^{3x-2}$

37. $f(x) = \dfrac{(x-1)^2}{e^x}$

38. $f(x) = (x+3)^2 e^x$

39. $f(x) = (x+5)^2 e^{2x-1}$

40. $f(x) = \dfrac{(x-3)^2}{e^{2x}}$

41. Let a and b be positive numbers: A curve whose equation is $y = e^{-ae^{-bx}}$ is called a *Gompertz growth curve*. These curves are used in biology to describe certain types of population growth. Compute the derivative of $y = e^{-2e^{-.01x}}$.

42. Find $\dfrac{dy}{dx}$ if $y = e^{-(1/10)e^{-x/2}}$.

43. Determine all solutions of the differential equation
$$y' = -4y.$$

44. Determine all solutions of the differential equation
$$y' = \tfrac{1}{3}y.$$

45. Determine all functions $y = f(x)$ such that
$$y' = -.5y \quad \text{and} \quad f(0) = 1.$$

46. Determine all functions $y = f(x)$ such that
$$y' = 3y \quad \text{and} \quad f(0) = \tfrac{1}{2}.$$

47. Verify the result (3). [*Hint:* Let $g(x) = f(x)e^{-kx}$. Show that $g'(x) = 0$.] You may assume that only a constant function has a zero derivative.

48. Let $f(x)$ be a function with the property that $f'(x) = 1/x$. Let $g(x) = f(e^x)$, and compute $g'(x)$.

Graph the following functions.

49. $y = e^{-x^2}$

50. $y = xe^{-x}$ for $x \geq 0$

SOLUTIONS TO PRACTICE PROBLEMS 3

1. We must use the general power rule. However, this is most easily done if we first use the laws of exponents to simplify the function inside the brackets.

$$e^{-3x}(1 + e^{6x}) = e^{-3x} + e^{-3x} \cdot e^{6x}$$
$$= e^{-3x} + e^{3x}.$$

Now

$$\frac{d}{dx}\left[e^{-3x} + e^{3x}\right]^{12} = 12 \cdot \left[e^{-3x} + e^{3x}\right]^{11} \cdot (-3e^{-3x} + 3e^{3x})$$

$$= 36 \cdot \left[e^{-3x} + e^{3x}\right]^{11} \cdot (-e^{-3x} + e^{3x}).$$

2. The differential equation $y' = -y/20$ is of the type $y' = ky$, where $k = -\tfrac{1}{20}$. Therefore, any solution has the form $f(x) = Ce^{-(1/20)x}$. Now $f(0) = Ce^{-(1/20)\cdot 0} = Ce^0 = C$, so that $f(0) = 2$ when $C = 2$. Therefore, the desired function is $f(x) = 2e^{-(1/20)x}$.

4.4 The Natural Logarithm Function

As a preparation for the definition of the natural logarithm, we shall make a geometrical digression. In Fig. 1 we have plotted several pairs of points. Observe how they are related to the line $y = x$.

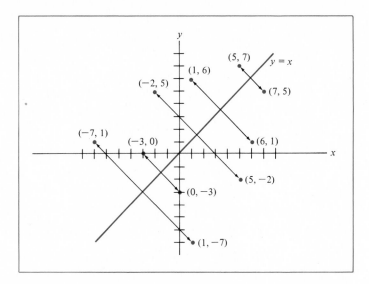

FIGURE 1

The points $(5, 7)$ and $(7, 5)$, for example, are the same distance from the line $y = x$. If we were to plot the point $(5, 7)$ with wet ink and then fold the page along the line $y = x$, the ink blot would produce a second blot at the point $(7, 5)$. If we think of the line $y = x$ as a mirror, then $(7, 5)$ is the mirror image of $(5, 7)$. We say that $(7, 5)$ is the *reflection* of $(5, 7)$ through the line $y = x$. Similarly, $(5, 7)$ is the reflection of $(7, 5)$ through the line $y = x$.

Now let us consider all points lying on the graph of the exponential function $y = e^x$ [see Fig. 2(a)]. If we reflect each such point through the line $y = x$, we obtain a new graph [see Fig. 2(b)]. For each positive x, there is exactly one value of y such that (x, y) is on the new graph. We call this value of y the *natural logarithm of* x, denoted $\ln x$. Thus the reflection of the graph of $y = e^x$ through the line $y = x$ is the graph of the natural logarithm function $y = \ln x$.

We may deduce some properties of the natural logarithm function from an inspection of its graph.

1. The point $(1, 0)$ is on the graph of $y = \ln x$ (because $(0, 1)$ is on the graph of $y = e^x$). In other words,

$$\ln 1 = 0. \tag{1}$$

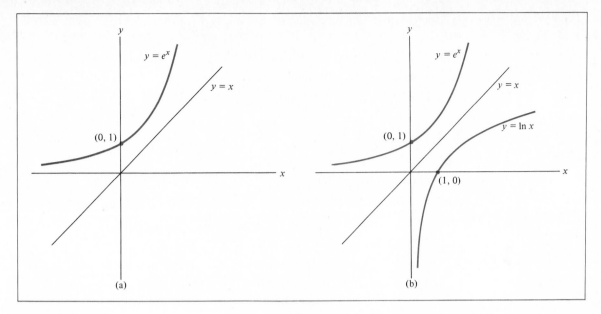

(a)   (b)

FIGURE 2

2. ln x is defined only for positive values of x.
3. ln x is negative for x between 0 and 1.
4. ln x is positive for x greater than 1.
5. ln x is an increasing function.

Let us study the relationship between the natural logarithm and exponential functions more closely. From the way in which the graph of ln x was obtained we know that (a, b) is on the graph of ln x if and only if (b, a) is on the graph of e^x. However, a typical point on the graph of ln x is of the form $(a, \ln a)$, $a > 0$. So for any positive value of a, the point $(\ln a, a)$ is on the graph of e^x. That is,

$$e^{\ln a} = a.$$

Since a was an arbitrary positive number, we have the following important relationship between the natural logarithm and exponential functions.

$$e^{\ln x} = x \qquad \text{for } x > 0. \tag{2}$$

Equation (2) can be put into verbal form.

For each positive number x, ln x is that exponent to which we must raise e in order to get x.

If b is any number, then e^b is positive and hence $\ln(e^b)$ makes sense. What is $\ln(e^b)$? Since (b, e^b) is on the graph of e^x, we know that (e^b, b) must be on the graph of $\ln x$. That is, $\ln(e^b) = b$. Thus we have shown that

$$\ln(e^x) = x \qquad \text{for any } x. \tag{3}$$

The identities (2) and (3) express the fact that the natural logarithm is the *inverse* of the exponential function. For instance, if we take a number x and compute e^x, then, by (3), we can undo the effect of the exponentiation by taking the natural logarithm; that is, the logarithm of e^x equals the original number x. Similarly, if we take a positive number x and compute $\ln x$, then, by (2), we can undo the effect of the logarithm by raising e to the $\ln x$ power; that is, $e^{\ln x}$ equals the original number x.

Scientific calculators have an "$\ln x$" key that will compute the natural logarithm of a number to as many as ten significant figures. For instance, entering the number 2 into the calculator and pressing the $\ln x$ key, one obtains $\ln 2 = .6931471806$ (to 10 significant figures). If a scientific calculator is unavailable, one may use a table of logarithms, such as Table 2 of the Appendix, which gives the values of $\ln x$ to five significant figures. For instance, this table shows that $\ln .8 = -.22314$ (to five significant figures).

The relationships (2) and (3) between e^x and $\ln x$ may be used to solve equations, as the next examples show.

EXAMPLE 1 Solve the equation $5e^{x-3} = 4$ for x.

Solution First divide each side by 5,

$$e^{x-3} = .8.$$

Taking the logarithm of each side and using (3), we have

$$\ln(e^{x-3}) = \ln .8$$

$$x - 3 = \ln .8$$

$$x = 3 + \ln .8.$$

[If desired, the numerical value of x can be obtained by using a scientific calculator or a natural logarithm table, namely, $x = 3 - .22314 = 2.77686$ (to five decimal places).]

EXAMPLE 2 Solve the equation $2 \ln x + 7 = 0$ for x.

Solution
$$2 \ln x = -7$$

$$\ln x = -3.5$$

$$e^{\ln x} = e^{-3.5}$$

$$x = e^{-3.5} \qquad [\text{by (2)}].$$

Other Exponential and Logarithm Functions   In our discussion of the exponential function, we mentioned that all exponential functions of the form b^x, where b is a fixed positive number, can be expressed in terms of *the* exponential function e^x. Now we can be quite explicit. For since $b = e^{\ln b}$, we see that

$$b^x = (e^{\ln b})^x = e^{(\ln b)x}.$$

Hence we have shown that

$$b^x = e^{kx}, \qquad \text{where } k = \ln b.$$

The natural logarithm function is sometimes called the *logarithm to the base e*, for it is the inverse of the exponential function e^x. If we reflect the graph of the function $y = 2^x$ through the line $y = x$, we obtain the graph of a function called the *logarithm to the base* 2, denoted by $\log_2 x$. Similarly, if we reflect the graph of $y = 10^x$ through the line $y = x$, we obtain the graph of a function called the *logarithm to the base* 10, denoted by $\log_{10} x$. (See Fig. 3.)

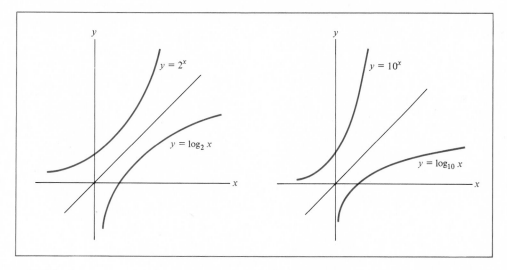

FIGURE 3

Logarithms to the base 10 are sometimes called *common* logarithms. Common logarithms are usually introduced into algebra courses for the purpose of simplifying certain arithmetic calculations. However, with the advent of the modern digital computer and the widespread availability of pocket electronic calculators, the need for common logarithms has diminished considerably. It can be shown that

$$\log_{10} x = \frac{1}{\ln 10} \cdot \ln x,$$

so that $\log_{10} x$ is simply a constant multiple of $\ln x$. However, we shall not need this fact.

The natural logarithm function is used in calculus because differentiation and integration formulas are simpler than for $\log_{10} x$ or $\log_2 x$, and so on. (Recall that we prefer the function e^x over the functions 10^x and 2^x for the same reason.) Also, $\ln x$ arises "naturally" in the process of solving certain differential equations that describe various growth processes.

PRACTICE PROBLEMS 4

1. Find $\ln e$.

2. Solve $e^{-3x} = 2$ using the natural logarithm function.

EXERCISES 4

1. Find $\ln(1/e)$.   2. Find $\ln(\sqrt{e})$.

3. If $e^{-x} = 1.7$, write x in terms of the natural logarithm.

4. If $e^x = 3.5$, write x in terms of the natural logarithm.

5. If $\ln x = 2.2$, write x using the exponential function.

6. If $\ln x = -5.7$, write x using the exponential function.

Simplify the following expressions.

7. $\ln e^2$ 8. $e^{\ln 1.37}$ 9. $e^{e^{\ln 1}}$

10. $\ln(e^{.73 \ln e})$ 11. $e^{5 \ln 1}$ 12. $\ln(\ln e)$

Solve the following equations for x.

13. $e^{2x} = 5$ 14. $e^{3x-1} = 4$ 15. $\ln(4 - x) = \frac{1}{2}$

16. $\ln 3x = 2$ 17. $\ln x^2 = 6$ 18. $e^{x^2} = 7$

19. $6e^{-.00012x} = 3$ 20. $2 - \ln x = 0$ 21. $\ln 5x = \ln 3$

22. $\ln(x^2 - 3) = 0$ 23. $\ln(\ln 2x) = 0$ 24. $3 \ln x = 8$

25. $2e^{x/3} - 9 = 0$ 26. $4 - 3e^{x+6} = 0$ 27. $300e^{.2x} = 1800$

28. $750e^{-.4x} = 375$ 29. $e^{5x} \cdot e^{\ln 5} = 2$ 30. $e^{x^2-5x+6} = 1$

31. $4e^x \cdot e^{-2x} = 6$ 32. $(e^x)^2 \cdot e^{2-3x} = 4$

In Exercises 33–36, find the coordinates of each extreme point of the given function and determine if the point is a maximum point or a minimum point.

33. $f(x) = e^{-x} + 3x$ 34. $f(x) = 5x - 2e^x$

35. $f(x) = \frac{1}{3}e^{2x} - x + \frac{1}{2} \ln \frac{3}{2}$ 36. $f(x) = 5 - \frac{1}{2}x - e^{-3x}$

37. When a drug or vitamin is administered intramuscularly (into a muscle), the concentration in the blood at time t after injection can be approximated by a function of the form $f(t) = c(e^{-k_1 t} - e^{-k_2 t})$. Sketch the graph of the function $f(t) = 5(e^{-.01t} - e^{-.51t})$ for $t \geq 0$ and discuss why this graph could be a reasonable model for the concentration of a drug injected intramuscularly.

38. Under certain geographic conditions, the wind velocity v at a height x centimeters above the ground is given by $v = K \ln(x/x_0)$, where K is a positive constant (depending on the air density, average wind velocity, etc.), and x_0 is a roughness parameter (depending on the roughness of the vegetation on the ground).* Suppose that $x_0 = .7$ cm (a value that applies to lawn grass 3 cm high) and $K = 300$ cm/sec.

   (a) At what height above the ground is the wind velocity zero?

   (b) At what height is the wind velocity 1200 cm/sec?

39. Use the tables of values of e^x and $\ln x$ to estimate $(1.6)^{10}$.

40. Find k such that $2^x = e^{kx}$ for all x.

SOLUTIONS TO PRACTICE PROBLEMS 4

1. Answer: 1. The number $\ln e$ is that exponent to which e must be raised in order to obtain e.

2. Take the logarithm of each side and use (3) to simplify the left side:

$$\ln e^{-3x} = \ln 2$$
$$-3x = \ln 2$$
$$x = -\frac{\ln 2}{3}.$$

4.5 The Derivative of ln x

Let us now compute the derivative of $\ln x$ for $x > 0$. Since $e^{\ln x} = x$, we have

$$\frac{d}{dx}(e^{\ln x}) = \frac{d}{dx}(x) = 1. \tag{1}$$

On the other hand, if we differentiate $e^{\ln x}$ by the chain rule, we find that

$$\frac{d}{dx}(e^{\ln x}) = e^{\ln x} \cdot \frac{d}{dx}(\ln x) = x \cdot \frac{d}{dx}(\ln x), \tag{2}$$

where the last equality used the fact that $e^{\ln x} = x$. By combining equations (1) (2) we obtain

$$x \cdot \frac{d}{dx}(\ln x) = 1.$$

---

* G. Cox, B. Collier, A. Johnson, and P. Miller, *Dynamic Ecology* (Englewood Cliffs, N.J.: Prentice-Hall, Inc., 1973), pp. 113–115.

In other words,

$$\frac{d}{dx}(\ln x) = \frac{1}{x}, \qquad x > 0. \tag{3}$$

By combining this differentiation formula with the chain rule, product rule, and quotient rule, we may differentiate many functions involving ln x.

EXAMPLE 1 Differentiate.

(a) $(\ln x)^5$ (b) $x \ln x$ (c) $\ln(x^3 + 5x^2 + 8)$

Solution (a) By the general power rule,

$$\frac{d}{dx}(\ln x)^5 = 5(\ln x)^4 \cdot \frac{d}{dx}(\ln x)$$

$$= 5(\ln x)^4 \cdot \frac{1}{x}$$

$$= \frac{5(\ln x)^4}{x}.$$

(b) By the product rule,

$$\frac{d}{dx}(x \ln x) = x \cdot \frac{d}{dx}(\ln x) + (\ln x) \cdot 1$$

$$= x \cdot \frac{1}{x} + \ln x$$

$$= 1 + \ln x.$$

(c) By the chain rule,

$$\frac{d}{dx}\ln(x^3 + 5x^2 + 8) = \frac{1}{x^3 + 5x^2 + 8} \cdot \frac{d}{dx}(x^3 + 5x^2 + 8)$$

$$= \frac{3x^2 + 10x}{x^3 + 5x^2 + 8}.$$

Let $g(x)$ be any differentiable function. For any value of x for which $g(x)$ is positive, the function $\ln(g(x))$ is defined. For such a value of x, the derivative is given by the chain rule as

$$\frac{d}{dx}[\ln g(x)] = \frac{1}{g(x)} \cdot \frac{d}{dx}g(x)$$

$$= \frac{g'(x)}{g(x)}.$$

Example 1(c) illustrates a special case of this formula.

EXAMPLE 2   The function $f(x) = (\ln x)/x$ has an extreme point for some $x > 0$. Find the point and determine whether it is a maximum or a minimum point.

Solution   By the quotient rule,

$$f'(x) = \frac{x \cdot \dfrac{1}{x} - \ln x \cdot 1}{x^2} = \frac{1 - \ln x}{x^2}$$

$$f''(x) = \frac{x^2 \cdot \left(-\dfrac{1}{x}\right) - (1 - \ln x)(2x)}{x^4} = \frac{2\ln x - 3}{x^3}.$$

If we set $f'(x) = 0$, then

$$1 - \ln x = 0$$

$$\ln x = 1$$

$$e^{\ln x} = e^1 = e$$

$$x = e.$$

Therefore, the only possible extreme point is at $x = e$. When $x = e$, $f(e) = (\ln e)/e = 1/e$. Furthermore,

$$f''(e) = \frac{2\ln e - 3}{e^3} = -\frac{1}{e^3} < 0,$$

which implies that the graph of $f(x)$ is concave down at $x = e$. Therefore, $(e, 1/e)$ is a maximum point of the graph of $f(x)$.

The next example introduces a function that will be needed later when we study integration.

EXAMPLE 3   The function $\ln|x|$ is defined for all nonzero values of x. Its graph is sketched in Fig. 1. Compute the derivative of $\ln|x|$.

Solution   If x is positive, then $|x| = x$, so

$$\frac{d}{dx}\ln|x| = \frac{d}{dx}\ln x = \frac{1}{x}.$$

If x is negative, then $|x| = -x$; and, by the chain rule,

$$\frac{d}{dx}\ln|x| = \frac{d}{dx}\ln(-x)$$

$$= \frac{1}{-x} \cdot \frac{d}{dx}(-x)$$

$$= \frac{1}{-x} \cdot (-1) = \frac{1}{x}.$$

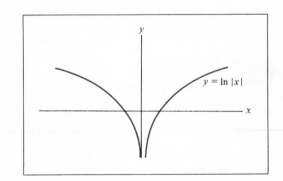

FIGURE 1

Therefore, we have established the following useful fact.

$$\frac{d}{dx}\ln|x| = \frac{1}{x}, \qquad x \neq 0.$$

PRACTICE PROBLEMS 5

Differentiate.

1. $f(x) = \dfrac{1}{\ln(x^4 + 5)}$

2. $f(x) = \ln(\ln x)$.

EXERCISES 5

Differentiate the following functions.

1. $\ln 2x$

2. $\ln x^2$

3. $\ln(x + 5)$

4. $x^2 \ln x$

5. $\dfrac{1}{x}\ln(x + 1)$

6. $\sqrt{\ln x}$

7. $e^{\ln x + x}$

8. $\ln\left(\dfrac{x}{x - 3}\right)$

9. $4 + \ln\left(\dfrac{x}{2}\right)$

10. $\ln \sqrt{x}$

11. $(\ln x)^2 + \ln x$

12. $\ln(x^3 + 2x + 1)$

13. $\ln(kx),\ k$ constant

14. $\dfrac{x}{\ln x}$

15. $\dfrac{x}{(\ln x)^2}$

16. $(\ln x)e^{-x}$

17. $e^{2x}\ln x$

18. $(\ln x + 1)^3$

19. $\ln(e^{5x} + 1)$

20. $\ln(e^{e^x})$

Find.

21. $\dfrac{d}{dt}(t^2 \ln 4)$

22. $\dfrac{d^2}{dx^2}\ln(1 + x^2)$

23. $\dfrac{d^2}{dt^2}(\ln t)^3$

24. Find the slope of the graph of $y = \ln|x|$ at $x = 3$ and $x = -3$.

25. Write the equation of the tangent line to the graph of $y = \ln(x^2 + e)$ at $x = 0$.

26. The function $f(x) = (\ln x + 1)/x$ has an extreme point for $x > 0$. Find the coordinates of the point. Is it a maximum point?

27. The function $f(x) = (\ln x)/\sqrt{x}$ has an extreme point for $x > 0$. Find the coordinates of the point. Is it a maximum point?

28. The function $f(x) = x/(\ln x + x)$ has an extreme point for $x > 1$. Find the coordinates of the point. Is it a minimum point?

29. Sketch the graph of the function $y = \ln x + (1/x) - \frac{1}{2}$.

30. Sketch the graph of $y = 1 + \ln(x^2 - 6x + 10)$.

31. If a cost function is $C(x) = (\ln x)/(40 - 3x)$, find the marginal cost when $x = 10$.

32. Suppose that the demand equation for a certain commodity is $p = 45/(\ln x)$. Determine the marginal revenue function for this commodity, and compute the marginal revenue when $x = 20$.

33. Suppose that the total revenue function for a manufacturer is $R(x) = 300 \ln(x + 1)$, so that the sale of x units of a product brings in about $R(x)$ dollars. Suppose also that the total cost of producing x units is $C(x)$ dollars, where $C(x) = 2x$. Find the value of x at which the profit function $R(x) - C(x)$ will be maximized. Show that the profit function has a maximum and not a minimum point at this value of x.

34. Evaluate $\displaystyle\lim_{h \to 0} \frac{\ln(7 + h) - \ln 7}{h}$.

35. Find the maximum area of a rectangle in the first quadrant with one corner at the origin, two sides on the coordinate axes, and one corner on the graph of $y = -\ln x$.

SOLUTIONS TO PRACTICE PROBLEMS 5

1. Here $f(x) = [\ln(x^4 + 5)]^{-1}$. By the chain rule,

$$f'(x) = -[\ln(x^4 - 5)]^{-2} \cdot \frac{d}{dx} \ln(x^4 + 5)$$

$$= -[\ln(x^4 + 5)]^{-2} \cdot \frac{4x^3}{x^4 + 5}.$$

2. $f'(x) = \dfrac{d}{dx} \ln(\ln x) = \dfrac{1}{\ln x} \cdot \dfrac{d}{dx} \ln x$

$$= \frac{1}{\ln x} \cdot \frac{1}{x} = \frac{1}{x \ln x}.$$

4.6 Properties of the Natural Logarithm Function

The natural logarithm function $\ln x$ possesses many of the familiar properties of logarithms to base 10 (or common logarithms) that are encountered in algebra.

Let x and y be positive numbers, b any number.

LI $\ln(xy) = \ln x + \ln y$.

LII $\ln\left(\dfrac{1}{x}\right) = -\ln x$.

LIII $\ln\left(\dfrac{x}{y}\right) = \ln x - \ln y$.

LIV $\ln(x^b) = b \ln x$.

Verification of LI   By equation (2) of Section 4 we have $e^{\ln(xy)} = xy$, $e^{\ln x} = x$, and $e^{\ln y} = y$. Therefore,

$$e^{\ln(xy)} = xy = e^{\ln x} \cdot e^{\ln y} = e^{\ln x + \ln y}.$$

By equating exponents, we get LI.

Verification of LII   Since $e^{\ln(1/x)} = 1/x$, we have

$$e^{\ln(1/x)} = \frac{1}{x} = \frac{1}{e^{\ln x}} = e^{-\ln x}.$$

By equating exponents, we get LII.

Verification of LIII   By LI and LII, we have

$$\ln\left(\frac{x}{y}\right) = \ln\left(x \cdot \frac{1}{y}\right)$$

$$= \ln x + \ln\left(\frac{1}{y}\right)$$

$$= \ln x - \ln y.$$

Verification of LIV   Since $e^{\ln(x^b)} = x^b$, we have

$$e^{\ln(x^b)} = x^b = (e^{\ln x})^b = e^{b \ln x}.$$

Equating exponents, we get LIV.

These properties of the natural logarithm should be learned thoroughly. You will find them useful in many calculations involving ln x and the exponential function.

EXAMPLE 1   Simplify $\ln 5 + 2 \ln 3$.

Solution

$$\ln 5 + 2 \ln 3 = \ln 5 + \ln 3^2 \qquad \text{(LIV)}$$

$$= \ln 5 + \ln 9$$

$$= \ln 45. \qquad \text{(LI)}$$

EXAMPLE 2   Simplify $\frac{1}{2} \ln(4t) - \ln(t^2 + 1)$.

Solution

$$\frac{1}{2} \ln(4t) - \ln(t^2 + 1) = \ln[(4t)^{1/2}] - \ln(t^2 + 1) \qquad \text{(LIV)}$$

$$= \ln(2\sqrt{t}) - \ln(t^2 + 1)$$

$$= \ln\left(\frac{2\sqrt{t}}{t^2 + 1}\right). \qquad \text{(LIII)}$$

EXAMPLE 3   Simplify $\ln x + \ln 3 + \ln y - \ln 5$.

Solution Use (LI) twice and (LIII) once.

$$(\ln x + \ln 3) + \ln y - \ln 5 = \ln 3x + \ln y - \ln 5$$

$$= \ln 3xy - \ln 5$$

$$= \ln\left(\frac{3xy}{5}\right).$$

EXAMPLE 4 Differentiate $f(x) = \ln[x(x + 1)(x + 2)]$.

Solution First rewrite $f(x)$, using (LI).

$$f(x) = \ln[x(x + 1)(x + 2)]$$

$$= \ln x + \ln(x + 1) + \ln(x + 2).$$

Then $f'(x)$ is easily calculated:

$$f'(x) = \frac{1}{x} + \frac{1}{x + 1} + \frac{1}{x + 2}.$$

The natural logarithm function can be used to simplify the task of differentiating products. Suppose, for example, that we wish to differentiate the function

$$g(x) = x(x + 1)(x + 2).$$

As we showed in Example 4,

$$\frac{d}{dx} \ln g(x) = \frac{1}{x} + \frac{1}{x + 1} + \frac{1}{x + 2}.$$

However,

$$\frac{d}{dx} \ln g(x) = \frac{g'(x)}{g(x)}.$$

Therefore, equating the two expressions for $\frac{d}{dx} \ln g(x)$, we have

$$\frac{g'(x)}{g(x)} = \frac{1}{x} + \frac{1}{x + 1} + \frac{1}{x + 2}.$$

Finally, we solve for $g'(x)$:

$$g'(x) = g(x) \cdot \left(\frac{1}{x} + \frac{1}{x + 1} + \frac{1}{x + 2}\right)$$

$$= x(x + 1)(x + 2)\left(\frac{1}{x} + \frac{1}{x + 1} + \frac{1}{x + 2}\right).$$

In a similar way, we differentiate the product of any number of terms by first taking natural logarithms, then differentiating, and finally solving for the desired derivative. This procedure is called *logarithmic differentiation*.

EXAMPLE 5 Differentiate the function $g(x) = (x^2 + 1)(x^3 - 3)(2x + 5)$ using logarithmic differentiation.

Solution Begin by taking the natural logarithm of both sides of the given equation:

$$\ln g(x) = \ln[(x^2 + 1)(x^3 - 3)(2x + 5)]$$
$$= \ln(x^2 + 1) + \ln(x^3 - 3) + \ln(2x + 5).$$

Now differentiate and solve for $g'(x)$:

$$\frac{g'(x)}{g(x)} = \frac{2x}{x^2 + 1} + \frac{3x^2}{x^3 - 3} + \frac{2}{2x + 5}$$

$$g'(x) = g(x)\left(\frac{2x}{x^2 + 1} + \frac{3x^2}{x^3 - 3} + \frac{2}{2x + 5}\right)$$

$$= (x^2 + 1)(x^3 - 3)(2x + 5)\left(\frac{2x}{x^2 + 1} + \frac{3x^2}{x^3 - 3} + \frac{2}{2x + 5}\right).$$

Let us now use logarithmic differentiation to finally establish the power rule:

$$\frac{d}{dx}(x^r) = rx^{r-1}.$$

Verification of the Power Rule Let $f(x) = x^r$. Then

$$\ln f(x) = \ln x^r = r \ln x.$$

Differentiation of this equation yields

$$\frac{f'(x)}{f(x)} = r \cdot \frac{1}{x}$$

$$f'(x) = r \cdot \frac{1}{x} \cdot f(x) = r \cdot \frac{1}{x} \cdot x^r = rx^{r-1}.$$

PRACTICE PROBLEMS 6

1. Differentiate $f(x) = \ln\left[\dfrac{e^x\sqrt{x}}{(x + 1)^6}\right]$.

2. Use logarithmic differentiation to differentiate $f(x) = (x + 1)^7(x + 2)^8(x + 3)^9$.

EXERCISES 6

Simplify the following expressions.

1. $\ln 5 + \ln x$
2. $\ln x^5 - \ln x^3$
3. $\frac{1}{2} \ln 9$
4. $3 \ln \frac{1}{2} + \ln 16$

5. $\ln 4 + \ln 6 - \ln 12$

6. $\ln 2 - \ln x + \ln 3$

7. $e^{2 \ln x}$

8. $\frac{3}{2} \ln 4 - 5 \ln 2$

9. $5 \ln x - \frac{1}{2} \ln y + 3 \ln z$

10. $e^{\ln x^2 + 3 \ln y}$

11. $\ln x - \ln x^2 + \ln x^4$

12. $\frac{1}{2} \ln xy + \frac{3}{2} \ln \dfrac{x}{y}$

13. Which is larger, $2 \ln 5$ or $3 \ln 3$?

14. Which is larger, $\frac{1}{2} \ln 16$ or $\frac{1}{3} \ln 27$?

Differentiate.

15. $\ln[(x + 5)(2x - 1)(4 - x)]$

16. $\ln[x^3(x + 1)^4]$

17. $\ln \left[\dfrac{(x + 1)(3x - 2)}{x + 2} \right]$

18. $\ln \left[\dfrac{x^2}{(3 - x)^3} \right]$

19. $\ln \left[\dfrac{\sqrt{x}}{x^2 + 1} \right]$

20. $\ln[e^{x^2}(x^4 + x^2 + 1)]$

Use logarithmic differentiation to differentiate the following functions.

21. $f(x) = (x + 1)^3(4x - 1)^2$

22. $f(x) = e^x(x - 4)^8$

23. $f(x) = (x - 2)^3(x - 3)^5(x + 2)^{-7}$

24. $f(x) = (x + 1)(2x + 1)(3x + 1)(4x + 1)$

25. $f(x) = x^x$

26. $f(x) = x^{1/x}$

27. $f(x) = e^x \sqrt{x^2 - 1}$

28. $f(x) = 2^x$

29. $f(x) = x^{\ln x}$

30. $f(x) = (2x - 1)^{2x}$

31. $f(x) = \dfrac{\sqrt{x - 1}(x - 2)}{x^2 - 3}$

32. $f(x) = \dfrac{xe^x}{\sqrt{3x^2 + 1}}$

33. There are substantial empirical data to show that if x and y measure the sizes of two organs of a particular animal, then x and y are related by an *allometric equation* of the form

$$\ln y - k \ln x = \ln c,$$

where k and c are positive constants that depend only on the type of parts or organs that are measured, and are constant among animals belonging to the same species.* Solve this equation for y in terms of x, k, and c.

34. In the study of epidemics, one finds the equation

$$\ln(1 - y) - \ln y = C - rt,$$

where y is the fraction of the population that has a specific disease at time t. Solve the equation for y in terms of t and the constants C and r.

---

* E. Batschelet, *Introduction to Mathematics for Life Scientists* (New York: Springer-Verlag, 1971), pp. 305–307.

35. Determine the values of h and k for which the graph of $y = he^{kx}$ passes through the points $(1, 6)$ and $(4, 48)$.

36. Find values of k and r for which the graph of $y = kx^r$ passes through the points $(2, 3)$ and $(4, 15)$.

SOLUTIONS TO PRACTICE PROBLEMS 6

1. Use the properties of the natural logarithm to express $f(x)$ as a sum of simple functions before differentiating.

$$f(x) = \ln\left[\frac{e^x \sqrt{x}}{(x+1)^6}\right]$$

$$= \ln e^x + \ln \sqrt{x} - \ln(x+1)^6$$

$$= x + \tfrac{1}{2}\ln x - 6\ln(x+1).$$

$$f'(x) = 1 + \frac{1}{2x} - \frac{6}{x+1}.$$

2.
$$f(x) = (x+1)^7(x+2)^8(x+3)^9.$$

$$\ln f(x) = 7\ln(x+1) + 8\ln(x+2) + 9\ln(x+3).$$

Now we differentiate both sides of the equation.

$$\frac{f'(x)}{f(x)} = \frac{7}{x+1} + \frac{8}{x+2} + \frac{9}{x+3}.$$

$$f'(x) = f(x)\left(\frac{7}{x+1} + \frac{8}{x+2} + \frac{9}{x+3}\right)$$

$$= (x+1)^7(x+2)^8(x+3)^9\left(\frac{7}{x+1} + \frac{8}{x+2} + \frac{9}{x+3}\right).$$

Chapter 4: CHECKLIST

☐ $b^x \cdot b^y = b^{x+y}$
☐ $b^{-x} = 1/b^x$
☐ $b^x/b^y = b^{x-y}$
☐ $(b^y)^x = b^{xy}$
☐ $a^x b^x = (ab)^x$
☐ $a^x/b^x = (a/b)^x$
☐ $b^0 = 1$
☐ Definition of e and e^x
☐ $\dfrac{d}{dx}(e^{kx}) = ke^{kx}$

☐ $\dfrac{d}{dx}(e^{g(x)}) = e^{g(x)}g'(x)$

- Graph of e^{kx}, k positive
- Graph of e^{kx}, k negative
- If $y = f(x)$ satisfies $y' = ky$, then $y = Ce^{kx}$ for some constant C.
- Reflection in the line $y = x$
- Definition of $\ln x$
- $\ln 1 = 0$
- $\ln e = 1$
- $e^{\ln x} = x, x > 0$
- $\ln e^x = x$
- $\dfrac{d}{dx}(\ln x) = \dfrac{1}{x}$
- $\ln(xy) = \ln x + \ln y$
- $\ln\left(\dfrac{x}{y}\right) = \ln x - \ln y$
- $\ln\left(\dfrac{1}{x}\right) = -\ln x$
- $\ln(x^b) = b \ln x$

Chapter 4: SUPPLEMENTARY EXERCISES

Calculate the following.

1. $27^{4/3}$
2. $4^{1.5}$
3. 5^{-2}
4. $16^{-.25}$

5. $(2^{5/7})^{14/5}$
6. $8^{1/2} \cdot 2^{1/2}$
7. $\dfrac{9^{5/2}}{9^{3/2}}$
8. $4^{.2} \cdot 4^{.3}$

Simplify the following.

9. $(e^{x^2})^3$
10. $e^{5x} \cdot e^{2x}$
11. $\dfrac{e^{3x}}{e^x}$

12. $2^x \cdot 3^x$
13. $(e^{8x} + 7e^{-2x})e^{3x}$
14. $\dfrac{e^{5x/2} - e^{3x}}{\sqrt{e^x}}$

Solve the following equations for x.

15. $e^{-3x} = e^{-12}$
16. $e^{x^2 - x} = e^2$

17. $(e^x \cdot e^2)^3 = e^{-9}$
18. $e^{-5x} \cdot e^4 = e$

Differentiate the following functions.

19. $10e^{7x}$
20. $e^{\sqrt{x}}$
21. xe^{x^2}
22. $\dfrac{e^x + 1}{x - 1}$

23. e^{e^x}
24. $(\sqrt{x} + 1)e^{-2x}$
25. $\dfrac{x^2 - x + 5}{e^{3x} + 3}$
26. x^e

27. Determine all solutions of the differential equation $y' = -y$.

28. Determine all functions $y = f(x)$ such that $y' = -1.5y$ and $f(0) = 2000$.

29. Determine all solutions of the differential equation $y' = 1.5y$ and $f(0) = 2$.

30. Determine all solutions of the differential equation $y' = \frac{1}{3}y$.

Graph the following functions.

31. $e^{-x} + x$

32. $e^x - x$

33. $e^{-(1/2)x^2}$

34. $100(x - 2)e^{-x}$ for $x \geq 2$

Simplify the following expressions.

35. $\dfrac{\ln x^2}{\ln x^3}$

36. $e^{2 \ln 2}$

37. $e^{-5 \ln 1}$

38. $\left[e^{\ln x}\right]^2$

39. $e^{(\ln 5)/2}$

40. $e^{\ln(x^2)}$

Solve the following equations for t.

41. $3e^{2t} = 15$

42. $3e^{t/2} - 12 = 0$

43. $2 \ln t = 5$

44. $2e^{-.3t} = 1$

45. $t^{\ln t} = e$

46. $\ln(\ln 3t) = 0$

Differentiate the following functions.

47. $\ln(5x - 7)$

48. $\ln(9x)$

49. $(\ln x)^2$

50. $(x \ln x)^3$

51. $\ln(x^6 + 3x^4 + 1)$

52. $\dfrac{x}{\ln x}$

53. $\ln\left(\dfrac{xe^x}{\sqrt{1 + x}}\right)$

54. $\ln\left[(x^2 + 3)^5(x^3 + 1)^{-4}\right]$

55. $\ln(\ln \sqrt{x})$

56. $\dfrac{1}{\ln x}$

57. $x \ln x - x$

58. $e^{2 \ln(x + 1)}$

59. $e^x \ln x$

60. $\ln(x^2 + e^x)$

Use logarithmic differentiation to differentiate the following functions.

61. $f(x) = (x^2 + 5)^6(x^3 + 7)^8(x^4 + 9)^{10}$

62. $f(x) = x^{1+x}$

63. $f(x) = 10^x$

64. $f(x) = \sqrt{x^2 + 5}\, e^{x^2}$

Sketch the following curves.

65. $y = x - \ln x$

66. $y = \ln(x^2 + 1)$

67. $y = (\ln x)^2$

68. $y = (\ln x)^3$

Applications of the Exponential and Natural Logarithm Functions

Chapter 5

Earlier, we introduced the exponential function e^x and the natural logarithm function $\ln x$ and studied their most important properties. From the way we introduced these functions, it is by no means clear that they have any substantial connection with the physical world. However, as this chapter will demonstrate, the exponential and natural logarithm functions intrude into the study of many physical problems, often in a very curious and unexpected way.

Here the most significant fact that we require is that the exponential function is uniquely characterized by its differential equation. In other words, we will constantly make use of the following fact, which was stated previously.

> The function* $y = Ce^{kt}$ satisfies the differential equation
> $$y' = ky.$$
> Conversely, if $y = f(t)$ satisfies the differential equation, then $y = Ce^{kt}$ for some constant C.

If $f(t) = Ce^{kt}$, then, by setting $t = 0$, we have
$$f(0) = Ce^0 = C.$$
Therefore, C is the value of $f(t)$ at $t = 0$.

5.1 Exponential Growth and Decay

In biology, chemistry, and economics it is often necessary to study the behavior of a quantity that is increasing as time passes. If, at every instant, the rate of increase of the quantity is proportional to the quantity at that instant, then we say

---

* Note that we use the variable t instead of x. The reason is that, in most applications, the variable of our exponential function is time. The variable t will be used throughout this chapter.

that the quantity is *growing exponentially* or is *exhibiting exponential growth*. A simple example of exponential growth is exhibited by the growth of bacteria in a culture. Under ideal laboratory conditions a bacteria culture grows at a rate proportional to the number of bacteria present. It does so because the growth of the culture is accounted for by the division of the bacteria. The more bacteria there are at a given instant, the greater the possibilities for division and hence the more rapid is the rate of growth.

Let us study the growth of a bacteria culture as a typical example of exponential growth. Suppose that $P(t)$ denotes the number of bacteria in a certain culture at time t. The rate of growth of the culture at time t is $P'(t)$. We assume that this rate of growth is proportional to the size of the culture at time t, so that

$$P'(t) = kP(t), \qquad (1)$$

where k is a positive constant of proportionality. If we let $y = P(t)$, then (1) can be written as

$$y' = ky.$$

Therefore, from our discussion at the beginning of this chapter we see that

$$y = P(t) = P_0 e^{kt}, \qquad (2)$$

where P_0 is the number of bacteria in the culture at time $t = 0$. The number k is called the *growth constant*.

EXAMPLE 1   Suppose that a certain bacteria culture grows at a rate proportional to its size. At time $t = 0$, approximately 20,000 bacteria are present. In 5 hours there are 400,000 bacteria. Determine a function that expresses the size of the culture as a function of time, measured in hours.

Solution   Let $P(t)$ be the number of bacteria present at time t. By assumption, $P(t)$ satisfies a differential equation of the form $y' = ky$, so $P(t)$ has the form

$$P(t) = P_0 e^{kt},$$

where the constants P_0 and k must be determined. The value of P_0 and k can be obtained from the data that give the population size at two different times. We are told that

$$P(0) = 20{,}000, \qquad P(5) = 400{,}000. \qquad (3)$$

The first condition immediately implies that $P_0 = 20{,}000$, so

$$P(t) = 20{,}000 e^{kt}.$$

Using the second condition in (3), we have

$$20{,}000 e^{k \cdot 5} = P(5) = 400{,}000$$

$$e^{5k} = 20$$

$$5k = \ln 20 \qquad (4)$$

$$k = \frac{\ln 20}{5} \approx .60.*$$

---

* Here and elsewhere in this chapter, we have used Table 2 of the Appendix and have carried out calculations to two significant figures.

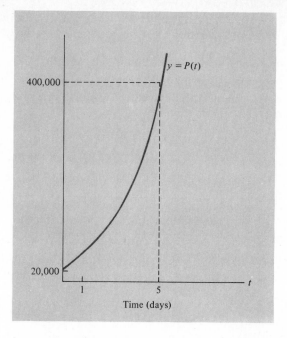

FIGURE 1   Graph of $y = 20{,}000e^{.6t}$.

So we may take

$$P(t) = 20{,}000e^{.6t}.$$

This function is a mathematical model of the growth of the bacteria culture. (See Fig. 1.)

EXAMPLE 2   Suppose that a colony of fruit flies is growing according to the exponential law $P(t) = P_0 e^{kt}$ and suppose that the size of the colony doubles in 12 days. Determine the growth constant k.

Solution   We do not know the initial size of the population at $t = 0$. However, we are told that $P(12) = 2P(0)$; that is,

$$P_0 e^{k \cdot 12} = 2P_0$$

$$e^{12k} = 2$$

$$12k = \ln 2$$

$$k = \tfrac{1}{12} \ln 2 \approx .058.$$

Notice that the initial size P_0 of the population was not given in Example 2. We were able to determine the growth constant because we were told the amount of time required for the colony to double in size. Thus the growth constant does not depend on the initial size of the population. This property is characteristic of exponential growth.

EXAMPLE 3   Suppose that the initial size of the colony in Example 2 was 300. At what time will the colony contain 1800 fruit flies?

Solution From Example 2 we have $P(t) = P_0 e^{.058t}$. Since $P(0) = 300$, we conclude that

$$P(t) = 300e^{.058t}.$$

Now that we have the explicit formula for the size of the colony, we can set $P(t) = 1800$ and solve for t:

$$300e^{.058t} = 1800$$

$$e^{.058t} = 6$$

$$.058t = \ln 6$$

$$t = \frac{\ln 6}{.058} \approx 31 \text{ days.}$$

The table below shows the growth of the colony in Example 3. Notice that 1800 is exactly halfway between 1200 ($t = 24$) and 2400 ($t = 36$). It is incorrect to guess that the population will equal 1800 when t is halfway between $t = 24$ and $t = 36$—that is, when $t = 30$. We saw in Example 3 that it takes approximately 31 days for the colony to reach 1800 fruit flies.

Population size	Day
300	0
600	12
1,200	24
2,400	36
4,800	48
9,600	60
19,200	72
$\vdots$	$\vdots$

Exponential Decay An example of negative exponential growth, or *exponential decay*, is given by the disintegration of a radioactive element such as uranium 235. It is known that, at any instant, the rate at which a radioactive substance is decaying is proportional to the amount of the substance that has not yet disintegrated. If $P(t)$ is the quantity present at time t, then $P'(t)$ is the rate of decay. Of course, $P'(t)$ must be negative, since $P(t)$ is decreasing. Thus we may write $P'(t) = kP(t)$ for some negative constant k. To emphasize the fact that the constant is negative, k is often replaced by $-\lambda*$, where λ is a positive constant. Then $P(t)$ satisfies the differential equation

$$P'(t) = -\lambda P(t). \tag{5}$$

The general solution of (5) has the form

$$P(t) = P_0 e^{-\lambda t}$$

for some positive number P_0. We call such a function an *exponential decay function*. The constant λ is called the *decay constant*.

---

* λ is the Greek lowercase letter lambda.

EXAMPLE 4   The decay constant for strontium 90 is $\lambda = .0244$, where the time is measured in years. How long will it take for a quantity P_0 of strontium 90 to decay to one-half its original size?

Solution   We have

$$P(t) = P_0 e^{-.0244t}.$$

Next, set $P(t)$ equal to $\frac{1}{2}P_0$ and solve for t:

$$P_0 e^{-.0244t} = \tfrac{1}{2}P_0$$

$$e^{-.0244t} = \tfrac{1}{2} = .5$$

$$-.0244t = \ln .5$$

$$t = \frac{\ln .5}{-.0244} \approx 28 \text{ years.}$$

The *half-life* of a radioactive element is the length of time required for a given quantity of that element to decay to one-half its original size. Thus strontium 90 has a half-life of about 28 years. (See Fig. 2.) Notice from Example 4 that the half-life does not depend on the initial amount P_0.

FIGURE 2   Half-life of radioactive strontium 90.

EXAMPLE 5   Radioactive carbon 14 has a half-life of about 5730 years. Find its decay constant.

Solution   If P_0 denotes the initial amount of carbon 14, then the amount after t years will be

$$P(t) = P_0 e^{-\lambda t}.$$

After 5730 years, $P(t)$ will equal $\frac{1}{2}P_0$. That is,

$$P_0 e^{-\lambda(5730)} = P(5730) = \tfrac{1}{2}P_0 = .5P_0.$$

Solving for λ gives

$$e^{-5730\lambda} = .5$$

$$-5730\lambda = \ln .5$$

$$\lambda = \frac{\ln .5}{-5730} \approx .00012.$$

One of the problems connected with aboveground nuclear explosions is the radioactive debris that falls on plants and grass, thereby contaminating the food supply of animals. Strontium 90 is one of the most dangerous components of "fallout" because it has a relatively long half-life and because it is chemically similar to calcium and is absorbed into the bone structure of animals (and humans) who eat contaminated food. Iodine 131 is also produced by nuclear explosions, but it presents less of a hazard because it has a half-life of 8 days.

EXAMPLE 6 If dairy cows eat hay containing too much iodine 131, their milk will be unfit to drink. Suppose that some hay contains 10 times the maximum allowable level of iodine 131. How many days should the hay be stored before it is fed to dairy cows?

Solution Let P_0 be the amount of iodine 131 present in the hay. Then the amount at time t is $P(t) = P_0 e^{-\lambda t}$ (t in days). The half-life of iodine 131 is 8 days, so

$$P_0 e^{-8\lambda} = .5P_0$$

$$e^{-8\lambda} = .5$$

$$-8\lambda = \ln .5$$

$$\lambda = \frac{\ln .5}{-8} \approx .087,$$

and

$$P(t) = P_0 e^{-.087t}.$$

Now that we have the formula for $P(t)$, we want to find t such that $P(t) = \frac{1}{10}P_0$. We have

$$P_0 e^{-.087t} = .1P_0,$$

so

$$e^{-.087t} = .1$$

$$-.087t = \ln .1$$

$$t = \frac{\ln .1}{-.087} \approx 26 \text{ days.}$$

Radiocarbon Dating   Knowledge about radioactive decay is valuable to social scientists who want to estimate the age of objects belonging to ancient civilizations. Several different substances are useful for radioactive-dating techniques; the most common is radiocarbon, C^{14}. Carbon 14 is produced in the upper atmosphere when cosmic rays react with atmospheric nitrogen. Because the C^{14} eventually decays, the concentration of C^{14} cannot rise above certain levels. An equilibrium is reached where C^{14} is produced at the same rate as it decays. Scientists usually assume that the total amount of C^{14} in the biosphere has remained constant over the past 50,000 years. Consequently, it is assumed that the *ratio* of C^{14} to ordinary nonradioactive carbon 12, C^{12}, has been constant during this same period. (The ratio is about one part C^{14} to 10^{12} parts of C^{12}.) Both C^{14} and C^{12} are in the atmosphere in the form of carbon dioxide. All living vegetation and most forms of animal life contain C^{14} and C^{12} in the same proportion as the atmosphere. The reason is that plants absorb carbon dioxide through photosynthesis. The C^{14} and C^{12} in plants is distributed through the various food chains to almost all animal life.

When an organism dies, it stops replacing its carbon, and therefore the amount of C^{14} begins to decrease through radioactive decay. (The C^{12} in the dead organism remains constant.) At a later date, the ratio of C^{14} to C^{12} can be measured in order to determine when the organism died. (See Fig. 3.)

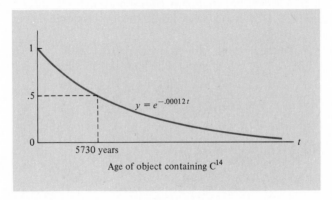

FIGURE 3 C^{14}–C^{12} ratio compared to the ratio in living plants.

EXAMPLE 7   A parchment fragment was discovered that had a C^{14}–C^{12} ratio of about .8 times the ratio found today in living matter. Estimate the age of the parchment.

Solution   We assume that the original C^{14}–C^{12} ratio in the parchment was the same as the ratio in living organisms today. Consequently, about eight-tenths of the original C^{14} remains. From Example 5 we obtain the formula for the amount of C^{14} present t years after the parchment was made from an animal skin:

$$P(t) = P_0 e^{-.00012t},$$

where P_0 = initial amount. We want to find t such that $P(t) = .8P_0$.

$$P_0 e^{-.00012t} = .8P_0$$

$$e^{-.00012t} = .8$$

$$-.00012t = \ln .8$$

$$t = \frac{\ln .8}{-.00012} \approx 1900 \text{ years old.}$$

A Sales Decay Curve   Marketing studies* have demonstrated that if advertising and other promotion of a particular product are stopped and if other market conditions remain fairly constant, then, at any time t, the sales of that product will be declining at a rate proportional to the amount of current sales at t. (See Fig. 4.) If S_0 is the number of sales in the last month during which advertising

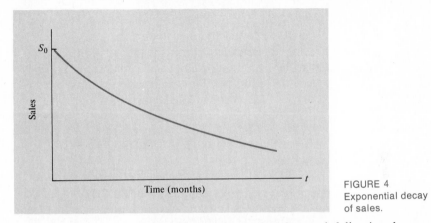

FIGURE 4
Exponential decay of sales.

occurred and if $S(t)$ is the number of sales in the tth month following the cessation of promotional effort, then a good mathematical model for $S(t)$ is

$$S(t) = S_0 e^{-\lambda t},$$

where λ is a positive number called the *sales decay constant*. The value of λ depends on many factors, such as the type of product, the number of years of prior advertising, the number of competing products, and other characteristics of the market.

The Time Constant   Consider an exponential decay function $y = Ce^{-\lambda t}$. In Fig. 5, we have drawn the tangent line to the decay curve when $t = 0$. The slope

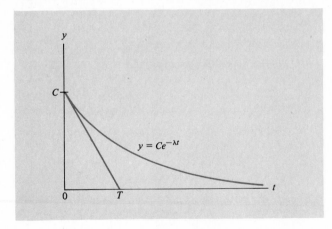

FIGURE 5   The time constant T in exponential decay: $T = 1/\lambda$.

* M. Vidale and H. Wolfe, "An Operations-Research Study of Sales Response to Advertising," *Operations Research*, 5 (1957), 370–381. Reprinted in F. Bass et al., *Mathematical Models and Methods in Marketing* (Homewood, Ill.: Richard D. Irwin, Inc., 1961).

there is the initial rate of decay. If the decay process were to continue at this rate, the decay curve would follow the tangent line and y would be zero at some time T. This time is called the *time constant* of the decay curve. It can be shown (see Exercise 29) that $T = 1/\lambda$ for the curve $y = Ce^{-\lambda t}$. Thus $\lambda = 1/T$, and the decay curve may be written in the form

$$y = Ce^{-t/T}.$$

If one has experimental data that tend to lie along an exponential decay curve, then the numerical constants for the curve may be obtained from Fig. 5. First, sketch the curve and estimate the y-intercept, C. Then sketch an approximate tangent line and from this estimate the time constant, T. This procedure is some-times used in biology and medicine.

PRACTICE PROBLEMS 1

1. (a) Solve the differential equation $P'(t) = -.6P(t)$, $P(0) = 50$.

    (b) Solve the differential equation $P'(t) = kP(t)$, $P(0) = 4000$, where k is some constant.

    (c) Find the value of k in part (b) for which $P(2) = 100P(0)$.

2. Under ideal conditions a colony of *Escherichia coli* bacteria can grow by a factor of 100 every 2 hours. If initially 4000 bacteria are present, how long will it take before there are 1,000,000 bacteria?

EXERCISES 1

1. Let $P(t)$ be the size of a certain insect population after t days, and suppose that $P(t)$ satisfies the differential equation

    $$P'(t) = .07P(t), \qquad P(0) = 400.$$

    Find the formula for $P(t)$.

2. Let $P(t)$ be the number of bacteria present in a culture after t minutes, and suppose that $P(t)$ satisfies the differential equation

    $$P'(t) = .55P(t).$$

    Find the formula for $P(t)$ if initially there are approximately 10,000 bacteria present.

3. Suppose that after t hours there are $P(t)$ cells present in a culture, where $P(t) = 5000e^{.2t}$.

    (a) How many cells were present initially?

    (b) When will 20,000 cells be present?

4. The size of a certain insect population is given by $P(t) = 300e^{.01t}$, where t is measured in days. At what time will the population equal 600? 1200?

5. Determine the growth constant of a population that is growing at a rate proportional to its size, where the population doubles in size every 40 days.

6. Determine the growth constant of a population that is growing at a rate proportional to its size, where the population triples in size every 5 hours.

7. The world's population was 4 billion on January 1, 1974, and was expected to reach 5 billion on January 1, 1985. Assume that at any time the population grows at a rate proportional to the population at that time.

   (a) Find the formula for $P(t)$, the world's population t years after January 1, 1974.

   (b) What will the population be on January 1, 2000?

   (c) In what year will the world's population reach 6 billion?

8. Mexico City is expected to become the most heavily populated city in the world by the end of this century. At the beginning of 1980, 14.5 million people lived in the metropolitan area of Mexico City, and the population was growing exponentially with growth constant .044. (Part of the growth is due to immigration.)

   (a) If this trend continues, how large will the population be in the year 2000?

   (b) In what year will the current population have doubled?

9. The growth rate of a certain bacteria culture is proportional to its size. If the bacteria culture doubles in size every 20 minutes, how long will it take for the culture to increase 12-fold?

10. The growth rate of a certain cell culture is proportional to its size. In 10 hours a population of 1 million cells grew to 9 million. How large will the cell culture be after 15 hours?

11. The weight in grams after t years $P(t)$ of a certain radioactive substance satisfies the differential equation

$$P'(t) = -.08P(t), \qquad P(0) = 30.$$

   (a) Find the formula for $P(t)$.

   (b) What is $P(10)$?

12. Five milligrams of a drug is injected into a patient, and the amount of drug present t hours after the injection satisfies the differential equation

$$P'(t) = -.09P(t).$$

   (a) Find the formula for $P(t)$.

   (b) Determine the amount of drug present 20 hours after the injection.

13. One hundred grams of a radioactive substance with decay constant .01 is buried in the ground. Assume that time is measured in years.

   (a) Give the formula for the amount remaining after t years.

   (b) How much will remain after 30 years?

   (c) What is the half-life of this radioactive substance?

14. The decay constant for cesium 137 is .023 when time is measured in years. Find the half-life of cesium 137.

15. Radioactive cobalt 60 has a half-life of 5.3 years.

    (a) Find the decay constant of cobalt 60.

    (b) If the initial amount of cobalt 60 is 10 grams, how much will be present after 2 years?

16. Five grams of a certain radioactive material decays to 3 grams in 1 year. After how many years will just 1 gram remain?

17. A 4500-year-old wooden chest was found in the tomb of the twenty-fifth century B.C. Chaldean king Meskalumdug of Ur. What percentage of the original C^{14} would you expect to find in the wooden chest? (Recall that the decay constant for C^{14} is .00012.)

18. In 1947, a cave with beautiful prehistoric wall paintings was discovered in Lascaux, France. Some charcoal found in the cave contained 20% of the C^{14} expected in living trees. How old are the Lascaux cave paintings?

19. Sandals woven from strands of tree bark were found recently in Fort Rock Creek Cave in Oregon. The bark contained 34% of the level of C^{14} found in living bark. Approximately how old are the sandals?

20. Many scientists believe there have been four ice ages in the past 1 million years. Before the technique of carbon dating was known, geologists erroneously believed that the retreat of the Fourth Ice Age began about 25,000 years ago. In 1950, logs from ancient spruce trees were found under glacial debris near Two Creeks, Wisconsin. Geologists determined that these trees had been crushed by the advance of ice during the Fourth Ice Age. Wood from the spruce trees contained 27% of the level of C^{14} found in living trees. Approximately how long ago did the Fourth Ice Age actually occur?

21. Let $f(t)$ be the value of the dollar t years after January 1, 1982, where the value is in terms of the purchasing power on January 1, 1982, and $f(0) = 1.00$. Suppose that the rate of decrease of $f(t)$ at any time t is proportional to $f(t)$. Give the formula for $f(t)$ if by January 1, 1984, the dollar had lost 15% of its purchasing power.

22. The consumer price index (CPI) gives a measure of the prices of commodities commonly purchased by consumers. For example, an increase of 10% in the CPI corresponds to a 10% average increase in the prices of consumer goods. Let $f(t)$ be the CPI at time t, where time is years since January 1, 1980. Suppose that $f(t)$ satisfies the differential equation

$$f'(t) = .12f(t), \qquad f(0) = 100.$$

(This means that at each time t, the CPI is rising at an annual rate of 12%, and the index is set equal to 100 on January 1, 1980.) How many years will it take for the CPI to double?

23. An island in the Pacific Ocean is contaminated by fallout from a nuclear explosion. If the strontium 90 is 100 times the level that scientists believe is "safe," how many years will it take for the island to once again be "safe" for human habitation? The half-life of strontium 90 is 28 years.

24. A common infection of the urinary tract in humans is caused by the bacterium *E. coli*. The infection is generally noticed when the bacteria colony reaches a population of about 10^8. The colony doubles in size about every 20 minutes. When a full bladder is emptied, about 90% of the bacteria are eliminated. Suppose that at the beginning of a certain time period, a person's bladder and urinary tract contain 10^8 *E. coli* bacteria. During an interval of T minutes the person drinks enough liquid to fill the bladder. Find the value of T such that if the bladder is emptied after T minutes, about 10^8 bacteria will still

remain. [*Note:* The average bladder holds about 1 liter of urine. It is seldom possible to eliminate an *E. coli* infection by diuresis without drugs—such as by drinking large amounts of water.]

25. By 1974 the United States had an estimated 80 million gallons of radioactive products from nuclear power plants and other nuclear reactors. These waste products were stored in various sorts of containers (made of such materials as stainless steel and cement), and the containers were buried in the ground and the ocean. Scientists feel that the waste products must be prevented from contaminating the rest of the earth until more than 99.99% of the radioactivity is gone (i.e., until the level is less than 0.0001 times the original level). If a storage cylinder contains waste products whose half-life is 1500 years, how many years must the container survive without leaking? [*Note:* Some of the containers are already leaking.]

26. In 1950 the world's population required 1×10^9 hectares* of arable land for food growth and in 1980 2×10^9 hectares were required. Current population trends indicate that if $A(t)$ denotes the amount of land needed t years after 1950, then

$$\frac{dA}{dt} = kA$$

for some constant k.

(a) Derive a formula for $A(t)$.

(b) The total amount of arable land on the earth's surface is estimated at 3.2×10^9 hectares. In what year will the earth exhaust its supply of land for growing food? [Data based on the Club of Rome's report, *The Limits to Growth* by D. H. Meadows, D. L. Meadows, J. Randers, and W. Behrens III (New York: Universe Books, 1972).]

27. A drought in the African veldt causes the death of much of the animal population. A typical herd of wildebeests suffers a death rate proportional to its size. The herd numbers 500 at the onset of the drought and only 200 remain 4 months later.

(a) Find a formula for the herd's population at time t months.

(b) How long will it take for the herd to diminish to one-tenth its original size?

28. In a laboratory experiment, 8 units of a sulfate were injected into a dog. After 50 minutes, only 4 units remained in the dog. Let $f(t)$ be the amount of sulfate present after t minutes. At any time, the rate of change of $f(t)$ is proportional to the value of $f(t)$. Find the formula for $f(t)$.

29. Let T be the time constant of the curve $y = Ce^{-\lambda t}$ as defined in Fig. 5. Show that $T = 1/\lambda$. [*Hint:* Express the slope of the tangent line in Fig. 5 in terms of C and T. Then set this slope equal to the slope of the curve $y = Ce^{-\lambda t}$ at $t = 0$.]

30. Suppose that a person is given an injection of 300 milligrams of penicillin at time $t = 0$, and let $f(t)$ be the amount (in milligrams) of penicillin present in the person's bloodstream t hours after the injection. Then the amount of penicillin present decays exponentially, and a typical formula for $f(t)$ is $f(t) = 300e^{-(2/3)t}$.

(a) What is the initial rate of decay of the penicillin?

(b) What is the time constant for this decay curve $y = f(t)$? (See the discussion accompanying Fig. 5.)

---

* A hectare equals 2.471 acres.

1. (a) Answer: $P(t) = 50e^{-.6t}$. Differential equations of the type $y' = ky$ have as their solution $P(t) = Ce^{kt}$, where C is $P(0)$.

    (b) Answer: $P(t) = 4000e^{kt}$. This problem is like the previous one except that the constant is not specified. Additional information is needed if one wants to determine a specific value for k.

    (c) Answer: $P(t) = 4000e^{2.3t}$. From the solution to part (b) we know that $P(t) = 4000e^{kt}$. We are given that $P(2) = 100P(0) = 100(4000) = 400,000$. So

    $$P(2) = 4000e^{k(2)} = 400,000$$

    $$e^{2k} = 100$$

    $$2k = \ln 100$$

    $$k = \frac{\ln 100}{2} \approx 2.3.$$

2. Let $P(t)$ be the number of bacteria present after t hours. We must first find an expression for $P(t)$ and then determine the value of t for which $P(t) = 1,000,000$. From the discussion at the beginning of the section we know that $P'(t) = k \cdot P(t)$. Also, we are given that $P(2)$ (the population after 2 hours) is $100P(0)$ (100 times the initial population). From Problem 1(c) we have an expression for $P(t)$:

    $$P(t) = 4000e^{2.3t}.$$

    Now we must solve $P(t) = 1,000,000$ for t.

    $$4000e^{2.3t} = 1,000,000$$

    $$e^{2.3t} = 250$$

    $$2.3t = \ln 250$$

    $$t = \frac{\ln 250}{2.3} \approx 2.4.$$

    Therefore, after 2.4 hours there will be 1,000,000 bacteria.

5.2 Compound Interest

When money is deposited in a savings account, interest is paid at stated intervals. If this interest is added to the account and thereafter earns interest itself, then the interest is called *compound interest*. The original amount deposited is called the *principal amount*. The principal amount plus the compound interest is called the *compound amount*. The interval between interest payments is referred to as the *interest period*. In formulas for compound interest, the interest rate is expressed as a decimal rather than a percent. Thus 6% is written as .06.

If \$1000 is deposited at 6% annual interest, compounded annually, the compound amount at the end of the first year will be

$$A_1 = \underset{\text{principal}}{1000} + \underset{\text{interest}}{1000(.06)} = 1000(1 + .06).$$

At the end of the second year the compound amount will be

$$A_2 = \underset{\substack{\text{compound} \\ \text{amount}}}{A_1} + \underset{\text{interest}}{A_1(.06)} = A_1(1 + .06)$$

$$= [1000(1 + .06)](1 + .06) = 1000(1 + .06)^2.$$

At the end of 3 years

$$A_3 = A_2 + A_2(.06) = A_2(1 + .06)$$
$$= [1000(1 + .06)^2](1 + .06) = 1000(1 + .06)^3.$$

After n years the compound amount will be

$$A = 1000(1 + .06)^n.$$

In this example the interest period was 1 year. The important point to note, however, is that at the end of each interest period the amount on deposit grew by a factor of $(1 + .06)$. In general, if the interest rate is i instead of .06, the compound amount will grow by a factor of $(1 + i)$ at the end of each interest period.

Suppose that a principal amount P is invested at a compound interest rate i per interest period, for a total of n interest periods. Then the compound amount A at the end of the nth period will be

$$A = P(1 + i)^n. \tag{1}$$

EXAMPLE 1 Suppose that \$5000 is invested at 8% per year, with interest compounded annually. What is the compound amount after 3 years?

Solution Substituting $P = 5000$, $i = .08$, and $n = 3$ into formula (1), we have

$$A = 5000(1 + .08)^3 = 5000(1.08)^3$$
$$= 5000(1.259712) = 6298.56 \text{ dollars.}$$

It is common practice to state the interest rate as a percent per year ("per annum"), even though each interest period is often shorter than 1 year. If the annual rate is r and if interest is paid and compounded m times per year, then the interest rate i for each period is given by

$$[\text{rate per period}] \quad i = \frac{r}{m} = \frac{[\text{annual interest rate}]}{[\text{periods per year}]}.$$

Many banks pay interest quarterly. If the stated annual rate is 5%, then $i = .05/4 = .0125$.

If interest is compounded for t years, with m interest periods each year, there will be a total of mt interest periods. If in formula (1) we replace n by mt and replace i by r/m, we obtain the following formula for the compound amount:

$$A = P\left(1 + \frac{r}{m}\right)^{mt},$$

where P = principal amount,
r = interest rate per annum,
m = number of interest periods per year,
t = number of years.

(2)

EXAMPLE 2 Suppose that $1000 is deposited in a savings account that pays 6% per annum, compounded quarterly. If no additional deposits or withdrawals are made, how much will be in the account at the end of 1 year?

Solution We use (2) with $P = 1000$, $r = .06$, $m = 4$, and $t = 1$.

$$A = 1000\left(1 + \frac{.06}{4}\right)^4 = 1000(1.015)^4$$

$$= 1000(1.06136355) \approx 1061.36 \text{ dollars.}$$

Note that the $1000 in Example 2 earned a total of $61.36 in (compound) interest. This is 6.136% of $1000. Savings institutions sometimes advertise this rate as the *effective* annual interest rate. That is, the savings institutions mean that *if* they paid interest only once a year, they would have to pay a rate of 6.136% in order to produce the same earnings as their 6% rate compounded quarterly. The stated annual rate of 6% is often called the *nominal rate*.

The effective annual rate can be increased by compounding the interest more often. Some savings institutions compound interest monthly or even daily.

EXAMPLE 3 Suppose that the interest in Example 2 were compounded monthly. How much would be in the account at the end of 1 year? What about the case when 6% annual interest is compounded daily?

Solution For monthly compounding, $m = 12$. From (2) we have

$$A = 1000\left(1 + \frac{.06}{12}\right)^{12} = 1000(1.005)^{12}$$

$$\approx 1000(1.06167781) \approx 1061.68 \text{ dollars.}$$

The effective rate in this case is 6.168%.

A "bank year" usually consists of 360 days (in order to simplify calculations). So, for daily compounding, we take $m = 360$. Then

$$A = 1000\left(1 + \frac{.06}{360}\right)^{360} \approx 1000(1.00016667)^{360}$$

$$\approx 1000(1.06183133) \approx 1061.83 \text{ dollars.}$$

With daily compounding, the effective rate is 6.183%.

What would happen if the interest in Example 3 were compounded more often than once a day? Would the total interest be much more than $61.83 if the interest were compounded every hour? Every minute? To answer these questions, we shall connect the notion of compound interest to the exponential function. Recall that the compound amount A is given by

$$A = P\left(1 + \frac{r}{m}\right)^{mt} = P\left(1 + \frac{r}{m}\right)^{(m/r) \cdot rt}.$$

If we set $h = r/m$, then $1/h = m/r$, and

$$A = P(1 + h)^{(1/h) \cdot rt}.$$

As the frequency of compounding is increased, m gets large and $h = r/m$ approaches 0. To determine what happens to the compound amount, we must therefore examine the limit

$$\lim_{h \to 0} P(1 + h)^{(1/h) \cdot rt}.$$

The following remarkable fact is proved in the appendix at the end of this section:

$$\lim_{h \to 0} (1 + h)^{1/h} = e.$$

Using this fact together with two limit theorems, we have

$$\lim_{h \to 0} P(1 + h)^{(1/h)rt} = P\left[\lim_{h \to 0} (1 + h)^{1/h}\right]^{rt} = Pe^{rt}.$$

These calculations show that the compound amount calculated from the formula $P\left(1 + \frac{r}{m}\right)^{mt}$ gets closer to Pe^{rt} as the number m of interest periods per year is increased. When the formula

$$A = Pe^{rt} \tag{3}$$

is used to calculate the compound amount, we say that the interest is *compounded continuously.*

We can now answer the question posed following Example 3. Suppose that $1000 is deposited for 1 year in an account paying 6% per annum. Then $P = 1000$, $r = .06$, and $t = 1$. If the interest is compounded continuously, the compound amount at the end of 1 year will be

$$1000e^{.06} \approx 1061.84 \text{ dollars.}$$

Recall from Example 3 that daily compounding of the 6% interest produced $1061.83. Consequently, more frequent compounding (such as every hour or every second) will produce at most 1 cent more.

In recent years, when banks have wanted to offer the maximum effective rate of interest permitted by law, many banks and savings institutions have advertised savings accounts that pay interest compounded continuously. However, as we have seen, the effect of compounding continuously is practically the same as compounding daily, unless the principal amount P is quite large.

When interest is compounded continuously, the compound amount $A(t)$ is an exponential function of the number of years t that interest is earned, $A(t) = Pe^{rt}$. Hence $A(t)$ satisfies the differential equation

$$\frac{dA}{dt} = rA.$$

The rate of growth of the compound amount is proportional to the amount of money present. Since the growth comes from the interest, we conclude that under continuous compounding, interest is earned continuously at a rate of growth proportional to the amount of money present.

The formula $A = Pe^{rt}$ contains four variables. (Remember that the letter e here represents a specific constant, $e = 2.718. \ldots$) In a typical problem, we are given values for three of these variables and must solve for the remaining variable.

EXAMPLE 4   How long is required for an investment of $1000 to double if the interest is 10%, compounded continuously?

Solution   Here $P = 1000$ and $r = .10$. For each time t, the value of the investment is $Pe^{rt} = 1000e^{.10t}$. We must find t such that this value is $2000. So we set

$$2000 = 1000e^{.10t}$$

and solve for t. We divide both sides by 1000 and then take logarithms of both sides to obtain

$$2 = e^{.10t}$$

$$\ln 2 = .10t,$$

and

$$t = 10 \ln 2 \approx 6.9 \text{ years.}$$

Remark The calculations in Example 4 would be essentially unchanged after the first step if the initial amount of the investment were changed from $1000 to any arbitrary amount P. When this investment doubles, the compound amount will be $2P$. So one sets $2P = Pe^{.10t}$ and solves for t as we did above, to conclude that any amount doubles in about 6.9 years.

If P dollars are invested today, the formula $A = Pe^{rt}$ gives the value of this investment after t years (assuming continuously compounded interest). We say that P is the *present value* of the amount A to be received in t years. If we solve for P in terms of A, we obtain

$$P = Ae^{-rt}. \tag{4}$$

The concept of the present value of money is an important theoretical tool in business and economics. Problems involving depreciation of equipment, for example, may be analyzed by calculus techniques when the present value of money is computed from (4) using continuously compounded interest.

EXAMPLE 5 Find the present value of $5000 to be received in 2 years if money can be invested at 12% compounded continuously.

Solution Use (4) with $A = 5000$, $r = .12$, and $t = 2$.

$$P = 5000e^{-(.12)(2)} = 5000e^{-.24}$$
$$\approx 5000(0.78663) = 3933.15 \text{ dollars.}$$

APPENDIX   A Limit Formula for *e*

For $h \neq 0$, we have

$$\ln(1 + h)^{1/h} = 1/h \ln(1 + h).$$

Taking the exponential of both sides, we find that

$$(1 + h)^{1/h} = e^{(1/h)\ln(1 + h)}.$$

Since the exponential function is continuous,

$$\lim_{h \to 0} (1 + h)^{1/h} = e^{\left[\lim_{h \to 0} (1/h)\ln(1 + h)\right]}. \tag{5}$$

To examine the limit inside the exponential function, we note that $\ln 1 = 0$, and hence

$$\lim_{h \to 0} (1/h)\ln(1 + h) = \lim_{h \to 0} \frac{\ln(1 + h) - \ln 1}{h}.$$

The limit on the right is a difference quotient of the type used to compute a derivative. In fact,

$$\lim_{h \to 0} \frac{\ln(1 + h) - \ln 1}{h} = \frac{d}{dx} \ln x \Big|_{x=1}$$

$$= \frac{1}{x} \Big|_{x=1} = 1.$$

Thus the limit inside the exponential function in (5) is 1. That is,

$$\lim_{h \to 0} (1 + h)^{1/h} = e^{[1]} = e.$$

PRACTICE PROBLEMS 2

1. One thousand dollars is to be invested in a bank for 4 years. Would 8% interest compounded semiannually be better than $7\frac{3}{4}\%$ interest compounded continuously?

2. A building was bought for $150,000 and sold 10 years later for $400,000. What interest rate (compounded continuously) was earned on the investment?

EXERCISES 2

1. Suppose that $1000 is deposited in a savings account at 10% interest compounded annually. What is the compound amount after 2 years?

2. Five thousand dollars is deposited in a savings account at 6% interest compounded monthly. Give the formula that describes the compound amount after 4 years.

3. Ten thousand dollars is invested at 8% interest compounded quarterly. Give the formula that describes the value of the investment after 3 years.

4. What is the effective annual rate of interest of a savings account paying 8% interest compounded semiannually?

5. One thousand dollars is invested at 14% interest compounded continuously. Compute the value of the investment at the end of 6 years.

6. A painting was purchased in 1983 for $100,000. If it appreciates at 12% compounded continuously, how much will it be worth in 1990?

7. Five hundred dollars is deposited in a savings account paying 7% interest compounded daily. *Estimate* the balance in the account at the end of 3 years.

8. Ten thousand dollars is deposited into a money market fund paying 18% interest compounded continuously. How much interest will be earned during the first half-year if this rate of 18% does not change?

9. An office building, built in 1975 at a cost of 10 million dollars, was appraised at 25 million dollars in 1983. At what annual rate of interest (compounded continuously) did the building appreciate?

10. One thousand dollars is deposited in a savings account at 6% interest compounded continuously. How many years are required for the balance in the account to reach $2500?

11. Ten thousand dollars is to be invested in a highly speculative venture for 1 year. Would you rather receive 40% interest compounded semiannually or 39% interest compounded continuously?

12. A lot purchased in 1964 for $5000 was appraised at $60,000 in 1983. If the lot continues to appreciate at the same rate, when will it be worth $100,000?

13. Ten thousand dollars is invested at 15% interest compounded continuously. When will the investment be worth $38,000?

14. How many years are required for an investment to double in value if it is appreciating at the rate of 13% compounded continuously?

15. A farm purchased in 1970 for $1,000,000 was valued at $3,000,000 in 1980. If the farm continues to appreciate at the same rate (with continuous compounding), when will it be worth $10,000,000?

16. Find the present value of $1000 payable at the end of 3 years, if money may be invested at 14% with interest compounded continuously.

17. Find the present value of $2000 to be received in 10 years, if money may be invested at 15% with interest compounded continuously.

18. A parcel of land bought in 1975 for $10,000 was worth $16,000 in 1980. If the land continues to appreciate at this rate, in what year will it be worth $45,000?

19. One hundred dollars is deposited in a savings account at 7% interest compounded continuously. What is the effective annual rate of return?

20. In a certain town, property values tripled from 1973 to 1984. If this trend continues, when will property values be at five times their 1973 level? (Use an exponential model for the property value at time t.)

21. How much money must you invest now at 12% interest compounded continuously, in order to have $10,000 at the end of 5 years?

22. Investment A is currently worth $70,200 and is growing at the rate of 13% per year compounded continuously. Investment B is currently worth $60,000 and is growing at the rate of 14% per year compounded continuously. After how many years will the two investments have the same value?

23. Suppose that the present value of $1000 to be received in 2 years is $559.90. What rate of interest, compounded continuously, was used to compute this present value?

24. Two thousand dollars is deposited in a savings account at 5% interest compounded continuously. Let $f(t)$ be the compound amount after t years. Find and interpret $f'(2)$.

SOLUTIONS TO PRACTICE PROBLEMS 2

1. Let us compute the balance after 4 years for each type of interest.

    8% compounded semiannually: Use formula (2). Here $P = 1000$, $r = .08$, $m = 2$ (semiannually means that there are two interest periods per year), and $t = 4$. Therefore,

    $$A = 1000\left(1 + \frac{.08}{2}\right)^{2 \cdot 4}$$

    $$= 1000(1.04)^8 = 1368.57.$$

    $7\frac{3}{4}\%$ *compounded continuously:* Use the formula $A = Pe^{rt}$, where $P = 1000$, $r = .0775$, and $t = 4$. Then

    $$A = 1000e^{(.0775) \cdot 4}$$

    $$= 1000e^{.31} = 1363.43.$$

    Therefore, 8% compounded semiannually is best.

2. If the $150,000 had been compounded continuously for 10 years at interest rate r, the balance would be $150,000e^{r \cdot 10}$. The question asks: For what value of r will the balance be 400,000? We need just solve an equation for r.

    $$150{,}000e^{r \cdot 10} = 400{,}000$$

    $$e^{r \cdot 10} \approx 2.67$$

    $$r \cdot 10 = \ln 2.67$$

    $$r = \frac{\ln 2.67}{10} \approx .098.$$

    Therefore, the investment earned 9.8% interest per year.

5.3 Applications of the Natural Logarithm Function to Economics

In this section we consider two applications of the natural logarithm to the field of economics. Our first application is concerned with relative rates of change and the second with elasticity of demand.

Relative Rates of Change The *logarithmic derivative* of a function $f(t)$ is defined by the equation

$$\frac{d}{dt} \ln f(t) = \frac{f'(t)}{f(t)} \qquad (1)$$

The quantity on either side of equation (1) is often called the *relative rate of change of $f(t)$ per unit change of t.* Indeed, this quantity compares the rate of change of $f(t)$ [namely $f'(t)$] with $f(t)$ itself. The *percentage rate of change* is the relative rate of change of $f(t)$ expressed as a percentage.

A simple example will illustrate these concepts. Suppose that $f(t)$ denotes the average price per pound of sirloin steak at time t and $g(t)$ denotes the average price of a new car (of a given make and model) at time t, where $f(t)$ and $g(t)$ are given in dollars and time is measured in years. Then the ordinary derivatives $f'(t)$ and $g'(t)$ may be interpreted as the rate of change of the price of a pound of sirloin steak and of a new car, respectively, where both are measured in dollars per year. Suppose that, at a given time t_0, we have $f(t_0) = \$5.25$ and $g(t_0) = \$12,000$. Moreover, suppose that $f'(t_0) = \$.75$ and $g'(t_0) = \$1500$. Then at time t_0 the price per pound of steak is increasing at a rate of $\$.75$ per year, while the price of a new car is increasing at a rate of $\$1500$ per year. Which price is increasing more quickly? It is not meaningful to say that the car price is increasing faster simply because $\$1500$ is larger than $\$.75$. We must take into account the vast difference between the actual cost of a car and the cost of steak. The usual basis of comparison of price increases is the percentage rate of increase. In other words, at $t = t_0$, the price of sirloin steak is increasing at the percentage rate

$$\frac{f'(t_0)}{f(t_0)} = \frac{.75}{5.25} \approx .143 = 14.3\%$$

per year, but at the same time the price of a new car is increasing at the percentage rate

$$\frac{g'(t_0)}{g(t_0)} = \frac{1500}{12,000} \approx .125 = 12.5\%$$

per year. Thus the price of sirloin steak is increasing at a faster percentage rate than the price of a new car.

Economists often use relative rates of change (or percentage rates of change) when discussing the growth of various economic quantities, such as national income or national debt, because such rates of change can be meaningfully compared.

EXAMPLE 1 Suppose that the Gross National Product of the United States at time t (measured in years from January 1, 1984) is predicted by a certain school of economists to be

$$f(t) = 3.4 + .04t + .13e^{-t},$$

where the Gross National Product is measured in trillions of dollars. What is the percentage rate of growth (or decline) of the economy at $t = 0$ and $t = 1$?

Solution Since

$$f'(t) = .04 - .13e^{-t},$$

we see that

$$\frac{f'(0)}{f(0)} = \frac{.04 - .13}{3.4 + .13} = -\frac{.09}{3.53} \approx -2.6\%.$$

$$\frac{f'(1)}{f(1)} = \frac{.04 - .13e^{-1}}{3.4 + .04 + .13e^{-1}} = -\frac{.00782}{3.4878} \approx -.2\%.$$

So on January 1, 1984, the economy is predicted to contract at a relative rate of 2.6% per year; on January 1, 1985, the economy is predicted to be still contracting but only at a relative rate of .2% per year.

EXAMPLE 2 Suppose that the value in dollars of a certain business investment at time t may be approximated empirically by the function $f(t) = 750,000e^{.6\sqrt{t}}$. Use a logarithmic derivative to describe how fast the value of the investment is increasing when $t = 5$ years.

Solution We have

$$\frac{f'(t)}{f(t)} = \frac{d}{dt} \ln f(t) = \frac{d}{dt}(\ln 750,000 + \ln e^{.6\sqrt{t}})$$

$$= \frac{d}{dt}(\ln 750,000 + .6\sqrt{t})$$

$$= (.6)\left(\frac{1}{2}\right)t^{-1/2} = \frac{.3}{\sqrt{t}}.$$

When $t = 5$,

$$\frac{f'(5)}{f(5)} = \frac{.3}{\sqrt{5}} \approx .1345 = 13.4\%.$$

Thus, when $t = 5$ years, the value of the investment is increasing at the relative rate of 13.4% per year.

In certain mathematical models, it is assumed that for a limited period of time, the percentage rate of change of a particular function is constant. The following example shows that such a function must be an exponential function.

EXAMPLE 3 Suppose that the function $f(t)$ has a constant relative rate of change k. Show that $f(t) = Ce^{kt}$ for some constant C.

Solution We are given that

$$\frac{d}{dt} \ln f(t) = k.$$

That is,

$$\frac{f'(t)}{f(t)} = k.$$

Hence $f'(t) = kf(t)$. But this is just the differential equation satisfied by the exponential function. Therefore, we must have $f(t) = Ce^{kt}$ for some constant C.

Elasticity of Demand In Section 2.7 we considered demand equations for monopolists and for entire industries. Recall that a demand equation expresses, for each quantity q to be produced, the market price which will generate a demand of exactly q. For instance, the demand equation

$$p = 150 - .01x \qquad (2)$$

says that in order to sell x units, the price must be set at $150 - .01x$ dollars. To be specific: In order to sell 6000 units, the price must be set at $150 - .01(6000) = \$90$ per unit.

Equation (2) may be solved for x in terms of p to yield

$$x = 100(150 - p). \qquad (3)$$

This last equation gives quantity in terms of price. If we let the letter q represent quantity, equation (3) becomes

$$q = 100(150 - p). \qquad (3')$$

This equation is of the form $q = f(p)$, where in this case $f(p)$ is the function $f(p) = 100(150 - p)$. In what follows it will be convenient to always write our demand functions so that the quantity q is expressed as a function $f(p)$ of the price p.

Usually, raising the price of a commodity lowers demand. Therefore, the typical demand function $q = f(p)$ is decreasing and has a negative slope everywhere.

A demand function $q = f(p)$ relates the quantity demanded to the price. Therefore, the derivative $f'(p)$ compares the change in quantity demanded with the change in price. By way of contrast, the concept of elasticity is designed to compare the *relative* rate of change of the quantity demanded with the *relative* rate of change of price.

Let us be more explicit. Consider a particular demand function $q = f(p)$ and a particular price p. Then at this price, the ratio of the relative rates of change of the quantity demanded and the price is given by

$$\frac{[\text{relative rate of change of quantity}]}{[\text{relative rate of change of price}]} = \frac{\dfrac{d}{dp} \ln f(p)}{\dfrac{d}{dp} \ln p}$$

$$= \frac{f'(p)/f(p)}{1/p}$$

$$= \frac{pf'(p)}{f(p)}.$$

Since $f'(p)$ is always negative for a typical demand function, the quantity $pf'(p)/f(p)$ will be negative for all values of p. For convenience, economists prefer to work with positive numbers and therefore the *elasticity of demand* is taken to be this quantity multiplied by -1.

The elasticity of demand $E(p)$ at price p for the demand function $q = f(p)$ is defined to be

$$E(p) = \frac{-pf'(p)}{f(p)}.$$

EXAMPLE 4 Suppose that the demand function for a certain metal is $q = 100 - 2p$, where p is the price per pound and q is the quantity demanded (in millions of pounds).

(a) What quantity can be sold at \$30 per pound?

(b) Determine the function $E(p)$.

(c) Determine and interpret the elasticity of demand at $p = 30$.

(d) Determine and interpret the elasticity of demand at $p = 20$.

Solution (a) In this case, $q = f(p)$, where $f(p) = 100 - 2p$. When $p = 30$, we have $q = f(30) = 100 - 2(30) = 40$. Therefore, 40 million pounds of the metal can be sold. We also say that the *demand* is 40 million pounds.

(b)
$$E(p) = \frac{-pf'(p)}{f(p)}$$

$$= \frac{-p(-2)}{100 - 2p}$$

$$= \frac{2p}{100 - 2p}.$$

(c) The elasticity of demand at price $p = 30$ is $E(30)$.

$$E(30) = \frac{2(30)}{100 - 2(30)}$$

$$= \frac{60}{40}$$

$$= \frac{3}{2}.$$

When the price is set at \$30 per pound, a small increase in price will result in a relative rate of decrease in quantity demanded of about $\frac{3}{2}$ times the relative rate of increase in price. For example, if the price is increased from \$30 by 1%, then the quantity demanded will decrease by about 1.5%.

(d) When $p = 20$, we have

$$E(20) = \frac{2(20)}{100 - 2(20)} = \frac{40}{60} = \frac{2}{3}.$$

When the price is set at \$20 per pound, a small increase in price will result in a relative rate of decrease in quantity demanded of only $\frac{2}{3}$ of the relative rate of increase of price. For example, if the price is increased from \$20 by 1%, the quantity demanded will decrease by $\frac{2}{3}$ of 1%.

Economists say that demand is *elastic* at price p_0 if $E(p_0) > 1$ and *inelastic* at price p_0 if $E(p_0) < 1$. In Example 4, the demand for the metal is elastic at \$30 per pound and inelastic at \$20 per pound.

The significance of the concept of elasticity may perhaps best be appreciated by studying how revenue, $R(p)$, responds to changes in price. Recall that

$$[\text{revenue}] = [\text{quantity}] \cdot [\text{price per unit}],$$

that is,

$$R(p) = f(p) \cdot p.$$

If we differentiate $R(p)$ using the product rule, we find that

$$R'(p) = \frac{d}{dp}[f(p) \cdot p] = f(p) \cdot 1 + p \cdot f'(p)$$

$$= f(p)\left[1 + \frac{pf'(p)}{f(p)}\right]$$

$$= f(p)[1 - E(p)]. \tag{4}$$

Now suppose that demand is elastic at some price p_0. Then $E(p_0) > 1$ and $1 - E(p_0)$ is negative. Since $f(p)$ is always positive, we see from (4) that $R'(p_0)$ is negative. Therefore, by the first derivative rule, $R(p)$ is decreasing at p_0. So an increase in price will result in a decrease in revenue, and a decrease in price will result in an increase in revenue. On the other hand, if demand is inelastic at p_0, then $1 - E(p_0)$ will be positive and hence $R'(p_0)$ will be positive. In this case an increase in price will result in an increase in revenue, and a decrease in price will result in an decrease in revenue. We may summarize this as follows:

The change in revenue is in the opposite direction of the change in price when demand is elastic and in the same direction when demand is inelastic.

PRACTICE PROBLEMS 3

The current toll for the use of a certain toll road is \$2.50. A study conducted by the state highway department determined that with a toll of p dollars, q cars will use the road each day, where $q = 60{,}000e^{-.5p}$.

1. Compute the elasticity of demand at $p = 2.5$.

2. Is demand elastic or inelastic at $p = 2.5$?

3. If the state increases the toll slightly, will the revenue increase or decrease?

EXERCISES 3

Determine the percentage rate of change of the functions at the points indicated.

1. $f(t) = t^2$ at $t = 10$ and $t = 50$

2. $f(t) = t^{10}$ at $t = 10$ and $t = 50$

3. $f(x) = e^{.3x}$ at $x = 10$ and $x = 20$

4. $f(x) = e^{-.05x}$ at $x = 1$ and $x = 10$

5. $f(t) = e^{.3t^2}$ at $t = 1$ and $t = 5$

6. $G(s) = e^{-.05s^2}$ at $s = 1$ and $s = 10$

7. $f(p) = 1/(p + 2)$ at $p = 2$ and $p = 8$

8. $g(p) = 5/(2p + 3)$ at $p = 1$ and $p = 11$

9. Suppose that the annual sales S (in dollars) of a company may be approximated empirically by the formula
$$S = 50,000\sqrt{e^{\sqrt{t}}},$$
where t is the number of years beyond some fixed reference date. Use a logarithmic derivative to determine the percentage rate of growth of sales at $t = 4$.

10. Suppose that the price of wheat per bushel at time t (in months) is approximated by
$$f(t) = 4 + .001t + .01e^{-t}.$$
What is the percentage rate of change of $f(t)$ at $t = 0$? $t = 1$? $t = 2$?

11. Suppose that an investment grows at a constant 12% rate per year. In how many years will the value of the investment double?

12. Suppose that the value of a piece of property is growing at a constant $r\%$ rate per year and that the value doubles in 3 years. Find r.

For each demand function, find $E(p)$ and determine if demand is elastic or inelastic (or neither) at the indicated price.

13. $q = 700 - 5p, p = 80$

14. $q = 600e^{-.2p}, p = 10$

15. $q = 400(116 - p^2), p = 6$

16. $q = (77/p^2) + 3, p = 1$

17. $q = p^2 e^{-(p+3)}, p = 4$

18. $q = 700/(p + 5), p = 15$

19. Currently, 1800 people ride a certain commuter train each day and pay $4 for a ticket. The number of people q willing to ride the train at price p is $q = 600(5 - \sqrt{p})$. The railroad would like to increase its revenue.

    (a) Is demand elastic or inelastic at $p = 4$?

    (b) Should the price of a ticket be raised or lowered?

20. A company can sell $q = 9000/(p + 60) - 50$ radios at a price of p dollars per radio. The current price is $30.

    (a) Is demand elastic or inelastic at $p = 30$?

    (b) If the price is lowered slightly, will revenue increase or decrease?

21. A movie theater has a seating capacity of 3000 people. The number of people attending a show at price p dollars per ticket is $q = (18,000/p) - 1500$. Currently, the price is $6 per ticket.

(a) Is demand elastic or inelastic at $p = 6$?

(b) If the price is lowered, will revenue increase or decrease?

22. A subway charges 65 cents per person and has 10,000 riders each day. The demand function for the subway is $q = 2000\sqrt{90 - p}$.

  (a) Is demand elastic or inelastic at $p = 65$?

  (b) Should the price of a ride be raised or lowered in order to increase the amount of money taken in by the subway?

23. A country which is the major supplier of a certain commodity wishes to improve its balance of trade position by lowering the price of the commodity. The demand function is $q = 1000/p^2$.

  (a) Compute $E(p)$.

  (b) Will the country succeed in raising its revenue?

24. Show that any demand function of the form $q = a/p^m$ has constant elasticity m.

A cost function $C(x)$ gives the total cost of producing x units of a product. The *elasticity of cost at quantity x* is defined to be

$$E_c(x) = \frac{\dfrac{d}{dx}\ln C(x)}{\dfrac{d}{dx}\ln x}.$$

25. Show that $E_c(x) = x \cdot C'(x)/C(x)$.

26. Show that E_c is equal to the marginal cost divided by the average cost.

27. Let $C(x) = (1/10)x^2 + 5x + 300$. Show that $E_c(50) < 1$. (Hence when producing 50 units, a small relative increase in production results in an even smaller relative increase in total cost. Also, the average cost of producing 50 units is greater than the marginal cost at $x = 50$.)

28. Let $C(x) = 1000e^{.02x}$. Determine and simplify the formula for $E_c(x)$. Show that $E_c(60) > 1$ and interpret this result.

SOLUTIONS TO PRACTICE PROBLEMS 3

1. The demand function is $f(p) = 60{,}000e^{-.5p}$.

$$f'(p) = -30{,}000e^{-.5p}.$$

$$E(p) = \frac{-pf'(p)}{f(p)} = \frac{-p(-30{,}000)e^{-.5p}}{60{,}000e^{-.5p}} = \frac{p}{2}.$$

$$E(2.5) = \frac{2.5}{2} = 1.25.$$

2. The demand is elastic, because $E(2.5) > 1$.

3. Since demand is elastic at $2.50, a slight change in price causes revenue to change in the *opposite* direction. Hence revenue will decrease.

5.4 Further Exponential Models

A skydiver, on jumping out of an airplane, falls at an increasing rate. However, the wind rushing past the skydiver's body creates an upward force that begins to counterbalance the downward force of gravity. This air friction finally becomes so great that the skydiver's velocity reaches a limiting speed called the *terminal velocity*. If we let $v(t)$ be the downward velocity of the skydiver after t seconds of free fall, then a good mathematical model for $v(t)$ is given by

$$v(t) = M(1 - e^{-kt}), \qquad (1)$$

where M is the terminal velocity and k is some positive constant (Fig. 1). When t is close to zero, e^{-kt} is close to one and the velocity is small. As t increases, e^{-kt} becomes small and so $v(t)$ approaches M.

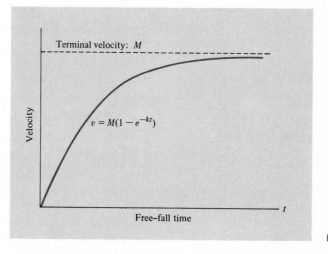

Terminal velocity: M

Velocity

$v = M(1 - e^{-kt})$

Free-fall time

t

FIGURE 1

EXAMPLE 1 Show that the velocity given in (1) satisfies the differential equation

$$\frac{dv}{dt} = k[M - v(t)], \qquad v(0) = 0. \qquad (2)$$

Solution From (1) we have $v(t) = M - Me^{-kt}$. Then

$$\frac{dv}{dt} = Mke^{-kt}.$$

However,

$$k[M - v(t)] = k[M - (M - Me^{-kt})] = kMe^{-kt},$$

so that the differential equation $\dfrac{dv}{dt} = k[M - v(t)]$ holds. Also,

$$v(0) = M - Me^0 = M - M = 0.$$

The differential equation (2) says that the rate of change in v is proportional to the difference between the terminal velocity M and the actual velocity v. It is not difficult to show that the only solution of (2) is given by the formula in (1).

The two equations (1) and (2) arise as mathematical models in a variety of situations. Some of these applications are described below.

The Learning Curve   Psychologists have found that in many learning situations a person's rate of learning is rapid at first and then slows down. Finally, as the task is mastered, the person's level of performance reaches a level above which it is almost physically impossible to rise. For example, within reasonable limits, each person seems to have a certain maximum capacity for memorizing a list of nonsense syllables. Suppose that a subject can memorize M syllables in a row if given sufficient time, say an hour, to study the list but cannot memorize $M + 1$ syllables in a row even if allowed several hours of study. By giving the subject different lists of syllables and varying lengths of time to study the lists, the psychologist can determine an empirical relationship between the number of nonsense syllables memorized accurately and the number of minutes of study time. It turns out that a good model for this situation is

$$y = M(1 - e^{-kt}) \tag{3}$$

for some appropriate positive constant k. (See Fig. 2.)

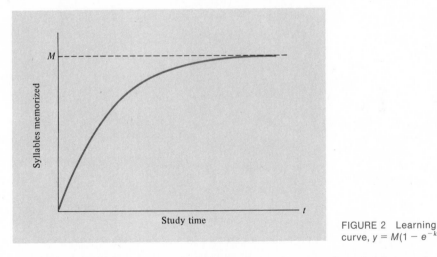

FIGURE 2   Learning curve, $y = M(1 - e^{-kt})$.

The *slope* of this learning curve at time t is approximately the number of additional syllables that can be memorized if the subject is given one more minute of study time. Thus the slope is a measure of the *rate of learning*. The differential equation satisfied by the function in (3) is

$$y' = k(M - y), \qquad f(0) = 0.$$

This equation says that if the subject is given a list of M nonsense syllables, then the rate of memorization is proportional to the number of syllables remaining to be memorized.

Diffusion of Information by Mass Media   Sociologists have found that the differential equation (2) provides a good model for the way information is spread (or "diffused") through a population when the information is being propagated constantly by mass media, such as television or magazines.* Given a fixed population P, let $f(t)$ be the number of people who have already heard a certain piece of information by time t. Then $P - f(t)$ is the number who have not yet heard the information. Also, $f'(t)$ is the rate of increase of the number of people who have heard the news (the "rate of diffusion" of the information). If the information is being publicized often by some mass media, then it is likely that the number of *newly informed* people per unit time is proportional to the number of people who have not yet heard the news. Therefore,

$$f'(t) = k[P - f(t)].$$

Assume that $f(0) = 0$ (i.e., there was a time $t = 0$ when nobody had heard the news). Then the remark following Example 1 shows that

$$f(t) = P(1 - e^{-kt}). \qquad (4)$$

(See Fig. 3.)

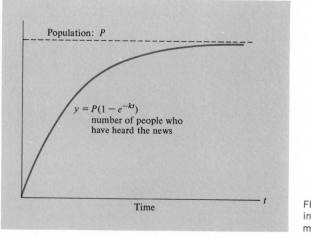

Population: P

$y = P(1 - e^{-kt})$
number of people who
have heard the news

Time

t

FIGURE 3   Diffusion of information by mass media.

EXAMPLE 2   Suppose that a certain piece of news (such as the resignation of a public official) is broadcast frequently by radio and television stations. Also suppose that one-half of the residents of a city have heard the news within 4 hours of its initial release. Use the exponential model (4) to estimate when 90% of the residents will have heard the news.

Solution   We must find the value of k in (4). If P is the number of residents, then the number who will have heard the news in the first four hours is given by (4) with $t = 4$.

---

* J. Coleman, *Introduction to Mathematical Sociology* (New York: The Free Press, 1964), p. 43.

By assumption, this number is half the population. So

$$\tfrac{1}{2}P = P(1 - e^{-k \cdot 4})$$

$$.5 = 1 - e^{-4k}$$

$$e^{-4k} = 1 - .5 = .5.$$

Solving for k, we find that $k \approx .17$. So the model for this particular situation is

$$f(t) = P(1 - e^{-.17t}).$$

Now we want to find t such that $f(t) = .90P$. We solve for t:

$$.90P = P(1 - e^{-.17t})$$

$$.90 = 1 - e^{-.17t}$$

$$e^{-.17t} = 1 - .90 = .10$$

$$-.17t = \ln .10$$

$$t = \frac{\ln .10}{-.17} \approx 14.$$

Therefore, 90% of the residents will hear the news within 14 hours of its initial release.

Intravenous Infusion of Glucose   The human body both manufactures and uses glucose ("blood sugar"). Usually, there is a balance in these two processes, so that the bloodstream has a certain "equilibrium level" of glucose. Suppose that a patient is given a single intravenous injection of glucose and let $A(t)$ be the amount of glucose (in milligrams) above the equilibrium level. Then the body will start using up the excess glucose at a rate proportional to the amount of excess glucose; that is,

$$A'(t) = -\lambda A(t), \tag{5}$$

where λ is a positive constant called the *velocity constant of elimination*. This constant depends on how fast the patient's metabolic processes eliminate the excess glucose from the blood. Equation (5) describes a simple exponential decay process.

Now suppose that, instead of a single shot, the patient receives a continuous intravenous infusion of glucose. A bottle of glucose solution is suspended above the patient, and a small tube carries the glucose down to a needle that runs into a vein. In this case, there are two influences on the amount of excess glucose in the blood: the glucose being added steadily from the bottle and the glucose being removed from the blood by metabolic processes. Let r be the rate of infusion of glucose (often from 10 to 100 milligrams per minute). If the body did not remove any glucose, the excess glucose would increase at a constant rate of r milligrams per minute; that is,

$$A'(t) = r. \tag{6}$$

Taking into account the two influences on $A'(t)$ described by (5) and (6), we can write

$$A'(t) = r - \lambda A(t). \tag{7}$$

If we let $M = r/\lambda$, then

$$A'(t) = \lambda(M - A(t)).$$

It can be shown that a solution of this differential equation is given by

$$A(t) = M(1 - e^{-\lambda t}) = \frac{r}{\lambda}(1 - e^{-\lambda t}). \tag{8}$$

Reasoning as in Example 1, we conclude that the amount of excess glucose rises until it reaches a stable level. (See Fig. 4).

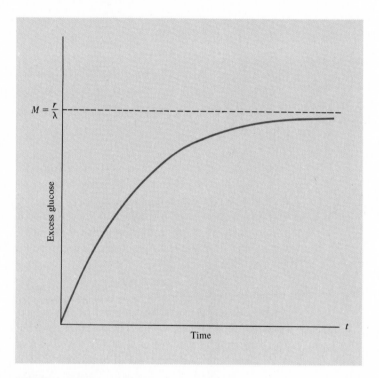

FIGURE 4   Continuous infusion of glucose.

$$y = \frac{r}{\lambda}(1 - e^{-\lambda t}).$$

The Logistic Growth Curve   The model for simple exponential growth discussed in Section 1 is adequate for describing the growth of many types of populations, but obviously a population cannot increase exponentially forever. The simple exponential growth model becomes inapplicable when the environment begins to inhibit the growth of the population. The logistic growth curve is an important exponential model that takes into account some of the effects of the environment on a population (Fig. 5). For small values of t, the curve has the same basic shape

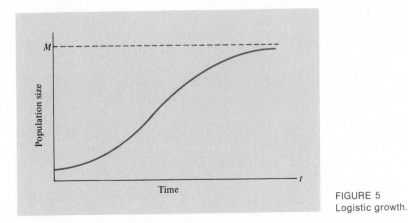

FIGURE 5
Logistic growth.

as an exponential growth curve. Then when the population begins to suffer from overcrowding or lack of food, the growth rate (the slope of the population curve) begins to slow down. Eventually, the growth rate tapers off to zero as the population reaches the maximum size that the environment will support. This latter part of the curve resembles the growth curves studied earlier in this section.

The equation for logistic growth has the general form

$$y = \frac{M}{1 + Be^{-Mkt}}, \tag{9}$$

where B, M, and k are positive constants. It can be shown that y satisfies the differential equation

$$y' = ky(M - y). \tag{10}$$

The factor y reflects the fact that the growth rate (y') depends in part on the size y of the population. The factor $M - y$ reflects the fact that the growth rate also depends on how close y is to the maximum level M.

The logistic curve is often used to fit experimental data that lie along an "S-shaped" curve. Examples are given by the growth of a fish population in a lake and the growth of a fruit fly population in a laboratory container. Also, certain enzyme reactions in animals follow a logistic law. One of the earliest applications of the logistic curve occurred in about 1840 when the Belgian sociologist P. Verhulst fit a logistic curve to six U.S. census figures, 1790 to 1840, and predicted the U.S. population for 1940. His prediction missed by less than 1 million persons (an error of about 1%).

EXAMPLE 3 Suppose that a lake is stocked with 100 fish. After 3 months there are 250 fish. A study of the ecology of the lake predicts that the lake can support 1000 fish. Find a formula for the number $P(t)$ of fish in the lake t months after it has been stocked.

Solution   The limiting population M is 1000. Therefore, we have

$$P(t) = \frac{1000}{1 + Be^{-1000kt}}.$$

At $t = 0$ there are 100 fish, so that

$$100 = P(0) = \frac{1000}{1 + Be^0} = \frac{1000}{1 + B}.$$

Thus $1 + B = 10$, or $B = 9$. Finally, since $P(3) = 250$, we have

$$250 = \frac{1000}{1 + 9e^{-3000k}}$$

$$1 + 9e^{-3000k} = 4$$

$$e^{-3000k} = \tfrac{1}{3}.$$

$$-3000k = \ln \tfrac{1}{3}$$

$$k \approx .00037.$$

Therefore,

$$P(t) = \frac{1000}{1 + 9e^{-.37t}}.$$

Several theoretical justifications can be given for using (9) and (10) in situations where the environment prevents a population from exceeding a certain size. A discussion of this topic may be found in *Mathematical Models and Applications* by D. Maki and M. Thompson (Englewood Cliffs, N.J.: Prentice-Hall, Inc., 1973), pp. 312–317.

An Epidemic Model   It will be instructive to actually "build" a mathematical model. Our example concerns the spread of a highly contagious disease. We begin by making several simplifying assumptions:

1. The population is a fixed number P and each member of the population is susceptible to the disease.
2. The duration of the disease is long, so that no cures occur during the time period under study.
3. All infected individuals are contagious and circulate freely among the population.
4. During each unit time period (such as 1 day or 1 week) each infected person makes c contacts, and each contact with an uninfected person results in transmission of the disease.

Consider a short period of time from t to $t + h$. Each infected person makes $c \cdot h$ contacts. How many of these contacts are with uninfected persons? If $f(t)$ is the number of infected persons at time t, then $P - f(t)$ is the number of uninfected persons, and $[P - f(t)]/P$ is the fraction of the population that is uninfected. Thus, of the $c \cdot h$ contacts made,

$$\left[\frac{P - f(t)}{P} \right] \cdot c \cdot h$$

will be with uninfected persons. This is the number of new infections produced by one infected person during the time period of length h. The total number of *new* infections during this period is

$$f(t) \left[\frac{P - f(t)}{P} \right] ch.$$

But this number must equal $f(t + h) - f(t)$, where $f(t + h)$ is the total number of infected persons at time $t + h$. So

$$f(t + h) - f(t) = f(t) \left[\frac{P - f(t)}{P} \right] ch.$$

Dividing by h, the length of the time period, we obtain the average number of new infections per unit time (during the small time period):

$$\frac{f(t + h) - f(t)}{h} = \frac{c}{P} f(t)[P - f(t)].$$

If we let h approach zero and let y stand for $f(t)$, the left-hand side approaches the rate of change in the number of infected persons and we derive the following equation:

$$\frac{dy}{dt} = \frac{c}{P} y(P - y). \tag{11}$$

This is the same type of equation as that used in (10) for logistic growth, although the two situations leading to this model appear to be quite dissimilar.

Comparing (11) with (10), we see that the number of infected individuals at time t is described by a logistic curve with $M = P$ and $k = c/P$. Therefore, by (9), we can write

$$f(t) = \frac{P}{1 + Be^{-ct}}.$$

B and c can be determined from the characteristics of the epidemic. (See Example 4 below.)

The logistic curve has an inflection point at that value of t for which $f(t) = P/2$. The position of this inflection point has great significance for applications of the logistic curve. From inspecting a graph of the logistic curve, we see that the inflection point is the point at which the curve has greatest slope. In other words, the inflection point corresponds to the instant of fastest growth of the logistic curve. This means, for example, that in the foregoing epidemic model the disease is spreading with the greatest rapidity precisely when half the population is infected. Any attempt at disease control (through immunization, for example) must strive to reduce the incidence of the disease to as low a point as possible, but in any case at least below the inflection point at $P/2$, at which point the epidemic is spreading fastest.

EXAMPLE 4 The Public Health Service monitors the spread of an epidemic of a particularly long-lasting strain of flu in a city of 500,000 people. At the beginning of the first week of monitoring, 200 cases have been reported; during the first week 300 new cases are reported. Estimate the number of infected individuals after 6 weeks.

Solution Here $P = 500,000$. If $f(t)$ denotes the number of cases at the end of t weeks, then

$$f(t) = \frac{P}{1 + Be^{-ct}}$$

$$= \frac{500,000}{1 + Be^{-ct}}.$$

Moreover, $f(0) = 200$, so that

$$200 = \frac{500,000}{1 + Be^0} = \frac{500,000}{1 + B},$$

and $B = 2499$. Consequently, since $f(1) = 300 + 200 = 500$, we have

$$500 = f(1) = \frac{500,000}{1 + 2499e^{-c}},$$

so that $e^{-c} \approx .4$ and $c \approx .92$. Finally,

$$f(t) = \frac{500,000}{1 + 2499e^{-.92t}}$$

and

$$f(6) = \frac{500,000}{1 + 2499e^{-.92(6)}} \approx 45,000.$$

After 6 weeks, about 45,000 individuals are infected.

This epidemic model is used by sociologists (where it is still called an epidemic model) to describe the spread of a rumor. In economics the model is used to describe the diffusion of knowledge about a product. An "infected person" represents an individual who possesses knowledge of the product. In both cases, it is assumed that the members of the population are themselves primarily responsible for the spread of the rumor or knowledge of the product. This situation is in contrast to the model described earlier where information was spread through a population by external sources, such as radio and television.

There are several limitations to this epidemic model. Each of the four simplifying assumptions made at the outset is unrealistic in varying degrees. More complicated models can be constructed that rectify one or more of these defects, but they require more advanced mathematical tools.

The Exponential Function in Lung Physiology

Let us conclude this section by deriving a useful model for the pressure in a person's lungs when the air is allowed to escape passively from the lungs with no use of the person's muscles. Let V be the volume of air in the lungs and let P be the relative pressure in the lungs when compared with the pressure in the mouth. The *total compliance* (of the respiratory system) is defined to be the derivative $\dfrac{dV}{dP}$. For normal values of V and P we may assume that the total compliance is a positive constant, say C. That is,

$$\frac{dV}{dP} = C. \tag{12}$$

We shall assume that the airflow during the passive respiration is smooth and not turbulent. Then Poiseuille's law of fluid flow says that the rate of change of volume as a function of time (i.e., the rate of air flow) satisfies

$$\frac{dV}{dt} = -\frac{P}{R}, \tag{13}$$

where R is a (positive) constant called the airway resistance. Under these conditions, we may derive a formula for P as a function of time. Since the volume is a function of the pressure, and the pressure is in turn a function of time, we may use the chain rule to write

$$\frac{dV}{dt} = \frac{dV}{dP} \cdot \frac{dP}{dt}.$$

From (12) and (13),

$$-\frac{P}{R} = C \cdot \frac{dP}{dt},$$

so

$$\frac{dP}{dt} = -\frac{1}{RC} \cdot P.$$

From this differential equation we conclude that P must be an exponential function of t. In fact,

$$P = P_0 e^{kt},$$

where P_0 is the initial pressure at time $t = 0$ and $k = -1/RC$. This relation between k and the product RC is useful to lung specialists, because they can experimentally compute k and the compliance C, and then use the formula $k = -1/RC$ to determine the airway resistance R.

PRACTICE PROBLEMS 4

1. A sociological study* was made to examine the process by which doctors decide to adopt a new drug. The doctors were divided into two groups. The doctors in group A had little interaction with other doctors and so received most of their information via mass media. The doctors in group B had extensive interaction with other doctors and so received most of their information via word of mouth. For each group, let $f(t)$ be the number who have learned about a new drug after t months. Examine the appropriate differential equations to explain why the two graphs were of the types shown below.

Group A   Group B

EXERCISES 4

1. Consider the function $f(x) = 5(1 - e^{-2x})$, $x \geq 0$.

   (a) Show that $f(x)$ is increasing and concave down for all $x \geq 0$.

   (b) Explain why $f(x)$ approaches 5 as x gets large.

   (c) Sketch the graph of $f(x)$, $x \geq 0$.

2. Consider the function $g(x) = 10 - 10e^{-.1x}$, $x \geq 0$.

   (a) Show that $g(x)$ is increasing and concave down for $x \geq 0$.

   (b) Explain why $g(x)$ approaches 10 as x gets large.

   (c) Sketch the graph of $g(x)$, $x \geq 0$.

3. Suppose that $y = 2(1 - e^{-x})$. Compute y' and show that $y' = 2 - y$.

4. Suppose that $y = 5(1 - e^{-2x})$. Compute y' and show that $y' = 10 - 2y$.

5. Suppose that $f(x) = 3(1 - e^{-10x})$. Show that $y = f(x)$ satisfies the differential equation

$$y' = 10(3 - y), \qquad f(0) = 0.$$

* James S. Coleman, Elihu Katz, and Herbert Menzel, "The Diffusion of an Innovation Among Physicians," *Sociometry*, 20 (1957), 253–270.

6. (*Ebbinghaus Model for Forgetting*) Suppose that a student learns a certain amount of material for some class. Let $f(t)$ denote the percentage of the material that the student can recall t weeks later. The psychologist Ebbinghaus found that this percent retention can be modeled by a function of the form

$$f(t) = (100 - a)e^{-\lambda t} + a,$$

where λ and a are positive constants and $0 < a < 100$. Sketch the graph of the function $f(t) = 85e^{-.5t} + 15, t \geq 0$.

7. When a grand jury indicted the mayor of a certain town for accepting bribes, the newspaper, radio, and television immediately began to publicize the news. Within an hour one-quarter of the citizens heard about the indictment. Estimate when three-quarters of the town heard the news.

8. Examine formula (8) for the amount $A(t)$ of excess glucose in the bloodstream of a patient at time t. Describe what would happen if the rate r of infusion of glucose were doubled.

9. Describe an experiment that a doctor could perform in order to determine the velocity constant of elimination of glucose for a particular patient.

10. Physiologists usually describe the continuous intravenous infusion of glucose in terms of the excess *concentration* of glucose, $C(t) = A(t)/V$, where V is the total volume of blood in the patient. In this case, the rate of increase in the concentration of glucose due to the continuous injection is r/V. Find a differential equation that gives a model for the rate of change of the excess concentration of glucose.

SOLUTIONS TO PRACTICE PROBLEMS 4

1. The difference between transmission of information via mass media and via word of mouth is that in the second case the rate of transmission depends not only on the number of people who have not yet received the information, but also on the number of people who know the information and therefore are capable of spreading it. Therefore, for group A, $f'(t) = k[P - f(t)]$, and for group B, $f'(t) = kf(t)[P - f(t)]$. Note that the spread of information by word of mouth follows the same pattern as the spread of an epidemic.

Chapter 5: CHECKLIST

- ☐ $y' = ky$ has the solution $y = Ce^{kt}$
- ☐ Exponential growth
- ☐ Exponential decay
- ☐ Half-life of a radioactive element
- ☐ Continuous compounding of interest
- ☐ Present value of money
- ☐ Percentage rate of change
- ☐ Elasticity of demand
- ☐ $y = M(1 - e^{-kt})$
- ☐ $y = M/(1 + Be^{-Mkt})$ (logistic growth)

Chapter 5: SUPPLEMENTARY EXERCISES

1. The atmospheric pressure (measured in inches of mercury) at height x miles above sea level $P(x)$ satisfies the differential equation $P'(x) = -.2P(x)$. Find the formula for $P(t)$ if the atmospheric pressure at sea level is 29.92.

2. The herring gull population in North America has been doubling every 13 years since 1900. Give a differential equation satisfied by $P(t)$, the population t years after 1900.

3. Find the present value of $10,000 payable at the end of 5 years if money may be invested at 12% with interest compounded continuously.

4. One thousand dollars is deposited in a savings account at 10% interest compounded continuously. How many years are required for the balance in the account to reach $3000?

5. The half-life of the radioactive element tritium is 12 years. Find its decay constant.

6. A piece of charcoal found at Stonehenge contained 63% of the level of C^{14} found in living trees. Approximately how old is the charcoal?

7. From 1970 to 1980, the population of Texas grew from 11.2 million to 14.2 million.

   (a) Give the formula for the population t years after 1970.

   (b) If this growth rate continues, how large will the population be in 1990?

   (c) In what year will the population reach 19 million?

8. A stock portfolio increased in value from $100,000 to $117,000 in 2 years. What rate of interest, compounded continuously, did this investment earn?

9. An investor initially invests $10,000 in a speculative venture. Suppose that the investment earns 20% interest compounded continuously for 5 years and 6% interest compounded continuously for 5 years thereafter.

   (a) How much does the $10,000 grow to after 10 years?

   (b) Suppose that the investor has the alternative of an investment paying 14% interest compounded continuously. Which investment is superior over a 10-year period, and by how much?

10. Two different bacteria colonies are growing near a pool of stagnant water. Suppose that the first colony initially has 1000 bacteria and doubles every 21 minutes. The second colony has 710,000 bacteria and doubles every 33 minutes. How much time will elapse before the first colony becomes as large as the second?

11. Find the percentage rate of change of the function $f(t) = 50e^{.2t^2}$ at $t = 10$.

12. Find $E(p)$ for the demand function $q = 4000 - 40p^2$, and determine if demand is elastic or inelastic at $p = 5$.

13. Suppose that for a certain demand function, $E(8) = 1.5$. If the price is increased to $8.16, estimate the percentage decrease in the quantity demanded. Will the revenue increase or decrease?

14. Find the percentage rate of change of the function $f(p) = \dfrac{1}{3p + 1}$ at $p = 1$.

15. A company can sell $q = 1000p^2 e^{-.02(p+5)}$ calculators at a price of p dollars per calculator. The current price is \$200. If the price is decreased, will the revenue increase or decrease?

16. Consider a demand function of the form $q = ae^{-bp}$, where a and b are positive numbers. Find $E(p)$ and show that the elasticity equals 1 when $p = 1/b$.

17. Refer to Practice Problems 4. Out of 100 doctors in group A, none knew about the drug at time $t = 0$, but 66 of them were familiar with the drug after 13 months. Find the formula for $f(t)$.

18. The growth of the yellow nutsedge weed is described by a formula $f(t)$ of type (9) in Section 4. A typical weed has length 8 centimeters after 9 days, length 48 centimeters after 25 days and reaches length 55 centimeters at maturity. Find the formula for $f(t)$.

Integration

Chapter 6

The derivative is one of the two fundamental concepts of calculus. The other is the *integral*, which we introduce in this chapter.

6.1 Antidifferentiation

We have developed several techniques for calculating the derivative $F'(x)$ of a function $F(x)$. In many applications, however, it is necessary to proceed in reverse. We are given the derivative $F'(x)$ and must determine the function $F(x)$. The process of determining $F(x)$ from $F'(x)$ is called *antidifferentiation*. The next example gives a typical application involving antidifferentiation.

EXAMPLE 1   For a number of years, the annual worldwide rate of oil consumption was growing exponentially with a growth constant of about .07. At the beginning of 1970, the rate was about 16.1 billion barrels of oil per year. Let $R(t)$ denote the rate of oil consumption at time t, where t is the number of years since the beginning of 1970. Then a reasonable model for $R(t)$ is given by

$$R(t) = 16.1e^{.07t}. \tag{1}$$

Use this formula for $R(t)$ to estimate the total amount of oil that will be consumed from 1970 to 1990.

Solution   Let $T(t)$ be the total amount of oil consumed from time 0 (1970) until time t. We wish to calculate $T(20)$, the amount of oil consumed from 1970 to 1990. We do this by first determining a formula for $T(t)$. Since $T(t)$ is the total oil consumed, the derivative $T'(t)$ is the *rate* of oil consumption, namely $R(t)$. Thus, although we do not yet have a formula for $T(t)$, we do know that

$$T'(t) = R(t).$$

Thus the problem of determining a formula for $T(t)$ has been reduced to a problem of antidifferentiation: Find a function whose derivative is $R(t)$. We shall solve this particular problem after developing some techniques for solving antidifferentiation problems in general.

Suppose that $f(x)$ is a given function and $F(x)$ is a function having $f(x)$ as its derivative—that is, $F'(x) = f(x)$. We call $F(x)$ an *antiderivative* of $f(x)$.

EXAMPLE 2 Find an antiderivative of $f(x) = x^2$.

Solution One such function is $F(x) = \frac{1}{3}x^3$, since

$$F'(x) = \frac{1}{3} \cdot 3x^2 = x^2.$$

Another antiderivative is $F(x) = \frac{1}{3}x^3 + 2$, since

$$\frac{d}{dx}\left(\frac{1}{3}x^3 + 2\right) = \frac{1}{3} \cdot 3x^2 + 0 = x^2.$$

In fact, if C is any constant, the function $F(x) = \frac{1}{3}x^3 + C$ is also an antiderivative of x^2, since

$$\frac{d}{dx}\left(\frac{1}{3}x^3 + C\right) = \frac{1}{3} \cdot 3x^2 + 0 = x^2.$$

(The derivative of a constant function is zero.)

EXAMPLE 3 Find an antiderivative of the function $f(x) = 2x - (1/x^2)$.

Solution Since

$$\frac{d}{dx}(x^2) = 2x \qquad \text{and} \qquad \frac{d}{dx}\left(\frac{1}{x}\right) = -\frac{1}{x^2},$$

we see that one antiderivative of $f(x)$ is given by

$$F(x) = x^2 + \frac{1}{x}.$$

However, any function of the form $x^2 + (1/x) + C$, C a constant, will do, since

$$\frac{d}{dx}\left(x^2 + \frac{1}{x} + C\right) = 2x - \frac{1}{x^2} + 0 = 2x - \frac{1}{x^2}.$$

Using the same reasoning as in Examples 2 and 3, we see that if $F(x)$ is an antiderivative of $f(x)$, then so is $F(x) + C$, where C is any constant. Thus if we know one antiderivative $F(x)$ of a function $f(x)$, we can write down an infinite number by adding all possible constants C to $F(x)$. It turns out that in this way we obtain all antidervatives of $f(x)$. That is, we have the following fundamental result.

> **Theorem I** If $F_1(x)$ and $F_2(x)$ are two antiderivatives of the same function $f(x)$, then $F_1(x)$ and $F_2(x)$ differ by a constant. In other words, there is a constant C such that
> $$F_2(x) = F_1(x) + C.$$

Our verification of this theorem will be based on the following fact, which is important in its own right.

> **Theorem II** If $F'(x) = 0$ for all x, then $F(x) = C$ for some constant C.

Verification of Theorem II If $F'(x) = 0$ for all x, then the curve $y = F(x)$ has slope equal to zero at every point. Thus the tangent line to $y = F(x)$ at any point is horizontal, which implies that the graph of $y = F(x)$ is a horizontal line. (Try to draw the graph of a function with horizontal tangent everywhere. There is no choice but to keep your pencil moving on a constant, horizontal line!) If the horizontal line is $y = C$, then $F(x) = C$ for all x.

Verification of Theorem I If $F_1(x)$ and $F_2(x)$ are two antiderivatives of $f(x)$, then the function $F(x) = F_2(x) - F_1(x)$ has derivative

$$F'(x) = F'_2(x) - F'_1(x)$$
$$= f(x) - f(x)$$
$$= 0.$$

So, by Theorem II, we know that $F(x) = C$ for some constant C. In other words, $F_2(x) - F_1(x) = C$, so that

$$F_2(x) = F_1(x) + C,$$

which is Theorem I.

Using Theorem I, we can find *all* antiderivatives of a given function once we know one antiderivative. For instance, since one antiderivative of x^2 is $\frac{1}{3}x^3$ (Example 2), all antiderivatives of x^2 have the form $\frac{1}{3}x^3 + C$, where C is a constant.

Suppose that $f(x)$ is a function whose antiderivatives are $F(x) + C$. The standard way to express this fact is to write

$$\int f(x)\,dx = F(x) + C.$$

The symbol $\int$ is called an *integral sign*. The entire notation $\int f(x)\,dx$ is called an *indefinite integral* and stands for antidifferentiation of the function $f(x)$. We always record the variable of interest by writing it, prefaced by the letter d. For example, if the variable of interest is t rather than x, then we write $\int f(t)\,dt$ for the antiderivative of $f(t)$.

EXAMPLE 4   Determine.

(a) $\int x^r \, dx, \quad r \neq -1$ (b) $\int e^{kx} \, dx, \quad k \neq 0$

Solution   (a) By the constant multiple and power rules,

$$\frac{d}{dx}\left(\frac{1}{r+1}\, x^{r+1}\right) = \frac{1}{r+1} \cdot \frac{d}{dx}\, x^{r+1} = \frac{1}{r+1} \cdot (r+1)x^r = x^r.$$

Thus $x^{r+1}/(r+1)$ is an antiderivative of x^r. Letting C represent any constant, we have

$$\int x^r \, dx = \frac{1}{r+1}\, x^{r+1} + C, \qquad r \neq -1. \tag{2}$$

(b) An antiderivative of e^{kx} is e^{kx}/k, since

$$\frac{d}{dx}\left(\frac{1}{k}\, e^{kx}\right) = \frac{1}{k} \cdot \frac{d}{dx}\, e^{kx} = \frac{1}{k}(ke^{kx}) = e^{kx}.$$

Hence

$$\int e^{kx} \, dx \quad \frac{1}{k}\, e^{kx} + C, \qquad k \neq 0. \tag{3}$$

Formula (2) does not give an antiderivative of x^{-1} because $1/(r+1)$ is undefined for $r = -1$. However, we know that for $x \neq 0$, the derivative of $\ln|x|$ is $1/x$. Hence $\ln|x|$ is an antiderivative of $1/x$, and we have

$$\int \frac{1}{x}\, dx = \ln|x| + C, \qquad x \neq 0. \tag{4}$$

Formulas (2), (3), and (4) each followed by "reversing" a familiar differentiation rule. In a similar fashion, one may use the sum rule and constant-multiple rule for derivatives to obtain corresponding rules for antiderivatives:

$$\int [f(x) + g(x)] \, dx = \int f(x)\, dx + \int g(x)\, dx \tag{5}$$

$$\int kf(x)\, dx = k \int f(x)\, dx, \qquad k \text{ a constant.} \tag{6}$$

In words, (5) says that a sum of functions may be antidifferentiated term by term, and (6) says that a constant multiple may be moved through the integral sign.

EXAMPLE 5   Compute:

$$\int \left(x^{-3} + 7e^{5x} + \frac{4}{x} \right) dx.$$

Solution   Using the rules given above, we have

$$\int \left(x^{-3} + 7e^{5x} + \frac{4}{x} \right) dx = \int x^{-3}\, dx + \int 7e^{5x}\, dx + \int \frac{4}{x}\, dx$$

$$= \int x^{-3}\, dx + 7 \int e^{5x}\, dx + 4 \int \frac{1}{x}\, dx$$

$$= \frac{1}{-2} x^{-2} + 7 \left(\frac{1}{5} e^{5x} \right) + 4 \ln|x| + C$$

$$= -\frac{1}{2} x^{-2} + \frac{7}{5} e^{5x} + 4 \ln|x| + C.$$

After some practice, most of the steps shown in the solution of Example 5 can be omitted.

A function $f(x)$ has infinitely many different antiderivatives, corresponding to the various choices of the constant C. In applications, it is often necessary to satisfy an additional condition, which then determines a specific value for C. As an illustration of this technique, consider the next example.

EXAMPLE 6   Find the antiderivative $F(x)$ of $3x^2 - 5$ for which $F(0) = 2$.

Solution   One antiderivative of $3x^2 - 5$ is $x^3 - 5x$. Therefore, by Theorem I, the antiderivatives are precisely the functions

$$F(x) = x^3 - 5x + C, \qquad C \text{ a constant.}$$

If $F(0) = 2$, then

$$2 = F(0) = 0^3 - 5 \cdot 0 + C = C,$$

so that $C = 2$, and the unique antiderivative satisfying the given condition is

$$F(x) = x^3 - 5x + 2.$$

Having introduced the basics of antidifferentiation, let us now solve the oil-consumption problem.

Solution of Example 1 (Continued)   The rate of oil consumption at time t is $R(t) = 16.1e^{.07t}$ billion barrels per year. Moreover, we observed that the total consumption $T(t)$, from time 0 to time t, is an antiderivative of $R(t)$. Using (3) and (6), we have

$$T(t) = \int 16.1e^{.07t}\, dt = \frac{16.1}{.07} e^{.07t} + C = 230e^{.07t} + C,$$

where C is a constant. However, in our particular example, $T(0) = 0$, since $T(0)$ is the amount of oil used from time 0 to time 0. Therefore, the constant C must satisfy

$$0 = T(0) = 230e^{.07(0)} + C = 230 + C,$$

$$C = -230.$$

Therefore,

$$T(t) = 230e^{.07t} - 230 = 230(e^{.07t} - 1).$$

The total oil that will be consumed from 1970 to 1990 is

$$T(20) = 230(e^{.07(20)} - 1) \approx 703 \text{ billion barrels}.$$

Antidifferentiation can be used to solve a variety of applied problems, of which the next two examples are typical.

EXAMPLE 7 A rocket is fired vertically into the air. Its velocity at t seconds after liftoff is $v(t) = 20t + 50$ meters per second. How far will the rocket travel during the first 100 seconds?

Solution If $s(t)$ denotes the height (in meters) of the rocket at time t seconds after liftoff, then $s'(t)$ is the rate (in meters per second) at which the height is changing at time t, which is just $v(t)$. In other words, $s(t)$ is an antiderivative of $v(t)$. Thus

$$s(t) = \int v(t)\,dt = \int (20t + 50)\,dt$$

$$= 10t^2 + 50t + C,$$

where C is a constant. At time 0 the height of the rocket is 0, so $s(0) = 0$ and

$$0 = s(0) = 10(0)^2 + 50(0) + C = C$$

$$C = 0$$

$$s(t) = 10t^2 + 50t.$$

At $t = 100$, the height of the rocket is

$$s(100) = 10(100)^2 + 50(100) = 105{,}000 \text{ meters}.$$

EXAMPLE 8 A factory's marginal cost function is $\frac{1}{100}x^2 - 2x + 120$, where x denotes the number of units produced per day. The factory has fixed costs of \$1000 per day. Find the cost of producing x units per day.

Solution Let $C(x)$ be the cost of producing x units per day. The derivative $C'(x)$ is just the marginal cost function. In other words, $C(x)$ is an antiderivative of the marginal cost function. Thus

$$C(x) = \int (\tfrac{1}{100}x^2 - 2x + 120)\,dx$$

$$= \tfrac{1}{300}x^3 - x^2 + 120x + C$$

for some constant C. However, the fixed costs are just the costs experienced when producing 0 units. That is, the fixed costs equal $C(0)$. So the given data imply that $C(0) = 1000$. Thus

$$1000 = C(0) = \tfrac{1}{300}(0)^3 - (0)^2 - 120(0) + C$$

$$C = 1000.$$

Therefore, the cost function $C(x)$ is given by

$$C(x) = \tfrac{1}{300}x^3 - x^2 + 120x + 1000.$$

PRACTICE PROBLEMS 1

1. Determine.

(a) $\int (x^3 + 4x)\,dx$ 　　　　　　(b) $\int t^{7/2}\,dt$

2. Let $R(x)$ be the revenue received from the sale of x units of a product. The marginal revenue function is $5 - 3x$. Find $R(x)$. [Note that the revenue is 0 if no units are sold. That is, $R(0) = 0$.]

EXERCISES 1

Find all antiderivatives of each of the following functions.

1. $f(x) = x$ 　　　　　　2. $f(x) = 9x^8$ 　　　　　　3. $f(x) = e^{3x}$

4. $f(x) = e^{-3x}$ 　　　　　5. $f(x) = 3$ 　　　　　　6. $f(x) = -4x$

In Exercises 7–22, find the value of k that makes the antidifferentiation formula true. [*Note:* You can check your answer without looking in the answer section. How?]

7. $\int x^{-5}\,dx = kx^{-4} + C$ 　　　　8. $\int x^{1/3}\,dx = kx^{4/3} + C$

9. $\int \sqrt{x}\,dx = kx^{3/2} + C$ 　　　　10. $\int \dfrac{6}{x^3}\,dx = \dfrac{k}{x^2} + C$

11. $\int \dfrac{10}{t^6}\,dt = kt^{-5} + C$ 　　　12. $\int \dfrac{3}{\sqrt{t}}\,dt = k\sqrt{t} + C$

13. $\int 5e^{-2t}\,dt = ke^{-2t} + C$ 　　　14. $\int 3e^{t/10}\,dt = ke^{t/10} + C$

15. $\int 2e^{4x-1}\,dx = ke^{4x-1} + C$ 　　16. $\int \dfrac{4}{e^{3x+1}}\,dx = \dfrac{k}{e^{3x+1}} + C$

17. $\int (x-7)^{-2}\,dx = k(x-7)^{-1} + C$ 　18. $\int \sqrt{x+1}\,dx = k(x+1)^{3/2} + C$

19. $\int (x+4)^{-1}\,dx = k\ln|x+4| + C$ 　20. $\int \dfrac{5}{(x-8)^4}\,dx = \dfrac{k}{(x-8)^3} + C$

21. $\int (3x+2)^4\,dx = k(3x+2)^5 + C$ 　22. $\int (2x-1)^3\,dx = k(2x-1)^4 + C$

Determine the following.

23. $\displaystyle\int (x^2 - x - 1)\,dx$

24. $\displaystyle\int (x^3 + 6x^2 - x)\,dx$

25. $\displaystyle\int \left(\frac{2}{\sqrt{x}} - 3\sqrt{x}\right) dx$

26. $\displaystyle\int \left[\frac{\sqrt{t}}{4} - 4(t-3)^{-2}\right] dt$

27. $\displaystyle\int \left(4 - 5e^{-5t} + \frac{e^{2t}}{3}\right) dt$

28. $\displaystyle\int (e^2 + 3t^2 - 2e^{3t})\,dt$

Find all functions $f(t)$ with the following property.

29. $f'(t) = t^{3/2}$

30. $f'(t) = \dfrac{4}{6 + t}$

31. $f'(t) = 0$

32. $f'(t) = t^2 - 5t - 7$

Find all functions $f(x)$ with the following properties.

33. $f'(x) = x,\ f(0) = 3$

34. $f'(x) = 8x^{1/3},\ f(1) = 4$

35. $f'(x) = \sqrt{x} + 1,\ f(4) = 0$

36. $f'(x) = x^2 + \sqrt{x},\ f(1) = 3$

37. $f'(x) = 2/x,\ f(1) = 2$

38. $f'(x) = 3,\ f(2) = 7$

39. A ball is thrown upward from a height of 256 feet above the ground, with an initial velocity of 96 feet per second. From physics it is known that the velocity at time t is $96 - 32t$ feet per second.

    (a) Find $s(t)$, the function giving the height of the ball at time t.

    (b) How long will it take for the ball to reach the ground?

    (c) How high will the ball go?

40. A rock is dropped from the top of a 400-foot cliff. Its velocity at time t seconds is $v(t) = -32t$ feet per second.

    (a) Find $s(t)$, the height of the rock above the ground at time t.

    (b) How long will it take to reach the ground?

    (c) What will be its velocity when it hits the ground?

41. Let $P(t)$ be the total output of a factory assembly line after t hours of work. Suppose that the rate of production at time t is $60 + 2t - \frac{1}{4}t^2$ units per hour. Find the formula for $P(t)$. [Hint: The rate of production is $P'(t)$ and $P(0) = 0$.]

42. After t hours of operation a coal mine is producing coal at the rate of $40 + 2t - \frac{1}{5}t^2$ tons of coal per hour. Find a formula for the total output of the coal mine after t hours of operation.

43. A package of frozen strawberries is taken from a freezer at $-5°C$ into a room at $20°C$. At time t the average temperature of the strawberries is increasing at the rate of $10e^{-.4t}$ degrees Celsius per hour. Find the temperature of the strawberries at time t.

44. A flu epidemic hits a town. Let $P(t)$ be the number of persons sick with the flu at time t, where time is measured in days from the beginning of the epidemic and $P(0) = 100$. Suppose that after t days the flu is spreading at the rate of $120t - 3t^2$ people per day. Find the formula for $P(t)$.

45. A small tie shop finds that at a sales level of x ties per day its marginal profit is $MP(x)$ dollars per tie, where $MP(x) = 1.30 + .06x - .0018x^2$. Also, the shop will lose $95 per day at a sales level of $x = 0$. Find the profit from operating the shop at a sales level of x ties per day.

46. A soap manufacturer estimates that its marginal cost of producing soap powder is $.2x + 1$ hundred dollars per ton at a production level of x tons per day. Fixed costs are $200 per day. Find the cost of producing x tons of soap powder per day.

47. The United States has been consuming iron ore at the rate of $R(t)$ million metric tons per year at time t, where $t = 0$ corresponds to 1980 and $R(t) = 94e^{.016t}$. Find a formula for the total U.S. consumption of iron ore from 1980 until time t.

48. The rate of production of natural and manufactured gas in the United States has been $R(t)$ quadrillion Btu per year at time t, with $t = 0$ corresponding to 1976 and $R(t) = 20e^{.02t}$. Find a formula for the total U.S. production of natural and manufactured gas from 1976 until time t.

SOLUTIONS TO PRACTICE PROBLEMS 1

1. (a) $\int (x^3 + 4x)\,dx = \frac{1}{4}x^4 + 2x^2 + C$

    (b) $\int t^{7/2}\,dt = \frac{1}{\frac{9}{2}}t^{9/2} + C = \frac{2}{9}t^{9/2} + C$

2. $R(x)$ is an antiderivative of the marginal revenue function. So

$$R(x) = \int (5 - 3x)\,dx = 5x - \tfrac{3}{2}x^2 + C$$

for some constant C. Since $R(0) = 0$,

$$0 = R(0) = 5(0) - \tfrac{3}{2}(0)^2 + C = C$$

$$C = 0.$$

Thus $R(x) = 5x - \tfrac{3}{2}x^2$.

6.2 Definite Integrals

In order to motivate the definite integral, let us return for a few moments to Example 1 of Section 1. Recall that we were given a formula $R(t)$ for the annual rate of oil consumption at time t. From this formula we determined the function $T(t)$, which gives the total oil consumed from time 0 to time t. Recall that $t = 0$ corresponds to 1970 and that $T(t)$ is an antiderivative of $R(t)$.

Consider the following related question: How much oil was consumed between 1975 and 1980? This is easy to determine using the function $T(t)$. The year 1975 corresponds to $t = 5$, and 1980 corresponds to $t = 10$. So the oil consumed between 1975 and 1980 is

$$T(10) - T(5). \tag{1}$$

[Indeed, $T(10)$ gives the oil consumed between 1970 and 1980 and $T(5)$ gives the oil consumed during the first five of those years.] The quantity $T(10) - T(5)$ is called the *net change* of $T(t)$ over the interval $t = 5$ to $t = 10$. More generally, if $F(x)$ is any function and a, b are numbers, then the *net change of $F(x)$ over the interval $x = a$ to $x = b$ is the number

$$F(b) - F(a).$$

The quantity $F(b) - F(a)$ is often abbreviated by the symbol $F(x)\big|_a^b$. For example, the net change (1) can be written in this notation as $T(t)\big|_5^{10}$.

Since $T(t)$ is an antiderivative of $R(t)$, the quantity (1) is the net change of an antiderivative of $R(t)$ over a certain interval. This suggests the following definition.

Definition of the Definite Integral Let $f(x)$ be a function and a, b be numbers. Then the definite integral of $f(x)$ over the interval from $x = a$ to $x = b$, denoted $\int_a^b f(x)\, dx$, is the net change of an antiderivative of $f(x)$ over that interval. Thus, if $F(x)$ is an antiderivative of $f(x)$, we have

$$\int_a^b f(x)\, dx = F(x)\Big|_a^b = F(b) - F(a).$$

For example, the quantity (1) can be expressed as the definite integral

$$\int_5^{10} R(t)\, dt.$$

EXAMPLE 1 Evaluate.

(a) $\displaystyle\int_2^5 3x^2\, dx$ (b) $\displaystyle\int_{-1}^2 e^{3x}\, dx$

Solution (a) An antiderivative of $3x^2$ is x^3. So

$$\int_2^5 3x^2\, dx = x^3\Big|_2^5 = 5^3 - 2^3 = 117.$$

(b) An antiderivative of e^{3x} is $\frac{1}{3}e^{3x}$. Therefore,

$$\int_{-1}^2 e^{3x}\, dx = \tfrac{1}{3}e^{3x}\Big|_{-1}^2 = \tfrac{1}{3}e^{3\cdot 2} - \tfrac{1}{3}e^{3\cdot(-1)} = \tfrac{1}{3}e^6 - \tfrac{1}{3}e^{-3}.$$

In computing the definite integral we may use *any* antiderivative of $f(x)$. The value of the definite integral does not depend on the choice of antiderivative. For instance, consider the definite integral $\int_2^5 3x^2\, dx$. The antiderivatives of $3x^2$ are all of the form $x^3 + C$, where C is any constant. Moreover,

$$(x^3 + C)\Big|_2^5 = (5^3 + C) - (2^3 + C) = 5^3 - 2^3 = 117.$$

So the value of the definite integral does not depend on the choice of C. The same argument works for a general function $f(x)$ in place of $3x^2$.

EXAMPLE 2   Refer to the oil-consumption data of Example 1, Section 1. How much oil will be consumed from 1978 to 1985?

Solution   As we saw above, the total consumption during a given time interval can be represented as a definite integral over that interval. Since $t = 8$ corresponds to 1978 and $t = 15$ to 1985, the desired quantity equals

$$\int_8^{15} R(t)\,dt = \int_8^{15} 16.1e^{.07t}\,dt.$$

Since an antiderivative of $16.1e^{.07t}$ is $230e^{.07t}$, the definite integral above equals

$$230e^{.07t}\Big|_8^{15} = 230e^{.07(15)} - 230e^{.07(8)}$$

$$\approx 255 \text{ billion barrels.}$$

EXAMPLE 3   Suppose that the velocity $v(t)$ of a rocket t seconds after liftoff is $v(t) = 20t + 50$ meters per second. Determine the distance it travels during the time interval from $t = 60$ to $t = 70$.

Solution   Let $s(t)$ denote the height of the rocket at time t. The desired distance is the net change $s(70) - s(60)$. Since $s(t)$ is an antiderivative of $v(t)$, this net change is a definite integral:

$$s(70) - s(60) = \int_{60}^{70} v(t)\,dt$$

$$= \int_{60}^{70} (20t + 50)\,dt$$

$$= (10t^2 + 50t)\Big|_{60}^{70}$$

$$= 13{,}500 \text{ meters.}$$

Thus the rocket travels 13,500 meters during the time interval from $t = 60$ to $t = 70$.

EXAMPLE 4   The marginal cost function for a certain factory is $\frac{1}{100}x^2 - 2x + 120$ dollars per unit, where x is the total daily production. Find the net increase in cost if production is raised from 100 to 105 units per day.

Solution   Let $C(x)$ denote the cost of producing x units per day. Then $C'(x) = \frac{1}{100}x^2 - 2x + 120$. Hence

$$C(105) - C(100) = \int_{100}^{105} C'(x)\,dx$$

$$= \int_{100}^{105} \left(\frac{1}{100}x^2 - 2x + 120 \right) dx$$

$$= (\tfrac{1}{300}x^3 - x^2 + 120x)\Big|_{100}^{105}$$

$$= 5433.75 - 5333.33 = 100.42.$$

Thus, if the factory is currently producing 100 units per day, it will cost an additional $100.42 to produce 5 additional units per day.

One of the most important applications of the definite integral is to the calculation of areas under curves.

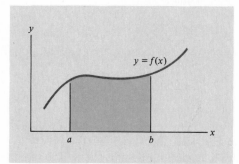

PROBLEM Let $f(x)$ be a function that is continuous and nonnegative for all x between a and b, where $a < b$. Determine the area of the region lying beneath the curve $y = f(x)$ and above the x-axis, from $x = a$ to $x = b$.

The region referred to is the shaded region in Fig. 1. We see that the region is bounded on the left by the vertical line $x = a$ and on the right by the vertical line $x = b$. We shall refer to the area of this region simply as the *area under the curve* $y = f(x)$ *from* $x = a$ *to* $x = b$.

This problem has a rather simple (and surprising) solution.

FIGURE 1

The Fundamental Theorem of Calculus Let $f(x)$ be continuous, and nonnegative for x between a and b. Then the area under the curve $y = f(x)$ from $x = a$ to $x = b$ is given by the definite integral

$$\int_a^b f(x)\,dx.$$

At first, this theorem may not seem very fundamental. (It may not even seem very interesting!) However, it allows us to accomplish several impressive goals. First of all, it permits us to apply everything we know about antidifferentiation to the calculation of areas. Although it is not at all apparent that antidifferentiation and calculation of areas are related problems, they indeed are, and the fundamental theorem of calculus provides the relationship. A second consequence of the fundamental theorem is that it allows us to interpret various quantities involving antiderivatives in terms of areas. Such interpretations are useful in many applications.

We shall explain why this theorem is true after we examine two examples of its application.

EXAMPLE 5 Find the area under the curve $y = x$ from $x = 1$ to $x = 2$.

Solution The region whose area we must calculate is the shaded region in Fig. 2(a). Here $f(x) = x$, $a = 1$, and $b = 2$. The value of the definite integral is then

$$\int_1^2 x\,dx = \tfrac{1}{2}x^2\Big|_1^2 = \tfrac{1}{2}\cdot 2^2 - \tfrac{1}{2}\cdot 1^2 = \tfrac{3}{2}.$$

Therefore, the area is $\tfrac{3}{2}$.

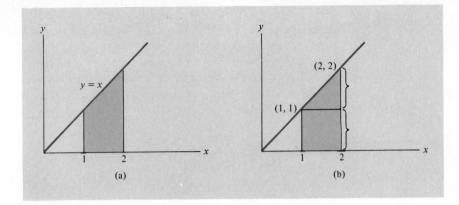

FIGURE 2

In Example 5 it is also possible to compute the area by elementary geometry. Let us break up the region into a triangle and a square, as pictured in Fig. 2(b). The area of the triangle is $\frac{1}{2} \cdot$ (base) $\cdot$ (height) $= \frac{1}{2} \cdot 1 \cdot 1 = \frac{1}{2}$. The area of the square is $1 \cdot 1 = 1$. Thus the area of the region is $\frac{1}{2} + 1 = \frac{3}{2}$, the same result that we obtained by using the definite integral.

EXAMPLE 6   Compute the area under the curve $y = x^2 - 3x + 5$ from $x = 1$ to $x = 4$.

Solution   Using our curve-sketching techniques ($y' = 2x - 3$, $y' = 0$ when $x = \frac{3}{2}$), we obtain the graph shown in Fig. 3. Since the function $f(x) = x^2 - 3x + 5$ is non-negative for all x between 1 and 4, the area under the curve is given by the integral

$$\int_1^4 (x^2 - 3x + 5)\,dx = \left(\frac{x^3}{3} - \frac{3}{2}x^2 + 5x \right)\Big|_1^4$$

$$= \left(\frac{64}{3} - \frac{48}{2} + 20 \right) - \left(\frac{1}{3} - \frac{3}{2} + 5 \right) = 13\frac{1}{2}.$$

FIGURE 3

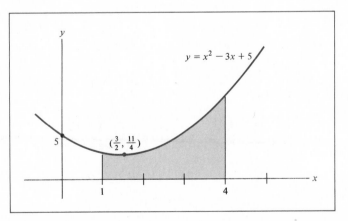

Historically, calculus centered about two problems: finding the slope of a curve at a point (differentiation) and finding the area of a region under the graph of a function. The fundamental theorem of calculus is so named because it connects these two problems. Computing an area can be accomplished by finding an antiderivative, the reverse process of differentiation.

To explain why the Fundamental Theorem of Calculus is true, we need the following theorem.

Theorem III Let $f(x)$ be a continuous nonnegative function for $a \leq x \leq b$. Let $A(x)$ be the area under the graph of the function from a to the number x. (See Fig. 4.) Then $A(x)$ is an antiderivative of $f(x)$.

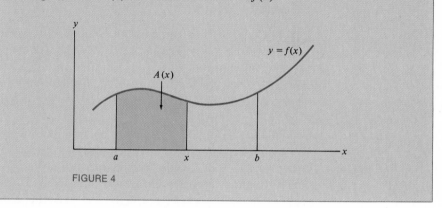

FIGURE 4

Verification of the Fundamental Theorem of Calculus

Note that the "area function" $A(x)$ in Theorem III has the property that $A(a)$ is zero and $A(b)$ equals the area under the curve $y = f(x)$ from a to b. Therefore, we may conclude from Theorem III that

$$\int_a^b f(x)\, dx = A(b) - A(a) = A(b) - 0$$

$$= [\text{area under } y = f(x) \text{ from } a \text{ to } b].$$

This is the Fundamental Theorem of Calculus.

It is not difficult to explain why Theorem III is true. Let h be a small positive number. Then $A(x + h) - A(x)$ is the area of the shaded region in Fig. 5. This shaded region is approximately a rectangle of width h, height $f(x)$, and area $h \cdot f(x)$. Thus

$$A(x + h) - A(x) \approx h \cdot f(x),$$

where the approximation becomes better as h approaches zero. Dividing by h, we have

$$\frac{A(x + h) - A(x)}{h} \approx f(x).$$

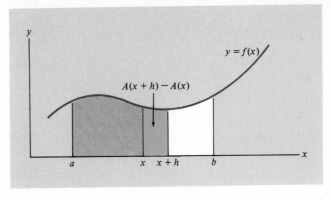

FIGURE 5

Since the approximation becomes more exact as h approaches zero, the quotient must approach $f(x)$. However, the limit definition of the derivative tells us that the quotient approaches $A'(x)$ as h approaches zero. Therefore, we have $A'(x) = f(x)$. Since x represented any number between a and b, this shows that $A(x)$ is an anti-derivative of $f(x)$.

Note The verification above applies only for $f(x)$ nonnegative for x between a and b. However, this restriction on $f(x)$ is easily removed.

PRACTICE PROBLEMS 2

1. Find the area under the curve $y = e^{x/2}$ from $x = -3$ to $x = 2$.

2. Let $C(x)$ be the cost of producing x units of a certain commodity. A common notation for the marginal cost function is $MC(x)$ (or just MC). What economic interpretation can be given to the area of the shaded region in Fig. 6?

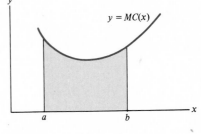

FIGURE 6

EXERCISES 2

Calculate the following definite integrals.

1. $\int_{-1}^{1} x \, dx$

2. $\int_{4}^{5} e^2 \, dx$

3. $\int_{1}^{2} 5 \, dx$

4. $\int_{-1}^{-1/2} \frac{1}{x^2}\,dx$

5. $\int_{1}^{2} 8x^3\,dx$

6. $\int_{0}^{1} e^{x/3}\,dx$

7. $\int_{0}^{1} 4e^{-3x}\,dx$

8. $\int_{1}^{3} \frac{5}{x}\,dx$

9. $\int_{1}^{4} 3\sqrt{x}\,dx$

10. $\int_{1}^{8} 2x^{1/3}\,dx$

11. $\int_{0}^{5} e^{.2t}\,dt$

12. $\int_{0}^{1} \frac{4}{e^{3t}}\,dt$

13. $\int_{3}^{6} x^{-1}\,dx$

14. $\int_{1}^{3} (5t-1)^3\,dt$

15. $\int_{-1}^{1} \frac{4}{(t+2)^3}\,dt$

16. $\int_{-3}^{0} \sqrt{25+3t}\,dt$

17. $\int_{2}^{3} (5-2t)^4\,dt$

18. $\int_{4}^{9} \frac{3}{t-2}\,dt$

19. $\int_{0}^{3} (x^3+x-7)\,dx$

20. $\int_{-5}^{5} (e^{x/10}-x^2-1)\,dx$

21. $\int_{2}^{4} \left(x^2+\frac{2}{x^2}-\frac{1}{x+5}\right)\,dx$

22. $\int_{1}^{2} (4x^3+3x^{-4}-5)\,dx$

23. Suppose that the velocity of a car at time t is $40+8/(t+1)^2$ kilometers per hour. How far does the car travel from $t=1$ to $t=9$ hours?

24. A helicopter is rising straight up in the air. Its velocity at time t is $2t+1$ feet per second for the first 5 seconds. How high does it rise in that time period?

25. After t hours of operation an assembly line is assembling power lawn mowers at the rate of $21-\frac{1}{5}t$ mowers per hour. How many mowers are assembled during the time from $t=2$ to $t=6$ hours?

26. Some food is placed in a freezer. After t hours the temperature of the food is dropping at a rate of $12+4/(t+3)^2$ degrees Fahrenheit per hour. How many degrees will the temperature of the food change during the first 2 hours?

27. Suppose that the marginal cost function of a handbag manufacturer is $\frac{3}{32}x^2-x+200$ dollars per unit at production level x (where x is measured in units of 100 handbags). Find the total cost of producing 6 additional units if 2 units are currently being produced.

28. If $C(x)$ is the cost of producing x units of some commodity, then $C(0)$ represents the fixed costs. Suppose that the marginal cost function is $C'(x)=.06x^2-8x+250$ dollars per unit. Find the total cost of producing the first 100 units, not counting the fixed costs.

29. Let $P(x)$ denote the profit earned from the sale of x units of some commodity. Then $P'(x)$ is the marginal profit. Let a and b be two values of x, with $a<b$, and give an economic interpretation to the area under the marginal profit curve $y=P'(x)$ from $x=a$ to $x=b$.

30. Suppose that the marginal profit function for a company is $100+50x-3x^2$. Find the total profit earned from the sale of 3 additional units if 5 units are currently being produced.

Find the area under each of the given curves.

31. $y=4x;\ x=2$ to $x=3$

32. $y=3x^2;\ x=-1$ to $x=1$

33. $y = e^{x/2}$; $x = 0$ to $x = 1$

34. $y = \sqrt{x}$; $x = 0$ to $x = 4$

35. $y = (x - 3)^4$; $x = 1$ to $x = 4$

36. $y = e^{3x}$; $x = -\frac{1}{3}$ to $x = 0$

Find the area under each of the given curves by antidifferentiation and use elementary geometry to check the answer.

37. $y = 5$; $x = -1$ to $x = 2$

38. $y = x + 1$; $x = 2$ to $x = 4$

39. $y = 2x$; $x = 0$ to $x = 3$

40. $y = 2x + 1$; $x = 0$ to $x = 3$

41. Verify the formula

$$\ln b = \int_1^b \frac{1}{t}\, dt, \qquad b > 0.$$

42. For $x > 1$, show how $\ln x$ may be interpreted as the area under an appropriate curve $y = f(x)$. (This interpretation of $\ln x$ is used in some texts as the *definition* of the natural logarithm function.)

43. For each positive number x, let $A(x)$ be the area under the curve $y = x^2 + 1$ from 0 to x. Find $A'(3)$.

44. For each number $x > 2$, let $A(x)$ be the area under the curve $y = x^3$ from 2 to x. Find $A'(6)$.

SOLUTIONS TO PRACTICE PROBLEMS 2

1. The desired area is

$$\int_{-3}^{2} e^{x/2}\, dx = 2e^{x/2}\Big|_{-3}^{2} = 2e - 2e^{-3/2} \approx 4.99.$$

2. The area shaded is the area under the curve MC from $x = a$ to $x = b$. This area can be computed as the definite integral $\int_a^b MC(x)\, dx$. Since $C(x)$ is an antiderivative of $MC(x)$,

$$\int_a^b MC(x)\, dx = C(x)\Big|_a^b = C(b) - C(a).$$

So the shaded area equals the additional cost of increasing production from a units to b units.

6.3 Areas in the xy-Plane

In Section 2 we introduced the concept of the definite integral and showed how the definite integral is used to compute the area of the region lying above the x-axis and below the graph of the nonnegative function $y = f(x)$. In this section we show how to use the definite integral to compute areas of more general regions. Three simple but important properties of the integral will be used repeatedly.

Let $f(x)$ and $g(x)$ be functions and a, b, k any constants. Then

$$\int_a^b f(x)\,dx + \int_a^b g(x)\,dx = \int_a^b [f(x) + g(x)]\,dx \qquad (1)$$

$$\int_a^b f(x)\,dx - \int_a^b g(x)\,dx = \int_a^b [f(x) - g(x)]\,dx \qquad (2)$$

$$\int_a^b kf(x)\,dx = k\int_a^b f(x)\,dx. \qquad (3)$$

Verification of (1) Let $F(x)$ and $G(x)$ be antiderivatives of $f(x)$ and $g(x)$, respectively. Then $F(x) + G(x)$ is an antiderivative of $f(x) + g(x)$, so

$$\int_a^b [f(x) + g(x)]\,dx = [F(x) + G(x)]\Big|_a^b$$

$$= [F(b) + G(b)] - [F(a) + G(a)]$$

$$= [F(b) - F(a)] + [G(b) - G(a)]$$

$$= \int_a^b f(x)\,dx + \int_a^b g(x)\,dx.$$

The verifications of (2) and (3) are similar and use the facts that $F(x) - G(x)$ is an antiderivative of $f(x) - g(x)$ and $kF(x)$ is an antiderivative of $kf(x)$.

Let us now consider regions that are bounded both above and below by graphs of functions. Referring to Fig. 1, we would like to find a simple expression for the area of the shaded region under the graph of $y = f(x)$ and above the graph of $y = g(x)$ from $x = a$ to $x = b$. It is the region under the graph of $f(x)$ with the region under the graph of $g(x)$ taken away. Therefore,

$$[\text{area of shaded region}] = [\text{area under } f(x)] - [\text{area under } g(x)]$$

$$= \int_a^b f(x)\,dx - \int_a^b g(x)\,dx$$

$$= \int_a^b [f(x) - g(x)]\,dx \qquad [\text{by property (2)}].$$

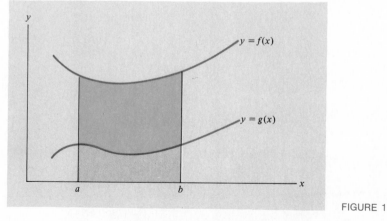

FIGURE 1

> *Area Between Two Curves*   If $y = f(x)$ lies above $y = g(x)$ from $x = a$ to $x = b$, the area of the region between $f(x)$ and $g(x)$ from $x = a$ to $x = b$ is
> $$\int_a^b [f(x) - g(x)]\, dx.$$

EXAMPLE 1   Find the area of the region between $y = 2x^2 - 4x + 6$ and $y = -x^2 + 2x + 1$ from $x = 1$ to $x = 2$.

Solution   Upon sketching the two graphs (Fig. 2), we see that $f(x) = 2x^2 - 4x + 6$ lies above $g(x) = -x^2 + 2x + 1$ for $1 \le x \le 2$. Therefore, our formula gives the area of the shaded region as

$$\int_1^2 [(2x^2 - 4x + 6) - (-x^2 + 2x + 1)]\, dx = \int_1^2 (3x^2 - 6x + 5)\, dx$$
$$= (x^3 - 3x^2 + 5x)\Big|_1^2$$
$$= 6 - 3 = 3.$$

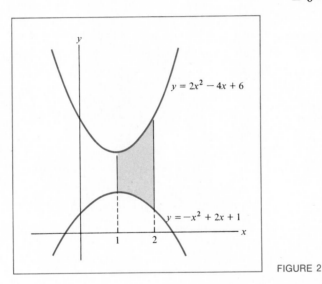

FIGURE 2

EXAMPLE 2   Find the area of the region between $y = x^2$ and $y = (x - 2)^2 = x^2 - 4x + 4$ from $x = 0$ to $x = 3$.

Solution   Upon sketching the graphs (Fig. 3), we see that the two graphs cross; and by setting $x^2 = x^2 - 4x + 4$, we find that they cross when $x = 1$. Thus one graph does not always lie above the other from $x = 0$ to $x = 3$, so that we cannot directly apply our rule for finding the area between two curves. However, the difficulty is easily surmounted if we break the region into two parts, namely the area from $x = 0$ to $x = 1$ and the area from $x = 1$ to $x = 3$. For from $x = 0$

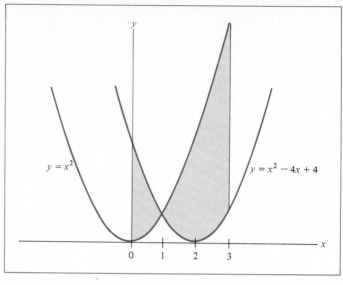

FIGURE 3

to $x = 1$, $y = x^2 - 4x + 4$ is on top; from $x = 1$ to $x = 3$, $y = x^2$ is on top. Consequently,

$$[\text{area from } x = 0 \text{ to } x = 1] = \int_0^1 [(x^2 - 4x + 4) - (x^2)]\,dx$$

$$= \int_0^1 (-4x + 4)\,dx$$

$$= (-2x^2 + 4x)\Big|_0^1$$

$$= 2 - 0 = 2.$$

$$[\text{area from } x = 1 \text{ to } x = 3] = \int_1^3 [(x^2) - (x^2 - 4x + 4)]\,dx$$

$$= \int_1^3 (4x - 4)\,dx$$

$$= (2x^2 - 4x)\Big|_1^3$$

$$= 6 - (-2) = 8.$$

Thus the total area is $2 + 8 = 10$.

In our derivation of the formula for the area between two curves, we examined functions that are nonnegative. However, the statement of the rule does not contain this stipulation, and rightfully so. Consider the case where $f(x)$ and $g(x)$ are not always positive. Let us determine the area of the shaded region in Fig. 4(a). Select some constant c such that the graphs of the functions $f(x) + c$ and $g(x) + c$ lie completely above the x-axis [Fig. 4(b)]. The region between them will have the same area as the original region. Using the rule as applied to

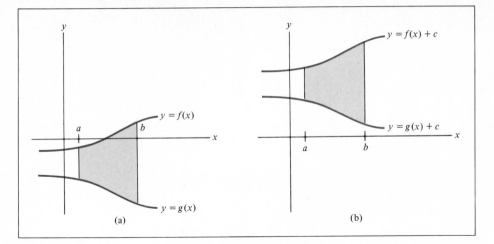

(a)   (b)

FIGURE 4

nonnegative functions, we have

$$[\text{area of the region}] = \int_a^b \left[(f(x) + c) - (g(x) + c)\right] dx$$

$$= \int_a^b \left[f(x) - g(x)\right] dx.$$

Therefore, we see that our rule is valid for any functions $f(x)$ and $g(x)$ as long as the graph of $f(x)$ lies above the graph of $g(x)$ for all x from $x = a$ to $x = b$.

EXAMPLE 3   Set up the integral that gives the area between the curves $y = x^2 - 2x$ and $y = -e^x$ from $x = -1$ to $x = 2$.

Solution   Since $y = x^2 - 2x$ lies above $y = -e^x$ (Fig. 5), the rule for finding the area between two curves can be applied directly. The area between the curves is

$$\int_{-1}^2 (x^2 - 2x + e^x)\,dx.$$

Sometimes we are asked to find the area between two curves without being given the values of a and b. In these cases there is a region that is completely enclosed by the two curves. As the next examples illustrate, we must first find the points of intersection of the two curves in order to obtain the values of a and b. In such problems careful curve sketching is especially important.

EXAMPLE 4   Set up the integral that gives the area bounded by the curves $y = x^2 + 2x + 3$ and $y = 2x + 4$.

Solution   The two curves are sketched in Fig. 6, and the region bounded by them is shaded. In order to find the points of intersection, we set $x^2 + 2x + 3 = 2x + 4$ and solve for x. We obtain $x^2 = 1$ or $x = -1$ and $x = +1$. When $x = -1, 2x + 4 = 2(-1) + 4 = 2$. When $x = 1, 2x + 4 = 2(1) + 4 = 6$. Thus the curves intersect at the points $(1, 6)$ and $(-1, 2)$.

FIGURE 5

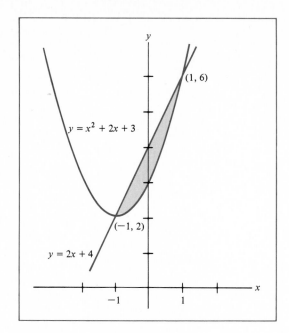

FIGURE 6

Since $y = 2x + 4$ lies above $y = x^2 + 2x + 3$ from $x = -1$ to $x = 1$, the area between the curves is given by

$$\int_{-1}^{1} \left[(2x + 4) - (x^2 + 2x + 3) \right] dx = \int_{-1}^{1} (1 - x^2) \, dx.$$

EXAMPLE 5 Set up the integral that gives the area bounded by the two curves $y = 2x^2$ and $y = x^3 - 3x$.

FIGURE 7

Solution First we make a rough sketch of the two curves as in Fig. 7. The curves intersect where $x^3 - 3x = 2x^2$ or $x^3 - 2x^2 - 3x = 0$. Note that

$$x^3 - 2x^2 - 3x$$
$$= x(x^2 - 2x - 3)$$
$$= x(x - 3)(x + 1).$$

So the solutions to $x^3 - 2x^2 - 3x = 0$ are $x = 0, 3, -1$, and the curves intersect at $(-1, 2)$, $(0, 0)$, and $(3, 18)$.

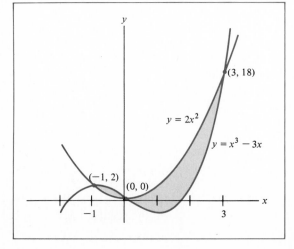

From $x = -1$ to $x = 0$, the curve $y = x^3 - 3x$ lies above $y = 2x^2$. But from $x = 0$ to $x = 3$, the reverse is true. Thus the area between the curves is given by

$$\int_{-1}^{0} (x^3 - 3x - 2x^2)\, dx + \int_{0}^{3} (2x^2 - x^3 + 3x)\, dx.$$

EXAMPLE 6 Beginning in 1974, with the advent of dramatically higher oil prices, the exponential rate of growth of world oil consumption slowed down from a growth constant of 7% to a growth constant of 4% per year. A fairly good model for the annual rate of oil consumption since 1974 is given by

$$R_1(t) = 21.3e^{.04(t-4)}, \qquad t \geq 4,$$

where $t = 0$ corresponds to 1970. Determine the total amount of oil saved between 1976 and 1980 by not consuming oil at the rate predicted by the model of Example 1, Section 1, namely

$$R(t) = 16.1e^{.07t}, \qquad t \geq 0.$$

Solution If oil consumption had continued to grow as it did prior to 1974, then the total oil consumed between 1976 and 1980 would have been

$$\int_{6}^{10} R(t)\, dt. \tag{4}$$

However, taking into account the slower increase in the rate of oil consumption since 1974, we find that the total oil consumed between 1976 and 1980 was approximately

$$\int_{6}^{10} R_1(t)\, dt. \tag{5}$$

The integrals in (4) and (5) may be interpreted as the areas under the curves $y = R(t)$ and $y = R_1(t)$, respectively, from $t = 6$ to $t = 10$. (See Fig. 8.) By

FIGURE 8

(a) Consumption rate predicted by pre-1974 data

(b) Consumption rate changed by 1974 oil price rise

superimposing the two curves we see that the area between them from $t = 6$ to $t = 10$ represents the total oil that was saved by consuming oil at the rate given by $R_1(t)$ instead of $R(t)$. (See Fig. 9.) The area between the two curves equals

$$\int_6^{10} [R(t) - R_1(t)]\, dt = \int_6^{10} [16.1 e^{.07t} - 21.3 e^{.04(t-4)}]\, dt$$

$$= \left(\frac{16.1}{.07} e^{.07t} - \frac{21.3}{.04} e^{.04(t-4)} \right) \Bigg|_6^{10}$$

$$\approx 13.02.$$

Thus about 13 billion barrels of oil were saved between 1976 and 1980.

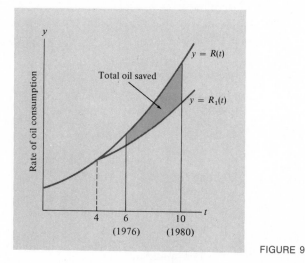

FIGURE 9

PRACTICE PROBLEMS 3

1. Find the area between the curves $y = x + 3$ and $y = x^2 + x - 13$ from $x = 1$ to $x = 3$.

2. A company plans to increase its production from 10 to 15 units per day. The present marginal cost function is $MC_1(x) = x^2 - 20x + 108$. By redesigning the production process and purchasing new equipment the company can change the marginal cost function to $MC_2(x) = \frac{1}{2}x^2 - 12x + 75$. Determine the area between the graphs of the two marginal cost curves from $x = 10$ to $x = 15$. Interpret this area in economic terms.

EXERCISES 3

Find the area of the region between the curves.

1. $y = 2x^2$ and $y = 8$ (a horizontal line) from $x = -2$ to $x = 2$

2. $y = 13 - 3x^2$ and $y = 1$ from $x = -2$ to $x = 2$

3. $y = x^2 - 6x + 12$ and $y = 1$ from $x = 0$ to $x = 4$

4. $y = x(2 - x)$ and $y = 4$ from $x = 0$ to $x = 2$

5. $y = x^2$ and $y = (x - 4)^2$ from $x = 0$ to $x = 3$

6. $y = x^2$ and $y = (x + 3)^2$ from $x = -3$ to $x = 0$

7. $y = 3x^2$ and $y = -3x^2$ from $x = -1$ to $x = 2$

8. $y = e^{2x}$ and $y = -e^{2x}$ from $x = -1$ to $x = 1$

Find the area of the region bounded by the curves.

9. $y = x^2 + x$ and $y = 3 - x$ 10. $y = 3x - x^2$ and $y = 4 - 2x$

11. $y = -x^2 + 6x - 5$ and $y = 2x - 5$ 12. $y = 2x^2 + x - 7$ and $y = x + 1$

13. Find the area between $y = x^2 - 3x$ and the x-axis

    (a) from $x = 0$ to $x = 3$.   (b) from $x = 0$ to $x = 4$.

    (c) from $x = -2$ to $x = 3$.

14. Find the area between $y = x^2$ and $y = 1/x^2$

    (a) from $x = 1$ to $x = 4$.   (b) from $x = \frac{1}{2}$ to $x = 4$.

15. Find the area bounded by $y = 1/x^2$, $y = x$, and $y = 8x$, for $x \geq 0$. (Make a careful sketch.)

16. Find the area bounded by $y = 1/x$, $y = 4x$, and $y = \frac{1}{2}x$, for $x \geq 0$.

17. Find the area bounded by $y = 12/x$, $y = \frac{3}{2}\sqrt{x}$, and $y = \frac{1}{3}x$, for $0 \leq x \leq 6$.

18. Find the area of the region in the first quadrant of the xy-plane that is bounded by $y = 12 - x^2$, $y = 4x$, and $y = x$.

19. Refer to the oil-consumption data in Example 1, Section 1. Suppose that in 1970 the growth constant for the annual rate of oil consumption had been held to .04. What effect would this action have had on oil consumption from 1970 to 1974?

20. The marginal profit function for a certain company is $MP_1(x) = -x^2 + 14x - 24$. The company expects the daily production level to rise from $x = 6$ to $x = 8$ units. The management is considering a plan that would have the effect of changing the marginal profit to $MP_2(x) = -x^2 + 12x - 20$. Should the company adopt the plan? Determine the area between the graphs of the two marginal cost functions from $x = 6$ to $x = 8$. Interpret this area in economic terms.

SOLUTIONS TO PRACTICE PROBLEMS 3

1. First graph the two curves, as shown in the accompanying figure. The curve $y = x + 3$ lies on top. So the area between the curves is

$$\int_1^3 [(x + 3) - (x^2 + x - 13)]\,dx = \int_1^3 [-x^2 + 16]\,dx$$

$$= (-\tfrac{1}{3}x^3 + 16x)\Big|_1^3$$

$$= 23\tfrac{1}{3}.$$

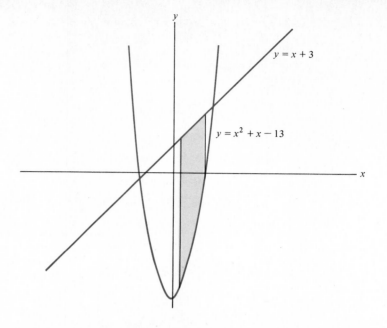

2. Graphing the two marginal cost functions yields the results shown in the accompanying figure. So the area between the curves equals

$$\int_{10}^{15} [MC_1(x) - MC_2(x)] \, dx = \int_{10}^{15} (x^2 - 20x + 108) - (\tfrac{1}{2}x^2 - 12x + 75)] \, dx$$

$$= \int_{10}^{15} [\tfrac{1}{2}x^2 - 8x + 33] \, dx$$

$$= (\tfrac{1}{6}x^3 - 4x^2 + 33x)\Big|_{10}^{15}$$

$$= 60\tfrac{5}{6}.$$

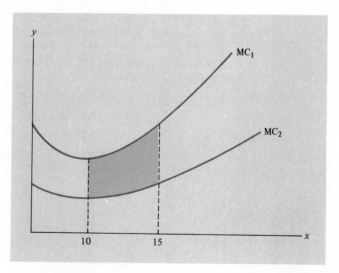

The area under a marginal cost curve is total cost. So the area between the curves represents the difference in the cost of increasing production from 10 to 15 units using the old production process versus the new production process. Using the new production process allows the company to perform the increased production for $60\frac{5}{6}$ dollars less than using the old process.

6.4 Riemann Sums

We have seen many applied problems which require us to calculate a definite integral $\int_a^b f(x)\,dx$. So far, our only technique for doing so requires us to determine an antiderivative of $f(x)$. However, as we shall now demonstrate, it is possible to numerically approximate a definite integral without first determining an antiderivative. The numerical approximation is in terms of so-called *Riemann sums*.

Let us begin by defining Riemann sums for a continuous function $f(x)$ over the interval from $x = a$ to $x = b$. Divide the interval on the x-axis from $x = a$ to $x = b$ into n equal subintervals (Fig. 1). (Here n may be any positive integer.) Since the entire interval is of length $b - a$, each subinterval has length $(1/n)$th of the total, or $(b - a)/n$. For brevity, let us set $\Delta x = (b - a)/n$. In each subinterval choose a point (any point will do). Label these points $x_1, x_2, x_3, \ldots, x_n$. (See Fig. 1.) Now form the sum

$$f(x_1)\,\Delta x + f(x_2)\,\Delta x + \cdots + f(x_n)\,\Delta x.$$

Such a sum is called a *Riemann sum* for the function $f(x)$ over the interval from $x = a$ to $x = b$.

FIGURE 1

The significance of Riemann sums is that they approximate the definite integral $\int_a^b f(x)\,dx$. More precisely, we can state the following fact.

Suppose that the interval from $x = a$ to $x = b$ is divided into n equal subintervals of length $\Delta x = (b - a)/n$. Let x_i be any point of the ith subinterval. Then

$$\int_a^b f(x)\,dx \approx f(x_1)\,\Delta x + f(x_2)\,\Delta x + \cdots + f(x_n)\,\Delta x, \qquad (1)$$

where the approximation becomes exact as n gets large.

Justification of Approximation (1)   We shall justify (1) only for the case when $f(x)$ is nonnegative for x between a and b. [Our argument can easily be modified to cover general $f(x)$.] Note that by the Fundamental Theorem of Calculus the left side of (1) equals the area under the curve $y = f(x)$ from $x = a$ to $x = b$.

Let us approximate this area by the area enclosed by n rectangles, as in Fig. 2. The first rectangle has height $f(x_1)$ and width Δx, so its area is $f(x_1)\Delta x$. The second rectangle has height $f(x_2)$ and width Δx, so its area is $f(x_2)\Delta x$. And so on. Each term on the right side of (1) corresponds to the area of one of the rectangles in Fig. 2. The sum of the areas of these rectangles closely approximates the area under the curve $y = f(x)$ from $x = a$ to $x = b$. This explains why (1) is true. Moreover, as n increases, the width of the rectangles decreases and the rectangular area approximates the area under the curve more closely. This completes the justification.

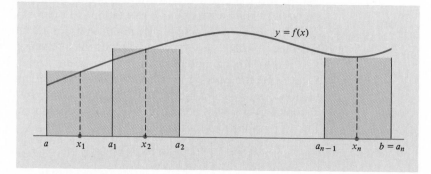

FIGURE 2

The Riemann sum approximation (1) is used in some texts to *define* the definite integral. Indeed, it is possible to prove that if f is continuous for $a \le x \le b$, then as Δx approaches 0, the limit of the Riemann sum is equal to the definite integral. Thus it is possible to define the definite integral as a limit of a Riemann sum. In this approach, our version of the fundamental theorem of calculus is a direct consequence of the definition of the definite integral. However, then the relationship of the integral with antidifferentiation must be proved.

The Riemann sum approximation may be used to explain the notation

$$\int f(x)\,dx$$

which is used for the integral. The symbol $\int$ is an old-fashioned letter "s" and stands for the Latin word "summum" (meaning sum). And, indeed, by (1) the integral is (approximately) a sum of quantities of the form $f(x)\Delta x$. Sir Isaac Newton, one of the inventors of calculus, denoted passage to the limit by changing the Δx to dx and therefore decided to write the integral as $\int f(x)\,dx$.

The approximation (1) may be used in a number of ways. First of all, we may apply it to estimate definite integrals. For this purpose, it is often convenient to choose $x_1, x_2, \ldots, x_n$ to be the midpoints of their respective subintervals, in which case approximation (1) is called the *rectangle rule*.

EXAMPLE 1   Approximate $\int_1^3 x^2\,dx$, using the rectangle rule with $n = 5$.

Note We know, of course, that

$$\int_1^3 x^2\,dx = \tfrac{1}{3}x^3 \Big|_1^3 = 9 - \tfrac{1}{3} = 8.6666\ldots,$$

so the rectangle rule is not really necessary here. But this example will give us some assurance that the rectangle rule works. Also, it will give us some idea of the accuracy we obtain for a given amount of calculation.

Solution   We have $\Delta x = (b - a)/n = (3 - 1)/5 = 2/5 = .4$. The endpoints of the five subintervals begin at $a = 1$ and are spaced .4 unit apart. The first midpoint of a subinterval is at $a + \dfrac{\Delta x}{2} = 1 + .2 = 1.2$. The midpoints are also spaced .4 unit apart.

Therefore, $x_1 = 1.2$, $x_2 = 1.6$, $x_3 = 2.0$, $x_4 = 2.4$, $x_5 = 2.8$, and

$$\int_1^3 x^2\,dx \approx (1.2)^2(.4) + (1.6)^2(.4) + (2.0)^2(.4) + (2.4)^2(.4) + (2.8)^2(.4)$$

$$= [1.44 + 2.56 + 4 + 5.76 + 7.84](.4)$$

$$= 8.64.$$

The error of this approximation is less than .03.

   Approximate techniques of integration become manageable for most examples by using one of the currently available calculators. (The examples in this book were done that way.) However, for very large n (such as $n = 100$), even the minicalculator must make way for a high-speed digital computer or a programmable calculator.
   The fundamental approximation (1) can be applied in a manner completely different from that described above. Namely it may be used to express certain quantities as definite integrals. The principle behind such applications of (1) is as follows: Suppose that we are interested in finding the value of a certain quantity, call it Q. We observe that we can obtain an approximation of Q by subdividing an interval from $x = a$ to $x = b$ into n equal subintervals and forming the sum $f(x_1)\Delta x + \cdots + f(x_n)\Delta x$, where $f(x)$ is some function (not given originally) and where x_i, Δx are as before. Also, we observe that this approximation to Q becomes exact as n gets large. We then invoke our fundamental formula to conclude that $Q = \int_a^b f(x)\,dx$.
   Examples 2–5 illustrate such applications of (1).

EXAMPLE 2   Revolve the region of Fig. 3(a) about the x-axis so that it sweeps out a solid [Fig. 3(b)]. Derive a formula for the volume of this *solid of revolution*.

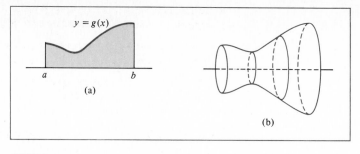

FIGURE 3

Solution Let us break the x-axis between a and b into a large number n of equal subintervals, each of length $\Delta x = (b - a)/n$. Using each subinterval as a base, we can divide the region into strips (see Fig. 4).

Let x_i be a point in the ith subinterval. Then the volume swept out by revolving the ith strip is approximately the same as the volume of the cylinder swept out by revolving the rectangle of height $g(x_i)$ and base Δx around the x-axis (Fig. 5). The volume of the cylinder is

$$[\text{area of circular side}] \cdot [\text{width}] = \pi[g(x_i)]^2 \cdot \Delta x.$$

FIGURE 4

FIGURE 5

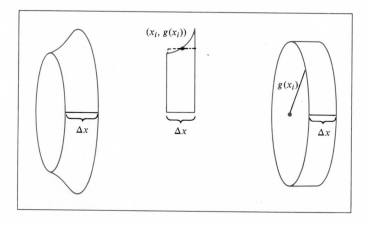

The total volume swept out by all the strips is approximated by the total volume swept out by the rectangles, which is

$$\text{volume} \approx \pi[g(x_1)]^2 \, \Delta x + \pi[g(x_2)]^2 \, \Delta x + \cdots + \pi[g(x_n)]^2 \, \Delta x.$$

As n gets larger and larger, this approximation becomes arbitrarily close to the true volume. The expression on the right is a Riemann sum with $f(x) = \pi[g(x)]^2$. Therefore, we conclude that

$$\text{volume} = \int_a^b \pi[g(x)]^2 \, dx.$$

EXAMPLE 3 Find the volume of the solid of revolution obtained by revolving the region of Fig. 6 about the x-axis.

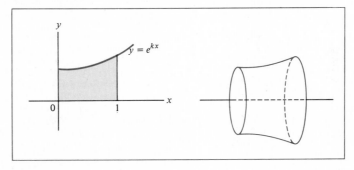

FIGURE 6

Solution   Using the formula derived in Example 2, we have

$$\text{volume} = \int_0^1 \pi(e^{kx})^2 \, dx$$

$$= \int_0^1 \pi e^{2kx} \, dx = \frac{\pi}{2k} e^{2kx} \Big|_0^1 = \frac{\pi}{2k}(e^{2k} - 1).$$

EXAMPLE 4 Find the volume of a right circular cone of radius r and height h.

Solution   The cone [Fig. 7(a)] is the solid of revolution swept out when the shaded region in Fig. 7(b) is revolved about the x-axis.

FIGURE 7

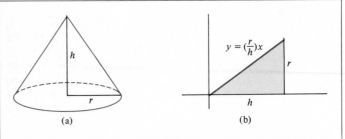

(a)   (b)

Using the formula developed in Example 2, the volume of the cone is

$$\int_0^h \pi \left(\frac{r}{h}x\right)^2 dx = \frac{\pi r^2}{h^2} \int_0^h x^2 \, dx = \frac{\pi r^2}{h^2} \frac{x^3}{3}\Big|_0^h = \frac{\pi r^2 h}{3}.$$

The next example provides an application of Riemann sums to economics.

EXAMPLE 5 (*Consumers' Surplus*) Using a demand curve of economics, derive a formula showing the amount that consumers benefit from an open system that has no price discrimination.

Solution   Figure 8(a) is a *demand curve* for a commodity. It is determined by complex economic factors and gives a relationship between the quantity sold and the unit price of a commodity. Specifically, it says that, in order to sell x units, the price must be set at $f(x)$ dollars per unit. Since, for most commodities, selling larger quantities requires a lowering of the price, demand functions are usually decreasing. Interactions between supply and demand determine the amount of a quantity available. Let A designate the amount of the commodity currently available and $B = f(A)$ the current selling price.

Divide the interval from 0 to A into n subintervals, each of length $\Delta x = (A - 0)/n$, and take x_i to be the right-hand endpoint of the ith interval. Consider the first subinterval, from 0 to x_1. [See Fig. 8(b).] Suppose that only x_1 units had been available. Then the price per unit could have been set at $f(x_1)$ dollars and these x_1 units sold. Of course, at this price we could not have sold any more units. However, those people who paid $f(x_1)$ dollars had a great demand for the commodity. It is extremely valuable to them, and there is no advantage to substituting another commodity at that price. They are actually paying what the commodity is worth to them. In theory, then, the first x_1 units of the commodity could be sold to these people at $f(x_1)$ dollars per unit, yielding (price per unit) · (number of units) $= f(x_1) \cdot (x_1) = f(x_1) \cdot \Delta x$ dollars.

After selling the first x_1 units, suppose that more units become available, so that now a total of x_2 units have been produced. Setting the price at $f(x_2)$, the remaining $x_2 - x_1 = \Delta x$ units can be sold, yielding $f(x_2) \cdot \Delta x$ dollars. Here, again,

FIGURE 8   Consumers' surplus.

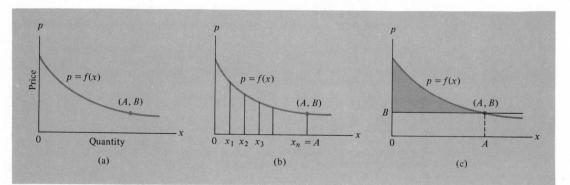

(a)   (b)   (c)

the second group of buyers would have paid as much for the commodity as it is worth to them. Continuing this process of price discrimination, the amount of money paid by the consumers would be

$$f(x_1)\,\Delta x + f(x_2)\,\Delta x + \cdots + f(x_n)\,\Delta x.$$

Taking n large, we see that this amount approaches $\int_0^A f(x)\,dx$.

Of course, in our open system, everyone pays the same price, B, and so the total amount paid is [price per unit] $\cdot$ [number of units] $= B \cdot A$. The amount of money saved by the consumers is

$$\int_0^A f(x)\,dx - B \cdot A = \int_0^A [f(x) - B]\,dx$$

and is called the *consumers' surplus*. Referring to Fig. 8(c), we see that the consumers' surplus is the area of the shaded region. It gives a numerical value to one benefit of a modern efficient economy.

The next example shows how the definite integral can be used to approximate the sum of a large number of terms.

EXAMPLE 6 Suppose that money is deposited daily into a saving account at an annual rate of $1000. The account pays 6% interest compounded continuously. Approximate the amount of money in the account at the end of 5 years.

Solution Divide the time interval from 0 to 5 years into daily subintervals. Each subinterval is then of duration $\Delta t = \frac{1}{365}$ years. Let $t_1, t_2, \ldots, t_n$ be points chosen from these subintervals. Since we deposit money at an annual rate of $1000, the amount deposited during one of the subintervals is $1000\,\Delta t$ dollars. If this amount is deposited at time t_i, the $1000\,\Delta t$ dollars will earn interest for the remaining $5 - t_i$ years. The total amount resulting from this one deposit at time t_i is then

$$1000\,\Delta t e^{.06(5-t_i)}.$$

Add up the effects of the deposits at times $t_1, t_2, \ldots, t_n$ to arrive at the total balance in the account:

$$A = 1000e^{.06(5-t_1)}\,\Delta t + 1000e^{.06(5-t_2)}\,\Delta t + \cdots + 1000e^{.06(5-t_n)}\,\Delta t.$$

This sum is a Riemann sum for the integral

$$\int_0^5 1000e^{.06(5-t)}\,dt = \frac{1000}{-.06}\,e^{.06(5-t)}\Big|_0^5$$

$$= \frac{1000}{-.06}\,(1 - e^{.3})$$

$$\approx 5831.$$

That is, the approximate balance in the account at the end of 5 years is $5831.

Average Value of a Function Let $f(x)$ be a function. The definite integral may be used to define the *average value of $f(x)$ from $x = a$ to $x = b$*. To calculate the average of a collection of numbers $y_1, y_2, \ldots, y_n$, we add up the numbers and divide by n to obtain

$$\frac{y_1 + y_2 + \cdots + y_n}{n}.$$

To determine the average value of $f(x)$ from $x = a$ to $x = b$ we proceed similarly. Choose n values of x, say $x_1, x_2, \ldots, x_n$, and calculate the corresponding function values $f(x_1), f(x_2), \ldots, f(x_n)$. The average of these values is

$$\frac{f(x_1) + f(x_2) + \cdots + f(x_n)}{n}. \tag{2}$$

If the points $x_1, x_2, \ldots, x_n$ are "evenly" spread out from a to b, we would expect this sum to closely approximate the average value of $f(x)$ from $x = a$ to $x = b$. In fact, as n becomes large the average (2) should approximate the average value of $f(x)$ to any arbitrary degree of accuracy. To guarantee that the points $x_1, x_2, \ldots, x_n$ are "evenly" spread out from a to b, let us divide the interval from $x = a$ to $x = b$ into n subintervals of equal length $\Delta x = (b - a)/n$ (Fig. 9). Then choose x_1 from

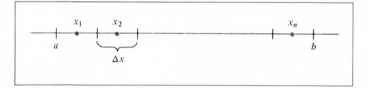

FIGURE 9

the first subinterval, x_2 from the second, and so forth. Thus, we see that for n large,

[average value of $f(x)$ from $x = a$ to $x = b$]

$$\approx f(x_1) \cdot \frac{1}{n} + f(x_2) \cdot \frac{1}{n} + \cdots + f(x_n) \cdot \frac{1}{n}$$

$$= \frac{1}{b - a}\left[f(x_1) \cdot \frac{b - a}{n} + f(x_2) \cdot \frac{b - a}{n} + \cdots + f(x_n) \cdot \frac{b - a}{n} \right]$$

$$= \frac{1}{b - a}\left[f(x_1)\Delta x + f(x_2)\Delta x + \cdots + f(x_n)\Delta x \right]$$

$$\approx \frac{1}{b - a} \int_a^b f(x)\,dx,$$

where the approximation becomes exact as n becomes large. Thus we have motivated the following important definition:

The *average value* of $f(x)$ from $x = a$ to $x = b$ is defined as the quantity

$$\frac{1}{b - a} \int_a^b f(x)\,dx. \tag{3}$$

EXAMPLE 7   Find the average value of $f(x) = \sqrt{x}$ over the interval from 0 to 9.

Solution   Using (2) with $a = 0$ and $b = 9$, the average value of $f(x) = \sqrt{x}$ from $x = 0$ to $x = 9$ is

$$\frac{1}{9-0} \int_0^9 \sqrt{x}\,dx.$$

Since $\sqrt{x} = x^{1/2}$, an antiderivative of $\sqrt{x}$ is $\frac{2}{3}x^{3/2}$. Therefore,

$$\frac{1}{9} \int_0^9 \sqrt{x}\,dx = \frac{1}{9}\left(\frac{2}{3}x^{3/2}\Big|_0^9\right) = \frac{1}{9}\left(\frac{2}{3}\cdot 9^{3/2} - 0\right) = \frac{1}{9}\left(\frac{2}{3}\cdot 27\right) = 2,$$

so that the average value of $\sqrt{x}$ over the interval from 0 to 9 is 2. The area of the shaded region is the same as the area of the pictured rectangle (Fig. 10).

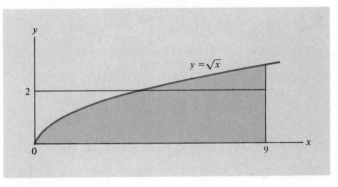

FIGURE 10

EXAMPLE 8   Suppose that the current world population is 4.2×10^9 and the population t years from now is given by the exponential growth law

$$P(t) = (4.2 \times 10^9)e^{.023t}.$$

Find the average population of the earth during the next 30 years. (This number is important in long-range planning for agricultural production and for the allocation of goods and services.)

Solution   The average value of the population $P(t)$ from $t = 0$ to $t = 30$ is

$$\frac{1}{30-0} \int_0^{30} P(t)\,dt = \frac{1}{30} \int_0^{30} (4.2 \times 10^9)e^{.023t}\,dt$$

$$= \left(\frac{1}{30}\right)\frac{4.2 \times 10^9}{.023} e^{.023t}\Big|_0^{30}$$

$$= \frac{4.2 \times 10^9}{.69}(e^{.69} - 1)$$

$$\approx 6 \times 10^9.$$

1. Approximate $\int_1^3 \frac{1}{x} dx$ using the rectangle rule with $n = 4$.

2. Suppose that the interval $0 \leq x \leq 2$ is divided into 100 subintervals, each of width Δx. Show that the sum,

$$[3(\tfrac{1}{100})^2 - \tfrac{1}{100}]\Delta x + [3(\tfrac{3}{100})^2 - \tfrac{3}{100}]\Delta x$$

$$+ [3(\tfrac{5}{100})^2 - \tfrac{5}{100}]\Delta x + \cdots + [3(\tfrac{199}{100})^2 - \tfrac{199}{100}]\Delta x,$$

is close to 6.

EXERCISES 4

Approximate the following integrals by the rectangle rule using the given value of n. Also, find the exact value of the integral. State your answers to five decimal places.

1. $\int_0^1 x^3 dx, n = 2, 4$

2. $\int_1^2 \frac{1}{(4x - 3)^2} dx; n = 4$

3. $\int_1^2 e^{-x} dx; n = 5$

4. $\int_0^4 (x^2 - 5) dx; n = 2, 4$

5. $\int_4^6 \frac{1}{x - 3} dx; n = 5$

6. $\int_2^5 \sqrt{x - 1} dx; n = 6$

Find the consumers' surplus for each of the following demand curves at the given sales level x.

7. $p = 3 - (x/10); x = 20$

8. $p = \frac{x^2}{200} - x + 50; x = 20$

9. $p = -.01x + 5; x = 100$

10. $p = \sqrt{16 - .02x}; x = 350$

Figure 11 shows a supply curve for a commodity. It gives the relationship between the selling price of the commodity and the quantity that producers will manufacture. At a higher selling price, a greater quantity will be produced. Therefore, the curve is increasing. If (A, B) is a point on the curve, then, in order to stimulate the production of A units of the commodity, the price per unit must be B dollars. Of course, some producers will be willing to produce the commodity even with a lower selling price. Since everyone receives the same price in an open efficient economy, most producers are receiving more than their minimal required price. The excess is called the *producers' surplus*. Using an argument analogous to that of the *consumers' surplus*, one can show that the total producers' surplus when the price is B is the area of the shaded region in Fig. 11. Find the producers' surplus for each of the following supply curves at the given sales level x.

11. $p = .01x + 3; x = 200$

12. $p = (x^2/9) + 1; x = 3$

13. $p = (x/2) + 7; x = 10$

14. $p = 1 + \tfrac{1}{2}\sqrt{x}; x = 36$

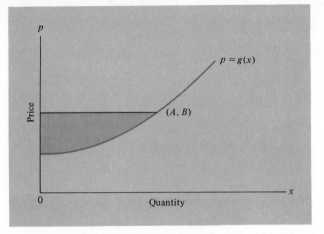

FIGURE 11 Producers' surplus.

For a particular commodity, the quantity produced and the unit price are given by the co-ordinates of the point where the supply and demand curves intersect. For each pair of supply and demand curves, determine the point of intersection (A, B) and the consumers' and pro-ducers' surplus. (See Fig. 12.)

15. Demand curve: $p = 12 - (x/50)$; supply curve: $p = (x/20) + 5$.

16. Demand curve: $p = \sqrt{25 - .1x}$; supply curve: $p = \sqrt{.1x + 9} - 2$.

17. Suppose that money is deposited daily into a savings account at an annual rate of $1000. If the account pays 5% interest compounded continuously, estimate the balance in the account at the end of 3 years.

18. Suppose that money is deposited daily into a savings account at an annual rate of $2000. If the account pays 6% interest compounded continuously, approximately how much will be in the account at the end of 2 years?

19. Suppose that money is deposited daily into a savings account at the rate of P dollars per year. The savings account pays 6% interest compounded continuously. Estimate the balance at the end of 4 years.

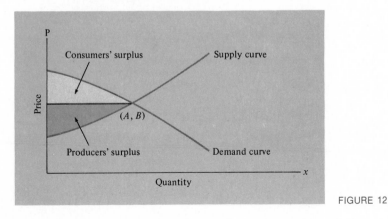

FIGURE 12

20. Suppose that money is deposited daily into a savings account at the rate of P dollars per year for T years. The account pays 5% interest compounded continuously. Find a definite integral that gives the approximate balance at the end of the T years.

Find the average value of $f(x)$ over the interval from $x = a$ to $x = b$, where:

21. $f(x) = x^2$; $a = 0, b = 3$

22. $f(x) = e^{x/3}$; $a = 0, b = 3$

23. $f(x) = x^3$; $a = -1, b = 1$

24. $f(x) = 5$; $a = 1, b = 10$

25. $f(x) = 1/x^2$; $a = \frac{1}{4}, b = \frac{1}{2}$

26. $f(x) = 2x - 6$; $a = 2, b = 4$

27. During a certain 12-hour period the temperature at time t (measured in hours from the start of the period) was $47 + 4t - \frac{1}{3}t^2$ degrees. What was the average temperature during that period?

28. Assuming that a country's population is now 3 million and growing exponentially with growth constant .02, what will be the average population during the next 50 years?

29. One hundred grams of radioactive radium having a half-life of 1690 years is placed in a concrete vault. What will be the average amount of radium in the vault during the next 1000 years?

30. One hundred dollars is deposited in the bank at 5% interest compounded continuously. What will be the average value of the money in the account during the next 20 years?

31. Suppose the interval $0 \le x \le 3$ is divided into 100 subintervals of width Δx. Let $x_1, \ldots, x_{100}$ be points in these subintervals. Suppose that in some application one needs to estimate the sum

$$(3 - x_1)^2 \, \Delta x + (3 - x_2)^2 \, \Delta x + \cdots + (3 - x_{100})^2 \, \Delta x.$$

Show that this sum is close to 9.

32. Suppose that the interval $0 \le t \le 3$ is divided into 1000 subintervals of width Δt. Let $t_1, t_2, \ldots, t_{1000}$ denote the right endpoints of these subintervals. Suppose that in some problem one needs to estimate the sum.

$$5000e^{-.1t_1} \, \Delta t + 5000e^{-.1t_2} \, \Delta t + \cdots + 5000e^{-.1t_{1000}} \, \Delta t.$$

Show that this sum is close to 13,000. [*Note:* A sum such as this would arise if one wanted to compute the present value of a continuous stream of income of $5000 per year for 3 years, with interest compounded continuously at 10%.]

33. Suppose that the interval $0 \le x \le 1$ is divided into 100 subintervals of width $\Delta x = \frac{1}{100}$. Show that the sum

$$\left[2 \cdot \tfrac{1}{100} + (\tfrac{1}{100})^3 \right] \Delta x + \left[2 \cdot \tfrac{2}{100} + (\tfrac{2}{100})^3 \right] \Delta x$$
$$+ \left[2 \cdot \tfrac{3}{100} + (\tfrac{3}{100})^3 \right] \Delta x + \cdots + \left[2 \cdot \tfrac{100}{100} + (\tfrac{100}{100})^3 \right] \Delta x$$

is close to $\frac{5}{4}$.

34. Suppose that water is flowing into a tank at a rate of $r(t)$ gallons per hour, where the rate depends on the time t according to the formula

$$r(t) = 20 - 4t, \qquad 0 \le t \le 5.$$

(a) Consider a brief period of time, say from t_1 to t_2. The length of this time period is $\Delta t = t_2 - t_1$. During this period the rate of flow does not change much and is approximately $20 - 4t_1$ (the rate at the beginning of the brief time interval). Approximately how much water flows into the tank during the time from t_1 to t_2?

(b) Explain why the total amount of water added to the tank during the time interval from $t = 0$ to $t = 5$ is given by $\int_0^5 r(t)\,dt$.

35. Suppose that the world consumes oil at the rate of $R_1(t)$ billion barrels per year at time t, where $t = 0$ corresponds to 1970 and $R_1(t) = 21.3e^{.04(t-4)}$ for $t \geq 4$. Use the rectangle rule with $n = 4$ to approximate $\int_6^{10} R_1(t)\,dt$. What interpretation can you give to each of the four terms in the approximation?

36. Find the average rate of oil consumption between 1976 and 1980, where the rate $R_1(t)$ is given in Exercise 35.

Find the volume of the solid of revolution generated by revolving about the x-axis the region under each of the following curves.

37. $y = \sqrt{r^2 - x^2}$ from $x = -r$ to $x = r$ (generates a sphere of radius r)

38. $y = r$ from $x = 0$ to $x = h$ (generates a cylinder)

39. $y = x^2$ from $x = 1$ to $x = 2$

40. $y = \dfrac{1}{x}$ from $x = 1$ to $x = 100$

41. $y = \sqrt{x}$ from $x = 0$ to $x = 4$ (The solid generated is called a *paraboloid*.)

42. $y = 2x - x^2$ from $x = 0$ to $x = 2$

43. $y = e^{-x}$ from $x = 0$ to $x = r$

44. $y = 2x + 1$ from $x = 0$ to $x = 1$ (The solid generated is called a *truncated cone*.)

SOLUTIONS TO PRACTICE PROBLEMS 4

1. Here $a = 1$, $b = 3$, $n = 4$, so $\Delta x = (3 - 1)/4 = \frac{1}{2}$. Moreover, by drawing the interval from $x = 1$ to $x = 3$, we see that $x_1 = 1.25, x_2 = 1.75, x_3 = 2.25, x_4 = 2.75$ (see Fig. 13). Thus

FIGURE 13

$$\int_1^3 \frac{1}{x}\,dx \approx \left[\frac{1}{1.25} + \frac{1}{1.75} + \frac{1}{2.25} + \frac{1}{2.75} \right] \cdot \frac{1}{2}$$

$$\approx 1.0898.$$

We note that

$$\int_1^3 \frac{1}{x}\,dx = \ln x \Big|_1^3 = \ln 3.$$

To four decimal places, $\ln 3 = 1.0986$, so the error from the rectangle rule is .0088. When $n = 10$, the error is .0015.

2. Here $\Delta x = (2 - 0)/100 = .02$. The subdivision is shown in Fig. 14, where the respective midpoints of the subintervals are seen to be $x_1 = .01, x_2 = .03, x_3 = .05, \ldots, x_{100} = 1.99$.

FIGURE 14

Therefore, the sum in question is

$$[3x_1^2 - x_1]\Delta x + [3x_2^2 - x_2]\Delta x + [3x_3^2 - x_3]\Delta x + \cdots + [3x_{100}^2 - x_{100}]\Delta x,$$

a Riemann sum for the function $f(x) = 3x^2 - x$ over the interval from $x = 0$ to $x = 2$. By (1), this sum is approximately equal to $\int_0^2 (3x^2 - x)\,dx$. Now,

$$\int_0^2 (3x^2 - x)\,dx = \left(x^3 - \frac{x^2}{2}\right)\Bigg|_0^2 = 6.$$

6.5 Techniques of Integration

This section presents two techniques of integration that will greatly expand our ability to find antiderivatives. Like the other rules of integration discussed in Section 1, these new methods are based on corresponding rules of differentiation: the chain rule and the product rule.

Integration by Substitution   Many integrals that appear complex have a simple form that becomes evident after making an appropriate change of variable. Integration by substitution is a technique for making such changes of variables.

Let $f(x)$ and $g(x)$ be two given functions and let $F(x)$ be an antiderivative of $f(x)$. The chain rule asserts that

$$\frac{d}{dx}[F(g(x))] = F'(g(x))g'(x)$$

$$= f(g(x))g'(x) \qquad [\text{since } F'(x) = f(x)].$$

Turning this formula into an integration formula, we have

$$\int f(g(x))g'(x)\,dx = F(g(x)) + C, \tag{1}$$

where C is any constant. One way to use this formula is illustrated in the following example.

EXAMPLE 1   Determine $\int (x^2 + 1)^3 \cdot 2x\,dx$.

Solution   If we set $f(x) = x^3$, $g(x) = x^2 + 1$, then $f(g(x)) = (x^2 + 1)^3$ and $g'(x) = 2x$. Therefore, we can apply formula (1). An antiderivative $F(x)$ of $f(x)$ is given by

$$F(x) = \tfrac{1}{4}x^4,$$

so that, by formula (1), we have

$$\int (x^2 + 1)^3 \cdot 2x \, dx = F(g(x)) + C$$
$$= \tfrac{1}{4}(x^2 + 1)^4 + C.$$

Formula (1) can be elevated from the status of a sometimes useful formula to a technique of integration by the introduction of a simple mnemonic device. Suppose that we are faced with integrating a function of the form $f(g(x))g'(x)$. Of course, we know the answer from formula (1). However, let us proceed somewhat differently. Replace the expression $g(x)$ by a new variable u and replace $g'(x) \, dx$ by du. Such a substitution has the advantage that it reduces the generally complex expression $f(g(x))$ to the simpler form $f(u)$. In terms of u, the integration problem may be written

$$\int f(g(x))g'(x) \, dx = \int f(u) \, du.$$

However, the integral on the right is easy to evaluate, since

$$\int f(u) \, du = F(u) + C.$$

Since $u = g(x)$, we then obtain

$$\int f(g(x))g'(x) \, dx = F(u) + C = F(g(x)) + C,$$

which is the correct answer by (1). Remember, however, that replacing $g'(x) \, dx$ by du only has status as a correct mathematical statement because doing so leads to the correct answers. We do not, in this book, seek to explain in any deeper way what this replacement means.

Let us rework Example 1 using this method.

Second Solution of Example 1   Set $u = x^2 + 1$. Then $du = \dfrac{d}{dx}(x^2 + 1) \, dx = 2x \, dx$, and

$$\int (x^2 + 1)^3 \cdot 2x \, dx = \int u^3 \, du$$
$$= \tfrac{1}{4}u^4 + C$$
$$= \tfrac{1}{4}(x^2 + 1)^4 + C \qquad (\text{since } u = x^2 + 1).$$

EXAMPLE 2   Evaluate $\int 2xe^{x^2} \, dx$.

Solution   Let $u = x^2$, so that $du = \dfrac{d}{dx}(x^2) \, dx = 2x \, dx$. Therefore,

$$\int 2xe^{x^2} \, dx = \int e^{x^2} \cdot 2x \, dx$$
$$= \int e^u \, du$$
$$= e^u + C$$
$$= e^{x^2} + C.$$

From Examples 1 and 2 we can deduce the following method for integration of functions of the form $f'(g(x))g'(x)$.

Integration by Substitution

1. Define a new variable $u = g(x)$, where $g(x)$ is chosen in such a way that, when written in terms of u, the integrand is simpler than when written in terms of x.

2. Transform the integral with respect to x into an integral with respect to u by replacing $g(x)$ everywhere by u and $g'(x)\,dx$ by du.

3. Integrate the resulting function of u.

4. Rewrite the answer in terms of x by replacing u by $g(x)$.

Let us try a few more examples.

EXAMPLE 3 Evaluate $\int 3x^2 \sqrt{x^3 + 1}\,dx$.

Solution The first problem facing us is to find an appropriate substitution that will simplify the integral. An immediate possibility is offered by setting $u = x^3 + 1$. Then $\sqrt{x^3 + 1}$ will become $\sqrt{u}$, a significant simplification. If $u = x^3 + 1$, then $du = \dfrac{d}{dx}(x^3 + 1)\,dx = 3x^2\,dx$, so that

$$\int 3x^2 \sqrt{x^3 + 1}\,dx = \int \sqrt{u}\,du$$
$$= \tfrac{2}{3}u^{3/2} + C$$
$$= \tfrac{2}{3}(x^3 + 1)^{3/2} + C.$$

Knowing the correct substitution to make is a skill that develops through practice. Basically, we look for an occurrence of function composition, $f(g(x))$, where $f(x)$ is a function that we know how to integrate and where $g'(x)$ also appears in the integrand. Sometimes $g'(x)$ does not appear exactly but can be obtained by multiplying by a constant. Such a shortcoming is easily remedied, as illustrated in Examples 4 and 5.

EXAMPLE 4 Find $\int x^2 e^{x^3}\,dx$.

Solution Let $u = x^3$; then $du = 3x^2\,dx$. A simple trick allows us to introduce the needed factor of 3 into the integrand. Note that

$$\int x^2 e^{x^3}\,dx = \int \frac{1}{3} \cdot 3x^2 e^{x^3}\,dx = \frac{1}{3} \int 3x^2 e^{x^3}\,dx.$$

Substituting, we obtain

$$\int x^2 e^{x^3}\, dx = \frac{1}{3} \int e^u\, du$$

$$= \frac{1}{3} e^u + C$$

$$= \frac{1}{3} e^{x^3} + C \qquad (\text{since } u = x^3).$$

EXAMPLE 5 Find $\int \dfrac{x}{x^2 + 1}\, dx$.

Solution If $u = x^2 + 1$, then $du = 2x\, dx$, and

$$\int \frac{x}{x^2 + 1}\, dx = \int \frac{1}{2} \cdot \frac{1}{x^2 + 1} \cdot 2x\, dx$$

$$= \frac{1}{2} \int \frac{1}{x^2 + 1} \cdot 2x\, dx = \frac{1}{2} \int \frac{1}{u}\, du$$

$$= \tfrac{1}{2} \ln|u| + C = \tfrac{1}{2} \ln|x^2 + 1| + C.$$

Integration by Parts Integration by parts is a technique of integration that exchanges a given integral for another (hopefully less complicated) integral. Let $f(x)$ and $g(x)$ be given functions and let $G(x)$ be an antiderivative of $g(x)$. The product rule asserts that

$$\frac{d}{dx}[f(x)G(x)] = f(x)G'(x) + G(x)f'(x)$$

$$= f(x)g(x) + f'(x)G(x) \qquad [\text{since } G'(x) = g(x)].$$

Therefore,

$$f(x)G(x) = \int f(x)g(x)\, dx + \int f'(x)G(x)\, dx.$$

This last formula may be rewritten in the following more useful form.

$$\int f(x)g(x)\, dx = f(x)G(x) - \int f'(x)G(x)\, dx. \qquad (2)$$

Equation (2) is the basis of *integration by parts*, one of the most important techniques of integration. The next three examples illustrate this technique.

EXAMPLE 6 Evaluate $\int xe^x\, dx$.

Solution   Set $f(x) = x$, $g(x) = e^x$. Then $f'(x) = 1$, $G(x) = e^x$, and (2) yields

$$\int xe^x \, dx = xe^x - \int 1 \cdot e^x \, dx$$

$$= xe^x - e^x + C.$$

The following principles underlie Example 6 and also illustrate general features of situations to which integration by parts may be applied:

1. The integrand is the product of two functions $f(x)$ and $g(x)$.
2. It is easy to compute $f'(x)$ and $G(x)$. That is, we can differentiate $f(x)$ and integrate $g(x)$.
3. The integral $\int f'(x)G(x) \, dx$ can be calculated.

Let us consider another example in order to see how these three principles work.

EXAMPLE 7   Evaluate $\int x(x + 5)^8 \, dx$.

Solution   Our calculations can be set up as follows:

$$f(x) = x, \qquad g(x) = (x + 5)^8$$

$$f'(x) = 1, \qquad G(x) = \frac{1}{9}(x + 5)^9.$$

Then

$$\int x(x + 5)^8 \, dx = x \cdot \frac{1}{9}(x + 5)^9 - \int 1 \cdot \frac{1}{9}(x + 5)^9 \, dx$$

$$= \frac{1}{9}x(x + 5)^9 - \frac{1}{9}\int (x + 5)^9 \, dx$$

$$= \frac{1}{9}x(x + 5)^9 - \frac{1}{9} \cdot \frac{1}{10}(x + 5)^{10} + C$$

$$= \frac{1}{9}x(x + 5)^9 - \frac{1}{90}(x + 5)^{10} + C.$$

We were led to try integration by parts because our integrand is the product of two functions. We were led to choose $f(x) = x$ [and not $(x + 5)^9$] because $f'(x) = 1$, so that the factor x in the integrand is made to disappear, thereby simplifying the integral.

EXAMPLE 8   Evaluate $\int x^2 \ln x \, dx$.

Solution   Set

$$f(x) = \ln x, \qquad g(x) = x^2$$

$$f'(x) = \frac{1}{x}, \qquad G(x) = \frac{x^3}{3}.$$

Then

$$\int x^2 \ln x \, dx = \frac{x^3}{3} \ln x - \int \frac{1}{x} \cdot \frac{x^3}{3} \, dx$$

$$= \frac{x^3}{3} \ln x - \frac{1}{3} \int x^2 \, dx$$

$$= \frac{x^3}{3} \ln x - \frac{1}{9} x^3 + C.$$

Let us illustrate how an integration technique, such as integration by parts, can arise in a fairly simple business situation. Recall from our discussion of compound interest that the present value of A dollars to be received t years from now is given by

$$P = Ae^{-rt},$$

where r is a specified annual rate of interest, with interest compounded continuously. Now suppose that a sum of money is to be received in a series of frequent payments over a period of time instead of in one lump sum at the end of the period. Such a series of payments is often viewed as a "continuous stream of income." If $K(t)$ is the annual rate of income at time t, and if the income is to be received over the next T years, then the *present value P of the stream of income* at interest rate r is defined by the integral

$$P = \int_0^T K(t)e^{-rt} \, dt. \tag{3}$$

A Riemann sum argument can be used to show that it takes a fund of P dollars available today in order to create a continuous stream of income of $K(t)$ dollars (annual rate) at time t.

The concept of the present value of a continuous stream of income is an important tool in management decision processes involving the selection or replacement of equipment. Even when $K(t)$ is a simple function, the evaluation of the integral in (3) usually requires special techniques such as integration by parts, as we see in the following example.

EXAMPLE 9   A printing company estimates that the rate of revenue generated by one of its printing presses at time t will be $80 - 2t$ thousand dollars per year. Find the present value of this continuous stream of income over the next 4 years at a 10% interest rate.

Solution   We use (3) with $K(t) = 80 - 2t$, $T = 4$, and $r = .1$. Our first task is to find an antiderivative of $K(t)e^{-rt} = (80 - 2t)e^{-.1t}$. We integrate by parts, with $f(t) = 80 - 2t$, $g(t) = e^{-.1t}$, and $G(t) = -10e^{-.1t}$.

$$\int (80 - 2t)e^{-.1t} \, dt = (80 - 2t)(-10e^{-.1t}) - \int 20e^{-.1t} \, dt$$

$$= -800e^{-.1t} + 20te^{-.1t} + 200e^{-.1t} + C$$

$$= 20te^{-.1t} - 600e^{-.1t} + C.$$

Then

$$P = \int_0^4 (80 - 2t)e^{-.1t}\, dt = (20te^{-.1t} - 600e^{-.1t})\Big|_0^4$$

$$= (80e^{-.4} - 600e^{-.4}) - (0 - 600e^0)$$
$$= 600 - 520e^{-.4}$$
$$\approx 251.$$

Thus the present value of the machine's earnings is approximately 251 thousand dollars.

PRACTICE PROBLEMS 5

Calculate the following integrals.

1. $\displaystyle\int \frac{e^{5x} + x^4}{(e^{5x} + x^5)^2}\, dx$ [*Hint:* Let $u = e^{5x} + x^5$.]

2. $\displaystyle\int \frac{2x - 1}{(x + 4)^{1/3}}\, dx$ [*Hint:* Let $f(x) = 2x - 1, g(x) = (x + 4)^{-1/3}$.]

EXERCISES 5

Calculate each of the following indefinite integrals.

1. $\displaystyle\int 2x(x^2 + 4)^5 dx$ 2. $\displaystyle\int 3x^2(x^3 + 1)^2\, dx$

3. $\displaystyle\int (x^2 - 5x)^3(2x - 5)\, dx$ 4. $\displaystyle\int 2x\sqrt{x^2 + 3}\, dx$

5. $\displaystyle\int 5e^{5x-3}\, dx$ 6. $\displaystyle\int 2xe^{-x^2}\, dx$

7. $\displaystyle\int \frac{3x^2}{x^3 - 1}\, dx$ 8. $\displaystyle\int \frac{2x + 1}{(x^2 + x + 3)^6}\, dx$

Calculate the following integrals, making the indicated substitutions.

9. $\displaystyle\int \frac{x^2}{\sqrt{x^3 - 1}}\, dx; u = x^3 - 1$ 10. $\displaystyle\int \frac{1}{\sqrt{2x + 1}}\, dx; u = 2x + 1$

11. $\displaystyle\int \frac{e^{1/x}}{x^2}\, dx; u = \frac{1}{x}$ 12. $\displaystyle\int \frac{e^{3x}}{e^{3x} + 1}\, dx; u = e^{3x} + 1$

13. $\displaystyle\int \frac{x^2 + 1}{x^3 + 3x + 2}\, dx; u = x^3 + 3x + 2$ 14. $\displaystyle\int \frac{\ln x}{x}\, dx; u = \ln x$

Determine the following integrals by making an appropriate substitution.

15. $\displaystyle\int (x^5 - 2x + 1)^{10}(5x^4 - 2)\, dx$ 16. $\displaystyle\int (x + 1)e^{x^2 + 2x + 4}\, dx$

17. $\displaystyle\int \frac{3}{2x-4}\,dx$

18. $\displaystyle\int \frac{e^{\sqrt{x}}}{\sqrt{x}}\,dx$

19. $\displaystyle\int \frac{x}{e^{x^2}}\,dx$

20. $\displaystyle\int \frac{3x-x^3}{x^4-6x^2+5}\,dx$

Use integration by parts to determine the following integrals.

21. $\displaystyle\int xe^{5x}\,dx$

22. $\displaystyle\int xe^{-x/2}\,dx$

23. $\displaystyle\int x(2x+1)^4\,dx$

24. $\displaystyle\int (x+1)e^x\,dx$

25. $\displaystyle\int x\sqrt{x+1}\,dx$

26. $\displaystyle\int x(x+5)^{-3}\,dx$

27. $\displaystyle\int \frac{3x}{e^x}\,dx$

28. $\displaystyle\int x^3 \ln x\,dx$

29. $\displaystyle\int \ln x\,dx$

30. $\displaystyle\int \frac{x}{\sqrt{3+2x}}\,dx$

$\qquad$ [Let $f(x) = \ln x,\ g(x) = 1$.]

$\qquad$ [Let $f(x) = x,\ g(x) = (3+2x)^{-1/2}$.]

Determine the following indefinite integrals.

31. $\displaystyle\int xe^{2x}\,dx$

32. $\displaystyle\int x\sqrt{x^2-1}\,dx$

33. $\displaystyle\int xe^{x^2}\,dx$

34. $\displaystyle\int x\sqrt{x-1}\,dx$

35. $\displaystyle\int \frac{2x-1}{\sqrt{3x-3}}\,dx$

36. $\displaystyle\int \frac{2x-1}{3x^2-3x+1}\,dx$

37. $\displaystyle\int (4x+3)(6x^2+9x)^{-7}\,dx$

38. $\displaystyle\int (4x+3)(6x+9)^{-7}\,dx$

39. $\displaystyle\int \frac{\ln x}{\sqrt{x}}\,dx$

40. $\displaystyle\int \frac{x+4}{e^{4x}}\,dx$

41. $\displaystyle\int \frac{1}{x\ln 5x}\,dx$

42. $\displaystyle\int x\ln 5x\,dx$

43. Suppose that an investment produces a continuous stream of income at the rate of $300t - 500$ dollars per year at time t. Find the present value of the investment income over the next 5 years, using a 10% interest rate.

44. Recompute the present value of the stream of income in Exercise 43 over 5 years using a 5% interest rate.

SOLUTIONS TO PRACTICE PROBLEMS 5

1. Let $u = e^{5x} + x^5$, $du = (5e^{5x} + 5x^4)\,dx$, and note that $5e^{5x} + 5x^4 = 5(e^{5x} + x^4)$. Then

$$\int \frac{e^{5x} + x^4}{(e^{5x} + x^5)^2}\,dx = \int \frac{1}{5} \cdot \frac{5e^{5x} + 5x^4}{(e^{5x} + x^5)^2}\,dx$$

$$= \frac{1}{5}\int \frac{1}{(e^{5x} + x^5)^2} \cdot (5e^{5x} + 5x^4)\,dx$$

$$= \frac{1}{5}\int \frac{1}{u^2}\,du$$

$$= \frac{1}{5}(-u^{-1}) + C$$

$$= -\frac{1}{5}(e^{5x} + x^5)^{-1} + C.$$

2. Use integration by parts with $f(x) = 2x - 1$, $g(x) = (x + 4)^{-1/3}$, and $G(x) = \frac{3}{2}(x + 4)^{2/3}$.

$$\int \frac{2x - 1}{(x + 4)^{1/3}}\,dx = \int (2x - 1)(x + 4)^{-1/3}\,dx$$

$$= (2x - 1) \cdot \tfrac{3}{2}(x + 4)^{2/3} - \int 2 \cdot \tfrac{3}{2}(x + 4)^{2/3}\,dx$$

$$= \tfrac{3}{2}(2x - 1)(x + 4)^{2/3} - 3\int (x + 4)^{2/3}\,dx$$

$$= \tfrac{3}{2}(2x - 1)(x + 4)^{2/3} - \tfrac{9}{5}(x + 4)^{5/3} + C.$$

6.6 Improper Integrals

In applications of calculus, especially to statistics, it is often necessary to consider the area of a region that extends infinitely far to the right or left along the x-axis. We have drawn several such regions in Fig. 1. The areas of such "infinite" regions may be computed using *improper integrals*.

FIGURE 1

(a)   (b)   (c)

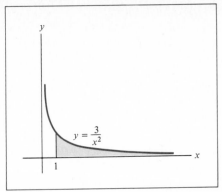

FIGURE 2

In order to motivate the idea of an improper integral, let us attempt to calculate the area under the curve $y = 3/x^2$ to the right of $x = 1$ (Fig. 2).

First, we shall compute the area under the graph of this function from $x = 1$ to $x = b$, where b is some number greater than 1. [See Fig. 3(a).] Then we shall examine how the area increases as we let b get larger, as in Fig. 3(b) and (c). The area from 1 to b is given by

$$\int_1^b \frac{3}{x^2}\, dx = -\frac{3}{x}\bigg|_1^b = \left(-\frac{3}{b}\right) - \left(-\frac{3}{1}\right) = 3 - \frac{3}{b}.$$

When b is large, $3/b$ is small and the integral nearly equals 3. That is, the area under the curve from 1 to b nearly equals 3. (See Table 1.) In fact, the area gets arbitrarily close to 3 as b gets larger. Thus it is reasonable to say that the region under the curve $y = 3/x^2$ for $x \geq 1$ has area 3.

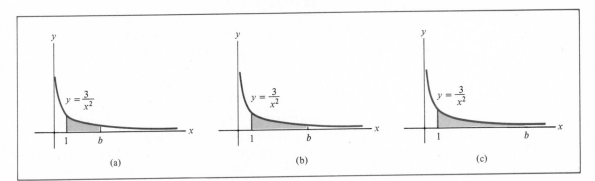

(a)   (b)   (c)

FIGURE 3

TABLE 1

b	$Area = \int_1^b \frac{3}{x^2}\, dx = 3 - \frac{3}{b}$
10	2.7000
100	2.9700
1,000	2.9970
10,000	2.9997

Recall from Chapter 1 that we write $b \to \infty$ as a shorthand for "b gets arbitrarily large, without bound." Then, to express the fact that the value of $\int_1^b \frac{3}{x^2}\, dx$ approaches 3 as $b \to \infty$, we write

$$\int_1^\infty \frac{3}{x^2}\, dx = 3.$$

FIGURE 4

We call $\int_1^\infty \dfrac{3}{x^2}\, dx$ an *improper* integral because the upper limit of the integral is ∞ (infinity) rather than a finite number.

☐ **Definition** Let a be fixed and suppose that $f(x)$ is a nonnegative function for all $x \geq a$. If $\lim\limits_{b \to \infty} \int_a^b f(x)\, dx = L$, then we define

$$\int_a^\infty f(x)\, dx = \lim_{b \to \infty} \int_a^b f(x)\, dx = L.$$

We say that the improper integral $\int_a^\infty f(x)\, dx$ is *convergent* and that the region under the curve $y = f(x)$ for $x \geq a$ has area L. (See Fig. 4.)

It is possible to consider improper integrals in which $f(x)$ is both positive and negative. However, we shall consider only nonnegative functions, since this is the case occurring in most applications.

EXAMPLE 1 Find the area under the curve $y = e^{-x}$ for $x \geq 0$ (Fig. 5).

Solution We must calculate the improper integral

$$\int_0^\infty e^{-x}\, dx.$$

We take $b > 0$ and compute

FIGURE 5

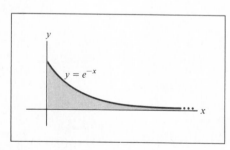

$$\int_0^b e^{-x}\, dx = -e^{-x}\Big|_0^b = (-e^{-b}) - (-e^0)$$

$$= 1 - e^{-b} = 1 - \frac{1}{e^b}.$$

We now consider the limit as $b \to \infty$ and note that $1/e^b$ approaches zero. Thus

$$\int_0^\infty e^{-x}\, dx = \lim_{b \to \infty} \int_0^b e^{-x}\, dx = \lim_{b \to \infty}\left(1 - \frac{1}{e^b}\right) = 1.$$

Therefore, the region in Fig. 5 has area 1.

EXAMPLE 2 Evaluate the improper integral $\int_7^\infty \dfrac{1}{(x-5)^2}\,dx.$

Solution

$$\int_7^b \frac{1}{(x-5)^2}\,dx = -\frac{1}{x-5}\bigg|_7^b = -\frac{1}{b-5} - \left(-\frac{1}{7-5}\right) = \frac{1}{2} - \frac{1}{b-5}.$$

As $b \to \infty$, the fraction $1/(b-5)$ approaches zero, so

$$\int_7^\infty \frac{1}{(x-5)^2}\,dx = \lim_{b\to\infty} \int_7^b \frac{1}{(x-5)^2}\,dx = \lim_{b\to\infty}\left(\frac{1}{2} - \frac{1}{b-5}\right) = \frac{1}{2}.$$

Not every improper integral is convergent. If the value of $\int_a^b f(x)\,dx$ does not have a limit as $b \to \infty$, then we cannot assign any numerical value to $\int_a^\infty f(x)\,dx$, and we say that the improper integral $\int_a^\infty f(x)\,dx$ is *divergent*.

EXAMPLE 3 Show that $\int_1^\infty \dfrac{1}{\sqrt{x}}\,dx$ is divergent.

Solution For $b > 1$ we have

$$\int_1^b \frac{1}{\sqrt{x}}\,dx = 2\sqrt{x}\,\bigg|_1^b = 2\sqrt{b} - 2. \qquad (1)$$

As $b \to \infty$, the quantity $2\sqrt{b} - 2$ increases without bound. That is, $2\sqrt{b} - 2$ can be made larger than any specific number. Therefore, $\int_1^b \dfrac{1}{\sqrt{x}}\,dx$ has no limit as $b \to \infty$, so $\int_1^\infty \dfrac{1}{\sqrt{x}}\,dx$ is divergent.

In some cases it is necessary to consider improper integrals of the form

$$\int_{-\infty}^b f(x)\,dx.$$

Let b be fixed and examine the value of $\int_a^b f(x)\,dx$ as $a \to -\infty$; that is, as a moves arbitrarily far to the left on the number line. If $\lim\limits_{a\to-\infty} \int_a^b f(x)\,dx = L$, we say that the improper integral $\int_{-\infty}^b f(x)\,dx$ is *convergent* and we write

$$\int_{-\infty}^b f(x)\,dx = L.$$

Otherwise, we say that the improper integral is *divergent*. An integral of the form $\int_{-\infty}^b f(x)\,dx$ may be used to compute the area of a region like that shown in Fig. 1(b).

EXAMPLE 4 Determine if $\int_{-\infty}^0 e^{5x}\,dx$ is convergent. If convergent, find its value.

$$\int_{-\infty}^{0} e^{5x}\,dx = \lim_{a \to -\infty} \int_{a}^{0} e^{5x}\,dx$$

$$= \lim_{a \to -\infty} \tfrac{1}{5}e^{5x}\Big|_{a}^{0}$$

$$= \lim_{a \to -\infty} (\tfrac{1}{5} - \tfrac{1}{5}e^{5a}).$$

As $a \to -\infty$, e^{5a} approaches 0 so that $\tfrac{1}{5} - \tfrac{1}{5}e^{5a}$ approaches $\tfrac{1}{5}$. Thus the improper integral converges and has value $\tfrac{1}{5}$.

Areas of regions that extend infinitely far to the left *and* right, such as the region in Fig. 1(c), are calculated using improper integrals of the form

$$\int_{-\infty}^{\infty} f(x)\,dx.$$

We define such an integral to have the value

$$\int_{-\infty}^{0} f(x)\,dx + \int_{0}^{\infty} f(x)\,dx,$$

provided that both of the latter improper integrals are convergent.

An important area that arises in probability theory is the area under the so-called *normal curve*, whose equation is

$$y = \frac{1}{\sqrt{2\pi}}e^{-x^2/2}.$$

(See Fig. 6.) It is of fundamental importance for probability theory that this area is 1. In terms of an improper integral, this fact may be written

$$\int_{-\infty}^{\infty} \frac{1}{\sqrt{2\pi}}e^{-x^2/2}\,dx = 1.$$

The proof of this result is beyond the scope of this book.

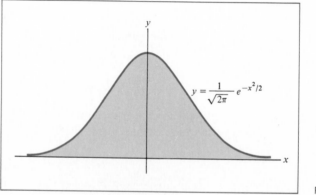

FIGURE 6

PRACTICE PROBLEMS 6

1. Does $1 - 2(1 - 3b)^{-4}$ approach a limit as $b \to \infty$?

2. Evaluate $\int_1^{\infty} \dfrac{x^2}{x^3 + 8}\,dx$.

3. Evaluate $\int_{-\infty}^{-2} \dfrac{1}{x^4}\,dx$.

EXERCISES 6

In Exercises 1–14, determine if the given expression approaches a limit as $b \to \infty$, and find that number when it exists.

1. $\dfrac{5}{b}$

2. b^2

3. $-3e^{2b}$

4. $\dfrac{1}{b} + \dfrac{1}{3}$

5. $\dfrac{1}{4} - \dfrac{1}{b^2}$

6. $\frac{1}{2}\sqrt{b}$

7. $2 - (b + 1)^{-1/2}$

8. $\dfrac{2}{b} - \dfrac{3}{b^{3/2}}$

9. $5 - \dfrac{1}{b - 1}$

10. $5(b^2 + 3)^{-1}$

11. $6 - 3b^{-2}$

12. $e^{-b/2} + 5$

13. $2(1 - e^{-3b})$

14. $4(1 - b^{-3/4})$

15. Find the area under the graph of $y = 1/x^2$ for $x \geq 2$.

16. Find the area under the graph of $y = (x + 1)^{-2}$ for $x \geq 0$.

17. Find the area under the graph of $y = e^{-x/2}$ for $x \geq 0$.

18. Find the area under the graph of $y = 4e^{-4x}$ for $x \geq 0$.

19. Find the area under the graph of $y = (x + 1)^{-3/2}$ for $x \geq 3$.

20. Find the area under the graph of $y = (3x + 1)^{-2}$ for $x \geq 1$.

21. Show that the region under the graph of $y = x^{-3/4}$ for $x \geq 1$ cannot be assigned any finite number as its area.

22. Show that the region under the graph of $y = (x - 1)^{-1/3}$ for $x \geq 2$ cannot be assigned any finite number as its area.

Evaluate the following improper integrals whenever they are convergent.

23. $\int_1^{\infty} \dfrac{1}{x^3}\,dx$

24. $\int_1^{\infty} \dfrac{2}{x^{3/2}}\,dx$

25. $\int_0^{\infty} e^{2x}\,dx$

26. $\int_0^{\infty} x^2 + 1\,dx$

27. $\int_0^{\infty} \dfrac{1}{(2x + 3)^2}\,dx$

28. $\int_0^{\infty} e^{-3x}\,dx$

29. $\int_2^{\infty} \dfrac{1}{(x - 1)^{5/2}}\,dx$

30. $\int_2^{\infty} e^{2-x}\,dx$

31. $\int_0^{\infty} .01e^{-.01x}\,dx$

32. $\int_0^{\infty} \dfrac{4}{(2x + 1)^3}\,dx$

33. $\int_0^{\infty} 6e^{1-3x}\,dx$

34. $\int_1^{\infty} e^{-.2x}\,dx$

35. $\displaystyle\int_3^\infty \frac{x^2}{\sqrt{x^3-1}}\,dx$ **36.** $\displaystyle\int_2^\infty \frac{1}{x\ln x}\,dx$ **37.** $\displaystyle\int_0^\infty xe^{-x^2}\,dx$

38. $\displaystyle\int_0^\infty \frac{x}{x^2+1}\,dx$ **39.** $\displaystyle\int_0^\infty 2x(x^2+1)^{-3/2}\,dx$ **40.** $\displaystyle\int_1^\infty (5x+1)^{-4}\,dx$

41. $\displaystyle\int_{-\infty}^0 e^{4x}\,dx$ **42.** $\displaystyle\int_{-\infty}^0 \frac{8}{(x-5)^2}\,dx$ **43.** $\displaystyle\int_{-\infty}^0 \frac{6}{(1-3x)^2}\,dx$

44. $\displaystyle\int_{-\infty}^0 \frac{1}{\sqrt{4-x}}\,dx$ **45.** $\displaystyle\int_0^\infty \frac{e^{-x}}{(e^{-x}+2)^2}\,dx$ **46.** $\displaystyle\int_{-\infty}^\infty \frac{e^{-x}}{(e^{-x}+2)^2}\,dx$

47. If $k>0$, show that $\displaystyle\int_0^\infty ke^{-kx}\,dx=1$.   **48.** If $k>0$, show that $\displaystyle\int_1^\infty \frac{k}{x^{k+1}}\,dx=1$.

49. The *capital value* of an asset such as a machine is sometimes defined as the present value of all future net earnings of the asset. (See Section 5.) The actual lifetime of the asset may not be known, and since some assets may last indefinitely, the capital value of the asset may be written in the form $\displaystyle\int_0^\infty K(t)e^{-rt}\,dt$, where $K(t)$ is the annual rate of earnings produced by the asset at time t, and where r is the annual rate of interest, compounded continuously. Find the capital value of an asset that generates income at the rate of $5000 per year, assuming an interest rate of 10%.

50. Find the capital value of an asset that at time t is producing income at the rate of $6000e^{.04t}$ dollars per year, assuming an interest rate of 16%.

SOLUTIONS TO PRACTICE PROBLEMS 6

1. The expression $1-2(1-3b)^{-4}$ may also be written in the form

$$1-\frac{2}{(1-3b)^4}.$$

When b is large, $(1-3b)^4$ is very large, and so $2/(1-3b)^4$ is very small. Thus $1-2(1-3b)^{-4}$ approaches 1 as $b\to\infty$.

2. The first step is to find an antiderivative of $x^2/(x^3+8)$. Using the substitution $u=x^3+8$, $du=3x^2\,dx$, we obtain

$$\int \frac{x^2}{x^3+8}\,dx=\frac{1}{3}\int \frac{1}{u}\,du=\frac{1}{3}\ln|u|+C=\frac{1}{3}\ln|x^3+8|+C.$$

Now,

$$\int_1^b \frac{x^2}{x^3+8}\,dx=\frac{1}{3}\ln|x^3+8|\;\Big|_1^b=\frac{1}{3}\ln(b^3+8)-\frac{1}{3}\ln 9.$$

Finally, we examine what happens as $b\to\infty$. Certainly, b^3+8 gets arbitrarily large, and so $\ln(b^3+8)$ must also get arbitrarily large. Hence

$$\int_1^b \frac{x^2}{x^3+8}\,dx$$

has no limit as $b \to \infty$, so the improper integral

$$\int_1^\infty \frac{x^2}{x^3 + 8}\, dx$$

is divergent.

3.

$$\int_a^{-2} \frac{1}{x^4}\, dx = \int_a^{-2} x^{-4}\, dx = \frac{x^{-3}}{-3}\Big|_a^{-2} = \frac{1}{-3x^3}\Big|_a^{-2}$$

$$= \frac{1}{-3(-2)^3} - \left(\frac{1}{-3 \cdot a^3}\right)$$

$$= \frac{1}{24} + \frac{1}{3a^3}.$$

$$\int_{-\infty}^{-2} \frac{1}{x^4}\, dx = \lim_{a \to -\infty} \int_a^{-2} \frac{1}{x^4}\, dx = \lim_{a \to -\infty} \left(\frac{1}{24} + \frac{1}{3a^3}\right) = \frac{1}{24}.$$

6.7 Applications of Calculus to Probability

Consider a cell population that is growing vigorously. Suppose that when a cell is 3 days old it divides and forms two new "daughter" cells. If the population is sufficiently large, it will contain cells of many different ages between 0 and 3, and it will turn out that the proportion of cells of various ages remains constant. That is, if a and b are any two numbers between 0 and 3, with $a < b$, then the proportion of cells whose ages lie between a and b will be essentially constant from one moment to the next, even though individual cells are aging and new cells are being formed all the time. In fact, biologists have found that under the ideal circumstances described, the proportion of cells whose ages are between a and b is given by the area under the graph of the function $f(x) = 2ke^{-kx}$ from $x = a$ to $x = b$, where $k = (\ln 2)/3$.* (See Fig. 1.)

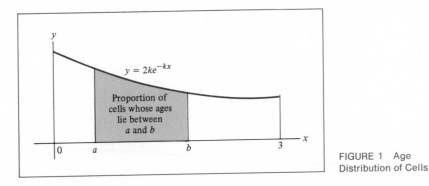

FIGURE 1   Age Distribution of Cells

---

* See J. R. Cook and T. W. James, "Age Distribution of Cells in Logarithmically Growing Cell Populations," in *Synchrony in Cell Division and Growth*, Erik Zeuthen, ed. (New York: John Wiley & Sons, 1964), pp. 485–495.

Now consider an experiment where we select a cell at random from the population and observe its age, X. Then the probability or likelihood* that X lies between a and b is given by the area under the graph of $f(x) = 2ke^{-kx}$ from a to b, as in Fig. 1. Let us denote this probability by $\Pr(a \leq X \leq b)$. Using the fact that the area under the graph of $f(x)$ is given by a definite integral, we have

$$\Pr(a \leq X \leq b) = \int_a^b f(x)\,dx = \int_a^b 2ke^{-kx}\,dx. \tag{1}$$

The function $f(x)$ that determines the probability in (1) for each a and b is called the *probability density function* of X (or of the experiment whose outcome is X).

The general situation we wish to describe in this section concerns an experiment whose outcome is a number X in a certain interval, say between A and B. For the cell population above, $A = 0$ and $B = 3$. Another typical experiment might consist of choosing a decimal X at random between $A = 5$ and $B = 6$. Or, one could select a random telephone call at some telephone switchboard and observe its duration, X. If we have no way of knowing how long a call might last, then X might be any nonnegative number. In this case it is convenient to say that X lies between 0 and ∞ and to take $A = 0$ and $B = \infty$. A similar situation arises in reliability studies where one measures the lifetime X of a transistor selected at random from a manufacturer's production line. Again, the possible values of X lie between 0 and ∞.

When we are dealing with experiments such as those described above, many questions can be reduced to calculating the probability that the outcome X lies between two specified numbers, say a and b. This probability, $\Pr(a \leq X \leq b)$, is a measure of the likelihood that an outcome of the experiment will lie between a and b. If the experiment is repeated many times, then the proportion of times X has a value between a and b will be close to $\Pr(a \leq X \leq b)$. In experiments of practical interest, it is often possible to find a density function $f(x)$ such that

$$\Pr(a \leq X \leq b) = \int_a^b f(x)\,dx, \tag{2}$$

for all a and b in the range of possible values of X.

Any function $f(x)$ with the following two properties is said to be a (*probability*) *density function:*

I. $f(x) \geq 0$, $A \leq x \leq B$
II. $\int_A^B f(x)\,dx = 1$

The additional condition (2) above is what relates a density function to the outcome X of a specific experiment. Graphically, properties I and II mean that for x between A and B, the graph of $f(x)$ must lie above or on the x-axis and the area under the graph must equal 1. Property II simply says that there is probability 1 (certainty) that X has a value between A and B. Of course, if $B = \infty$, then the integral in property II is an improper integral.

---

* For our purposes, it is sufficient to think of probability in the following intuitive terms. Suppose that an experiment with observed outcome X is repeated very often. Then the probability $\Pr(a \leq X \leq b)$ is given (approximately) as the fraction of repetitions in which X was between a and b.

EXAMPLE 1 Consider the cell population described earlier. Let $f(x) = 2ke^{-kx}$, where $k = (\ln 2)/3$. Show that $f(x)$ is indeed a probability density function on the interval from $x = 0$ to $x = 3$.

Solution Clearly $f(x) \geq 0$, since the exponential function is never negative. Thus property I is satisfied. For property II, we check that

$$\int_0^3 f(x)\,dx = \int_0^3 2ke^{-kx}\,dx = -2e^{-kx}\Big|_0^3 = -2e^{-k\cdot 3} + 2e^0$$

$$= 2 - 2e^{-[(\ln 2)/3]3} = 2 - 2e^{-\ln 2}$$

$$= 2 - 2(e^{\ln 2})^{-1} = 2 - 2(2)^{-1} = 2 - 1 = 1.$$

The simplest probability density function is that which assumes a constant value for $A \leq x \leq B$. In order for property II to hold, this constant value must be $1/(B - A)$; that is,

$$f(x) = \frac{1}{B - A}, \qquad A \leq x \leq B.$$

FIGURE 2 A Uniform Density Function.

For in this case the area under the graph of $f(x)$ in Fig. 2 is 1. Such a probability density function is said to be *uniform*. Experiments having uniform density functions generalize those experiments with a finite number of equally likely outcomes.

EXAMPLE 2 Suppose that a subway train leaves the station every 15 minutes. A person who arrives at a random time during the day must wait between 0 and 15 minutes for the next train. What are the chances the person must wait at least 10 minutes?

Solution Let X be the number of minutes the person must wait. The statement that the person arrives at a "random" time is usually interpreted to mean that X has a uniform probability density function. Since the possible values of X lie between 0 and 15, we take $f(x) = \frac{1}{15}$. Then the probability that the person waits at least 10 minutes is given by

$$\Pr(10 \leq X \leq 15) = \int_{10}^{15} \frac{1}{15}\,dx = \frac{1}{15}x\Big|_{10}^{15} = \frac{15}{15} - \frac{10}{15} = \frac{1}{3}.$$

(See Fig. 3.)

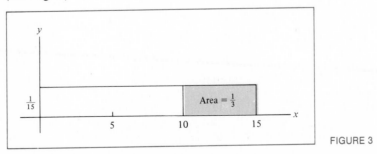

FIGURE 3

As we mentioned earlier, for some experiments the possible outcomes are the numbers between 0 and ∞. In such a case the probability density function $f(x)$ is defined for all $x \geq 0$, and property II is written as

$$\int_0^\infty f(x)\,dx = 1.$$

The most important function of this type has the form $f(x) = \lambda e^{-\lambda x}$, where λ is a positive constant. Just as in Example 1 of Section 6, one easily verifies that

$$\int_0^\infty \lambda e^{-\lambda x}\,dx = 1.$$

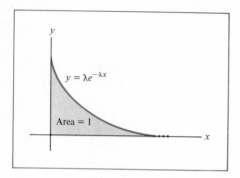

(See Fig. 4.) If the outcome X of an experiment has such a probability density function, the experiment is said to be *exponential* (or *exponentially distributed*). It can be shown that the constant λ may be interpreted as

$$\lambda = \frac{1}{a}, \qquad \text{where } a = \text{average value of } X.$$

Typical uses of exponential probability density functions are given in the next two examples.

FIGURE 4   Exponential Density Function.

EXAMPLE 3 Experiment has shown that the lifetime of a light bulb is exponentially distributed. Let X be the lifetime of a light bulb selected at random from the production line of a light bulb manufacturer. For simplicity, let us measure the lifetime in years, rather than hours, and suppose that the average light bulb produced by the manufacturer burns out in $\frac{1}{4}$ year of continuous use.

(a) What proportion of the light bulbs will burn out within $\frac{1}{2}$ year?

(b) What proportion will continue to burn for at least 1 year?

Solution The average value of X is $\frac{1}{4}$, so we let $\lambda = 4$ and $f(x) = 4e^{-4x}$.

(a) The proportion of light bulbs where X is less than or equal to $\frac{1}{2}$ is

$$\Pr(0 \leq X \leq \tfrac{1}{2}) = \int_0^{1/2} 4e^{-4x}\,dx = -e^{-4x}\Big|_0^{1/2} = -e^{-2} + 1$$

$$= 1 - .13534 = .86466.$$

(b) The proportion of light bulbs that do not burn out for at least 1 year is

$$\Pr(1 \leq X < \infty) = \int_1^\infty 4e^{-4x}\,dx.$$

To evaluate this improper integral, we compute

$$\int_1^b 4e^{-4x}\,dx = -e^{-4x}\Big|_1^b = -e^{-4b} + e^{-4} \to e^{-4}$$

as $b \to \infty$. Hence $\Pr(1 \leq X \leq \infty) = e^{-4} = .01832.$

EXAMPLE 4 A company makes a survey of the duration of telephone calls made by its employees. It finds that the lengths of calls are exponentially distributed, with the average call lasting 5 minutes. What is the probability that a randomly chosen call will last between 5 and 10 minutes?

Solution Let X be the length of the call. Since the average value of X is 5, we take $\lambda = \frac{1}{5} = .2$ and $f(x) = .2e^{-.2x}$. The desired probability is

$$\Pr(5 \leq X \leq 10) = \int_5^{10} .2e^{-.2x}\, dx = -e^{-.2x}\Big|_5^{10}$$

$$= -e^{-2} + e^{-1}$$

$$= -.13534 + .36788 = .23254.$$

PRACTICE PROBLEMS 7

1. An experimenter determines that the probability density function for a certain experiment with outcomes between 0 and 1 is given by $f(x) = x$. Why might you doubt his conclusion?

2. An experiment with outcomes between 0 and 1 has probability density function $f(x) = 4x^3$. What is the probability of an outcome larger than $\frac{1}{2}$?

EXERCISES 7

1. An experiment has the probability density function $f(x) = 6(x - x^2)$ and outcomes lying between 0 and 1. Determine the probability that an outcome

   (a) lies between $\frac{1}{4}$ and $\frac{1}{2}$ (b) lies between 0 and $\frac{1}{3}$

   (c) is at least $\frac{1}{4}$ (d) is at most $\frac{3}{4}$

2. Suppose that the outcome X of an experiment lies between 0 and 4, and the probability density function for X is $f(x) = \frac{1}{8}x$. Find.

   (a) $\Pr(X \leq 1)$ (b) $\Pr(2 \leq X \leq 2.5)$ (c) $\Pr(3.5 \leq X)$

3. If $f(x) = kx^2$, determine the value of k that makes $f(x)$ a probability density function on $0 \leq x \leq 2$.

4. If $f(x) = k/\sqrt{x}$, determine the value of k that makes $f(x)$ a probability density function on $1 \leq x \leq 4$.

5. Suppose that the outcomes X of an experiment lie between 0 and ∞, and X has an exponential density function $f(x) = 2e^{-2x}$. Find.

   (a) $\Pr(X \leq .1)$ (b) $\Pr(.1 \leq X \leq .5)$ (c) $\Pr(1 \leq X)$ (d) the average value of X

6. Suppose that the outcomes X of an experiment are exponentially distributed, with density function $f(x) = .25e^{-.25x}$. Find.

   (a) $\Pr(1 \leq X \leq 2)$ (b) $\Pr(X \leq 3)$ (c) $\Pr(4 \leq X)$ (d) the average value of X

7. An automated machine produces an automobile part every 3 minutes. An inspector arrives at a random time and must wait X minutes for a part.

   (a) Find the probability density function for X.

   (b) Find the probability that the inspector must wait at least 1 minute.

   (c) Find the probability that the inspector must wait no more than 1 minute.

8. The annual income of the households in a certain community ranges between 5 and 25 thousand dollars. Let X represent the annual income (in thousands of dollars) of a household chosen at random in this community, and suppose that the probability density function for X is $f(x) = kx$, $5 \leq x \leq 25$.

   (a) Find the value of k that makes $f(x)$ a density function.

   (b) Find the fraction of the households that have an annual income between 5 and 10 thousand dollars.

   (c) What fraction of the households have an income exceeding $20,000?

9. Suppose that in a certain farming region, and in a certain year, the number X of bushels of wheat produced on a given acre has a probability density function $f(x) = (x - 30)/50$, $30 \leq x \leq 40$.

   (a) What is the probability that an acre selected at random produced less than 35 bushels of wheat?

   (b) If the farming region had 20,000 acres of wheat, how many acres produced less than 35 bushels of wheat?

10. The parent corporation for a franchised chain of fast-food restaurants claims that the fraction X of their new restaurants that make a profit during their first year of operation has the probability density $f(x) = 12x^2 - 12x^3$, $0 \leq x \leq 1$.

    (a) What is the likelihood that less than 40% of the restaurants opened this year will make a profit during their first year of operation?

    (b) What is the likelihood that more than 50% of the restaurants will make a profit during their first year of operation?

11. Suppose that at a certain supermarket, the amount of time X one must wait at the express lane has the probability density function $f(x) = \frac{11}{10}(x + 1)^{-2}$, $0 \leq x \leq 10$. Find the probability of having to wait less than 4 minutes at the express lane.

12. Suppose that in a certain cell population, cells divide every 10 days and the age X of a cell selected at random has the probability density function $f(x) = 2ke^{-kx}$, $0 \leq x \leq 10$, where $k = (.1) \ln 2$.

    (a) Find the probability that a cell is at most 5 days old.

    (b) Upon examining a slide, a microbiologist finds that 10% of the cells are undergoing mitosis (a change in the cell leading to division). Compute the length of time required for mitosis; that is, find the number M such that

    $$\int_{10-M}^{10} 2ke^{-kx}\,dx = .10.$$

13. At a certain gas station, it takes an average of 4 minutes to get serviced. Suppose that the service time X for a car has an exponential probability density function.

(a) What fraction of the cars are serviced within 2 minutes?

(b) What is the probability that a car will have to wait at least 4 minutes?

14. The emergency flasher on an automobile is guaranteed for the first 12,000 miles that the car is driven. On the average, the flashers last about 50,000 miles. Let X be the time of failure of the flasher (measured in thousands of miles), and suppose X has an exponential probability density function. What percentage of the emergency flashers will have to be replaced during the warranty period?

15. Let X be the number of seconds between successive cars at a toll booth on the Ohio Turnpike on a typical Saturday afternoon. It can be shown that X has an exponential density function. If the average interarrival time is 2 seconds, find the probability that X is at least 3 seconds.

16. Let X be the relief time (in minutes) of an arthritic patient who has taken an analgesic for pain. Suppose that a certain analgesic provides relief within 4 minutes for 75% of a large group of patients, and suppose the density function for X is $f(x) = ke^{-kx}$. (This model has been used by some medical researchers.) Then one estimates that $\Pr(X \le 4) = .75$. Use this estimate to find an approximate value for k. [*Hint:* First show that $\Pr(X \le 4) = 1 - e^{-4k}$.]

SOLUTIONS TO PRACTICE PROBLEMS 7

1. Property II is not satisfied:

$$\int_0^1 f(x)\,dx = \int_0^1 x\,dx = \frac{x^2}{2}\Big|_0^1 = \frac{1}{2}.$$

Property II says that this integral should be 1.

2. $\int_{1/2}^1 4x^3\,dx = x^4\Big|_{1/2}^1 = 1 - \frac{1}{16} = \frac{15}{16}.$

6.8 Integral Tables

Extensive tables of integrals have been compiled to assist in the determination of complicated integrals. Table 3 in the Appendix gives a brief sample of the types of integrals that are found in tables.* We shall use Table 3 to illustrate some techniques of using an integral table.

Table 3 is organized like the more complete tables, with the integrals grouped into categories by the type of function that appears in the integrand. For instance, there are "Forms Involving $ax + b$," "Forms Involving $\sqrt{a^2 - x^2}$," and so on. When an integrand involves more than one type of function, it may be necessary to look through two or more categories in order to find the appropriate formula.

_____

* A comprehensive table may be found in *C.R.C. Standard Mathematical Tables*, 16th ed., Chemical Rubber Publishing Company, Cleveland, Ohio, 1981.

EXAMPLE 1   Use Table 3 to determine $\int \dfrac{1}{\sqrt{x^2 + 36}}\, dx.$

Solution   The integrand involves $\sqrt{x^2 + 36}$, which has the general form $\sqrt{x^2 + a^2}$. In Table 3, in the category "Forms Involving $\sqrt{x^2 \pm a^2}$," we find

$$8. \quad \int \frac{1}{\sqrt{x^2 \pm a^2}}\, dx = \ln|x + \sqrt{x^2 \pm a^2}| + C.$$

This formula does double duty since it relates to integrands of the form $\dfrac{1}{\sqrt{x^2 + a^2}}$

and $\dfrac{1}{\sqrt{x^2 - a^2}}$. Using the upper sign of the "$\pm$" signs, with $a = 6$, we obtain

$$\int \frac{1}{\sqrt{x^2 + 36}}\, dx = \ln|x + \sqrt{x^2 + 36}| + C.$$

EXAMPLE 2   Use Table 3 to determine $\int xe^{-x}\, dx.$

Solution   The category "Forms Involving e^x, $\ln x$" contains the formula

$$11. \quad \int x^m e^{kx}\, dx = \frac{x^m e^{kx}}{k} - \frac{m}{k} \int x^{m-1} e^{kx}\, dx.$$

Set $m = 1$ and $k = -1$ to obtain

$$\int xe^{-x}\, dx = -xe^{-x} + \int e^{-x}\, dx.$$

We know, of course, that

$$\int e^{-x}\, dx = -e^{-x} + C.$$

Hence

$$\int xe^{-x}\, dx = -xe^{-x} - e^{-x} + C.$$

Formulas of the type that appear in Example 2 are called *reduction formulas*. The integral is reduced to the sum of a function and a simpler integral (often, as here, involving a smaller exponent.) Sometimes reduction formulas must be applied several times before the integral on the right can be easily derived.

EXAMPLE 3   Use Table 3 to determine $\int \dfrac{e^{2x}}{\sqrt{4e^x + 3}}\, dx.$

Solution   This integral does not appear in Table 3. However, the substitution $u = e^x$, $du = e^x\, dx$ converts it into one that is in the table.

$$\int \frac{e^{2x}}{\sqrt{4e^x + 3}}\, dx = \int \frac{e^x}{\sqrt{4e^x + 3}} \cdot e^x\, dx = \int \frac{u}{\sqrt{4u + 3}}\, du.$$

The category "Forms Involving $\sqrt{ax + b}$" contains the formula,

$$\textbf{5.} \quad \int \frac{x}{\sqrt{ax + b}}\, dx = \frac{2ax - 4b}{3a^2} \sqrt{ax + b} + C.$$

With $a = 4$, $b = 3$, and u in place of x, we obtain

$$\int \frac{u}{\sqrt{4u + 3}}\, du = \frac{2(4)u - 4(3)}{3(16)} \sqrt{4u + 3} + C$$

$$= \frac{2u - 3}{12} \sqrt{4u + 3} + C.$$

Hence

$$\int \frac{e^{2x}}{\sqrt{4e^x + 3}}\, dx = \frac{2e^x - 3}{12} \sqrt{4e^x + 3} + C.$$

Sometimes more than one formula is needed to find an antiderivative.

EXAMPLE 4   Use Table 3 to determine

$$\int \frac{x + 10}{x^2 - 25}\, dx.$$

Solution   Observe that

$$x^2 - 25 = (x - 5)(x + 5) = (ax + b)(cx + d),$$

where $a = c = 1$, $b = -5$, and $d = 5$. Expressions of this type are found in formulas 3 and 4:

$$\textbf{3.} \quad \int \frac{1}{(ax + b)(cx + d)}\, dx = \frac{1}{ad - bc} \ln \left| \frac{ax + b}{cx + d} \right| + C$$

$$\textbf{4.} \quad \int \frac{x}{(ax + b)(cx + d)}\, dx = \frac{1}{ad - bc} \left[\frac{d}{c} \ln|cx + d| - \frac{b}{a} \ln|ax + b| \right] + C.$$

In these formulas we use $ad - bc = (1)(5) - (-5)(1) = 5 + 5 = 10$,

$$\int \frac{x}{(x - 5)(x + 5)}\, dx = \frac{1}{10} \left[\frac{5}{1} \ln|x + 5| - \frac{-5}{1} \ln|x - 5| \right] + C$$

$$= \frac{1}{2} \ln|x + 5| + \frac{1}{2} \ln|x - 5| + C,$$

and

$$\int \frac{1}{(x - 5)(x + 5)}\, dx = \frac{1}{10} \ln \left| \frac{x - 5}{x + 5} \right| + C$$

$$= \frac{1}{10} \ln|x - 5| - \frac{1}{10} \ln|x + 5| + C.$$

Combining these results, we have

$$\int \frac{x+10}{x^2-25}\, dx = \int \frac{x}{x^2-25}\, dx + 10 \int \frac{1}{x^2-25}\, dx$$

$$= \frac{1}{2} \ln|x+5| + \frac{1}{2} \ln|x-5|$$

$$+ 10\left[\frac{1}{10} \ln|x-5| - \frac{1}{10} \ln|x+5|\right] + C$$

$$= \frac{3}{2} \ln|x-5| - \frac{1}{2} \ln|x+5| + C.$$

Other integral tables may use different notation than used in Table 3. For instance, $\ln x$ might be written $\log(x)$, u might be used instead of x, and the "$+C$" might be missing from the end of the formulas. Also, the "dx" might be moved into the numerator of a quotient. For example, a table might list

$$\int \frac{dx}{\sqrt{x^2 \pm a^2}} \quad \text{instead of} \quad \int \frac{1}{\sqrt{x^2 \pm a^2}}\, dx.$$

Standard integral tables contain functions that we have not studied; primarily the trigonometric functions ($\sin x$, $\cos x$, $\tan x$, etc. . . .), the inverse trigonometric functions ($\sin^{-1} x$, $\cos^{-1} x$, $\tan^{-1} x$, etc.), and the hyperbolic functions ($\sinh x$, $\cosh x$, $\tanh x$, etc.). The hyperbolic functions can be defined in terms of exponential functions. For example, $\sinh x = (e^x - e^{-x})/2$. These functions, along with the trigonometric and inverse trigonometric functions, are treated in engineering calculus texts.

PRACTICE PROBLEMS 8

1. Use Table 3 to determine $\int 2\sqrt{x^2-5}\, dx$.
    [Hint: Any positive number can be written in the form a^2 for some suitable choice of a.]

2. Use Table 3 to determine $\int \frac{1}{x\sqrt{1-4x^2}}\, dx$

EXERCISES 8

In Exercises 1–6, find the appropriate formula in Table 3 and give the values of the constants a and b.

1. $\int \sqrt{x^2-25}\, dx$
2. $\int \frac{1}{x\sqrt{3x+1}}\, dx$
3. $\int \frac{x}{5-2x}\, dx$

4. $\int \frac{1}{3x^2+4x}\, dx$
5. $\int \frac{\sqrt{8-x^2}}{x}\, dx$
6. $\int \frac{1}{\sqrt{x^2+5}}\, dx$

In Exercises 7–36, use Table 3 to determine the antiderivative.

7. $\displaystyle\int \frac{1}{x\sqrt{9-x^2}}\,dx$

8. $\displaystyle\int \frac{x}{3x-2}\,dx$

9. $\displaystyle\int \frac{1}{4-e^{2x}}\,dx$

10. $\displaystyle\int \frac{\sqrt{16-x^2}}{x}\,dx$

11. $\displaystyle\int \frac{x}{\sqrt{x+5}}\,dx$

12. $\displaystyle\int \sqrt{x^2+15}\,dx$

13. $\displaystyle\int \frac{1}{x^2-3x-4}\,dx$

14. $\displaystyle\int \frac{1}{e^x-3}\,dx$

15. $\displaystyle\int \frac{1}{x^2-x}\,dx$

16. $\displaystyle\int x^5 \ln 2x\,dx$

17. $\displaystyle\int \frac{4}{x\sqrt{3-2x}}\,dx$

18. $\displaystyle\int \frac{x}{5\sqrt{5x-1}}\,dx$

19. $\displaystyle\int x^2 e^{3x}\,dx$

20. $\displaystyle\int \frac{1}{(x+2)(x+3)}\,dx$

21. $\displaystyle\int \sqrt{(3x)^2+1}\,dx$

22. $\displaystyle\int \frac{\sqrt{1-(x^2/4)}}{x}\,dx$

23. $\displaystyle\int \frac{1}{\sqrt{3x^2-1}}\,dx$

24. $\displaystyle\int \frac{x^2}{e^{x^3}-1}\,dx$

25. $\displaystyle\int \frac{x^3}{\sqrt{1+x^8}}\,dx$

26. $\displaystyle\int x^2 \ln(x^3-5)\,dx$

27. $\displaystyle\int \frac{x}{x^2+3x-10}\,dx$

28. $\displaystyle\int \frac{1}{x\sqrt{1-x^6}}\,dx$

29. $\displaystyle\int \frac{\ln x}{x\sqrt{1+5\ln x}}\,dx$

30. $\displaystyle\int x\sqrt{e^x}\,dx$

SOLUTIONS TO PRACTICE PROBLEMS 8

1. Think of 5 as $(\sqrt{5})^2$ and note that $\sqrt{x^2-5}=\sqrt{x^2-a^2}$ where $a=\sqrt{5}$. In the category "Forms Involving $\sqrt{x^2 \pm a^2}$," we find

$$7.\quad \int \sqrt{x^2 \pm a^2}\,dx = \frac{x}{2}\sqrt{x^2 \pm a^2} \pm \frac{a^2}{2}\ln\left|x+\sqrt{x^2 \pm a^2}\right| + C.$$

Using the lower signs of the "$\pm$" signs, we obtain

$$\int 2\sqrt{x^2-5}\,dx = 2\int \sqrt{x^2-5}\,dx$$

$$= 2\left[\frac{x}{2}\sqrt{x^2-5} - \frac{5}{2}\ln\left|x+\sqrt{x^2-5}\right|\right] + C$$

$$= x\sqrt{x^2-5} - 5\ln\left|x+\sqrt{x^2-5}\right| + C.$$

2. The given integral seems related to the formula

$$9.\quad \int \frac{1}{x\sqrt{a^2-x^2}}\,dx = -\frac{1}{a}\ln\left|\frac{a+\sqrt{a^2-x^2}}{x}\right| + C.$$

To use this formula we need the coefficient of x^2 to be 1 instead of 4. Note that

$$\sqrt{1-4x^2} = \sqrt{4(\tfrac{1}{4}-x^2)} = 2\sqrt{\tfrac{1}{4}-x^2} = 2\sqrt{a^2-x^2},$$

where $a = \frac{1}{2}$. Hence

$$\int \frac{1}{x\sqrt{1-4x^2}}\,dx = \int \frac{1}{2x\sqrt{\frac{1}{4}-x^2}}\,dx = \frac{1}{2}\int \frac{1}{x\sqrt{\frac{1}{4}-x^2}}\,dx$$

$$= \frac{1}{2}\left[-\frac{1}{\frac{1}{2}}\ln\left|\frac{\frac{1}{2}+\sqrt{\frac{1}{4}-x^2}}{x}\right|\right] + C$$

$$= -\ln\left|\frac{1+\sqrt{1-4x^2}}{2x}\right| + C.$$

In the last step, we multiplied the numerator and the denominator of the quotient inside the absolute value sign by 2.

Another way to solve this problem is to note that $\sqrt{1-4x^2} = \sqrt{1-(2x)^2}$, and then make the substitution $u = 2x$, $du = 2\,dx$. Then $\sqrt{1-4x^2} = \sqrt{1-u^2}$, and

$$\int \frac{1}{x\sqrt{1-4x^2}}\,dx = \int \frac{1}{2x\sqrt{1-4x^2}}\cdot 2\,dx = \int \frac{1}{u\sqrt{1-u^2}}\,du.$$

Using formula 9 displayed above, with $a = 1$ and u in place of x, we find that

$$\int \frac{1}{u\sqrt{1-u^2}}\,du = -\ln\left|\frac{1+\sqrt{1-u^2}}{u}\right| + C.$$

Hence

$$\int \frac{1}{x\sqrt{1-4x^2}}\,dx = -\ln\left|\frac{1+\sqrt{1-4x^2}}{2x}\right| + C.$$

Chapter 6: CHECKLIST

☐ Antiderivative

☐ Indefinite integral

☐ $\int x^r\,dx = \dfrac{1}{r+1}x^{r+1} + C, \qquad r \neq -1$

☐ $\int e^{kx}\,dx = \dfrac{1}{k}e^{kx} + C, \qquad k \neq 0$

☐ $\int \dfrac{1}{x}\,dx = \ln|x| + C, \qquad x \neq 0$

☐ $\int [f(x) + g(x)]\,dx = \int f(x)\,dx + \int g(x)\,dx$

☐ $\int kf(x)\,dx = k\int f(x)\,dx$

☐ Net change: $F(x)\big|_a^b$

☐ $\int_a^b f(x)\,dx = F(b) - F(a),\ F'(x) = f(x)$

☐ Fundamental Theorem of Calculus

- Area between two curves: $\int_a^b [f(x) - g(x)]\, dx$

- Riemann sum approximation to $\int_a^b f(x)\, dx$

- Rectangle rule

- Average value of $f(x)$: $\dfrac{1}{b-a} \int_a^b f(x)\, dx$

- $\int f(g(x))g'(x)\, dx = \int f(u)\, du, \qquad u = g(x)$

- $\int f(x)g(x)\, dx = f(x)G(x) - \int f'(x)G(x)\, dx, \qquad G'(x) = g(x)$

- Improper integral (convergent and divergent)
- Probability density function
- Uniform density function
- Exponential density function
- Integral tables

Chapter 6: SUPPLEMENTARY EXERCISES

Calculate each of the following indefinite integrals.

1. $\int e^{-x/2}\, dx$

2. $\int \dfrac{5}{\sqrt{x-7}}\, dx$

3. $\int \dfrac{x^2 - 1}{(x^3 - 3x + 2)^2}\, dx$

4. $\int x^4 e^{-x^5}\, dx$

5. $\int (2x + 3)^7\, dx$

6. $\int \left(9 - 4e^x + \dfrac{1}{x^4}\right) dx$

7. $\int (e^x + 4)^3 \cdot e^x\, dx$

8. $\int \dfrac{x - e^{-2x}}{x^2 + e^{-2x}}\, dx$

9. $\int (2x + 1)e^{-x/2}\, dx$

10. $\int \dfrac{5x}{\sqrt{x-7}}\, dx$

11. $\int \dfrac{x^2 - 1}{x^3 - 3x + 2}\, dx$

12. $\int x^3 e^{x^2}\, dx$

13. $\int x(2x + 3)^7\, dx$

14. $\int x^{-2} \ln x\, dx$

15. Find the function $f(x)$ for which $f'(x) = (x - 5)^2$, $f(8) = 2$.

16. Find the function $f(x)$ for which $f'(x) = e^{-5x}$, $f(0) = 1$.

Calculate the following definite integrals.

17. $\int_1^4 \dfrac{1}{x^2}\, dx$

18. $\int_3^6 e^{2 - (x/3)}\, dx$

19. $\int_{-1}^1 x^4 e^{x^5 - 1}\, dx$

20. $\int_0^5 (5 + 3x)^{-1}\, dx$

21. $\int_0^1 \dfrac{e^x - e^{-x}}{e^x + e^{-x}}\,dx$

22. $\int_2^3 \dfrac{x - 1}{(x^2 - 2x + 1)^2}\,dx$

23. Find the area under the curve $y = 1 + \sqrt{x}$ from $x = 1$ to $x = 9$.

24. Find the area under the curve $y = (3x - 2)^{-3}$ from $x = 1$ to $x = 2$.

25. Find the area of the region bounded by the curves $y = 16 - x^2$ and $y = 10 - x$.

26. Find the area of the region bounded by the curves $y = x^3 - 3x + 1$ and $y = x + 1$.

27. Find the area of the region between the curves $y = 5x + (1/x)$ and $y = 2x + (1/x)$ from $x = 2$ to $x = 3$.

28. Find the area of the region between the curves $y = 2x^2 + x$ and $y = x^2 + 2$ from $x = 0$ to $x = 2$.

29. Use the rectangle rule with $n = 2$ to approximate $\int_2^4 \dfrac{1}{x + 2}\,dx$. Then find the exact value to five decimal places.

30. Use the rectangle rule with $n = 5$ to approximate $\int_0^1 e^{x^2}\,dx$.

31. Find the consumers' surplus for the demand curve $p = \sqrt{25 - .04x}$ at the sales level $x = 400$.

32. Three thousand dollars is deposited in the bank at 6% interest compounded continuously. What will be the average value of the money in the account during the next 10 years?

33. Find the average value of $f(x) = 1/x^3$ from $x = \tfrac{1}{3}$ to $x = \tfrac{1}{2}$.

34. Suppose that the interval $0 \le x \le 1$ is divided into 100 subintervals of width $\Delta x = .01$. Show that the sum $[3e^{-.01}]\Delta x + [3e^{-.02}]\Delta x + [3e^{-.03}]\Delta x + \cdots + [3e^{-1}]\Delta x$ is close to $3(1 - e^{-1})$.

35. An airplane tire plant finds that its marginal cost of producing tires is $.04x + 150$ dollars at a production level of x tires per day. If fixed costs are \$500 per day, find the cost of producing x tires per day.

36. Suppose the marginal revenue function for a company is $400 - 3x^2$. Find the additional revenue received from doubling production if currently 10 units are being produced.

37. Let $y = f(t)$ be a function such that $y' = kty$ for some constant k. Show that $y = Ce^{kt^2/2}$ for some constant C. [*Hint:* Use the product rule to evaluate $\dfrac{d}{dt}[f(t)e^{-kt^2/2}]$, and then apply Theorem II of Section 1.]

38. The rate of production of copper ore in North America has been $2.6e^{.034t}$ million metric tons per year at time t, where $t = 0$ corresponds to 1975. Find a formula for the total production from 1980 to time t.

39. The known U.S. natural gas reserves are decreasing at the rate of $R(t)$ quadrillion Btu per year at time t, where $t = 0$ corresponds to 1980 and $R(t) = 22.3e^{-.14t}$.

  (a) If the known reserves in 1980 are 159 quadrillion Btu, what will be the reserves in 1986, assuming no new gas sources are discovered?

  (b) Compute $\int_0^{\infty} R(t)\,dt$ and interpret this quantity.

40. The United States has been consuming sulfur at the rate of $R(t)$ million long tons per year at time t, where $t = 0$ corresponds to 1980 and $R(t) = 12.4e^{.035t}$.

   (a) How much sulfur will be consumed in the United States between 1981 and 1985?

   (b) How much sulfur was consumed in the United States between 1975 and 1980?

41. A drug is injected into a patient at the rate of $f(t)$ cubic centimeters per minute at time t. What does the area under the graph of $y = f(t)$ from $t = 0$ to $t = 4$ represent?

42. Find the area under the graph of $y = e^x/(1 + e^x)^2$ for $x \geq 0$.

Evaluate the following improper integrals whenever they are convergent.

43. $\displaystyle\int_0^\infty e^{6-3x}\,dx$

44. $\displaystyle\int_1^\infty x^{-2/3}\,dx$

45. $\displaystyle\int_{-\infty}^0 \frac{8}{(5-2x)^3}\,dx$

46. $\displaystyle\int_0^\infty x^2 e^{-x^3}\,dx$

47. Show that the region under the graph of $y = x^{-1/4}$ for $x \geq 1$ cannot be assigned any finite number as its area.

48. Can the region under the graph of $y = 1/x$ for $x \geq 1$ be assigned a finite number as its area?

49. Suppose that the average lifetime of an electronic component is 72 months, and the lifetimes are exponentially distributed.

   (a) Find the probability that a component lasts for more than 24 months.

   (b) The *reliability function* $r(t)$ gives the probability that a component will last for more than t months. Compute $r(t)$ in this case.

50. Verify that for any number A, the function $f(x) = e^{A-x}$, $x \geq A$, is a probability density function.

Functions
of Several Variables

Chapter 7

7.1. Examples of Functions of Several Variables

In Chapters 1 through 6 we were concerned with calculus for functions of one variable. In the present chapter we show briefly how calculus can be extended to functions of several variables.

A function $f(x, y)$ of the two variables x and y is a rule that associates to each pair of values for the variables a number. Examples of functions of two variables are

$$f(x, y) = x^2 + 2y$$

$$g(x, y) = e^x(x + y).$$

An example of a function of three variables is

$$f(x, y, z) = 5xyz.$$

It is possible to consider functions of any number of variables.

EXAMPLE 1 Let $f(x, y) = xy + 5y - x^2$.

(a) Calculate $f(2, 3)$. That is, calculate the value of $f(x, y)$ when $x = 2$ and $y = 3$.

(b) Calculate $f(1, 0)$ and $f(0, 1)$.

(c) Calculate $f(x, y)$ when $(x, y) = (-4, 6)$—that is, when $x = -4$ and $y = 6$.

Solution (a) $f(2, 3) = 2 \cdot 3 + 5 \cdot 3 - (2)^2 = 17$.

(b) $f(1, 0) = 1 \cdot 0 + 5 \cdot 0 - 1^2 = -1$.
    $f(0, 1) = 0 \cdot 1 + 5 \cdot 1 - 0^2 = 5$.

(c) $f(-4, 6) = (-4) \cdot 6 + 5 \cdot 6 - (-4)^2 = -10$.

A function $f(x, y)$ of two variables may be graphed in a manner analogous to that for functions of one variable. It is necessary to use a three-dimensional coordinate system, where each point is identified by three coordinates (x, y, z). For each choice of x, y, the graph of $f(x, y)$ includes the point $(x, y, f(x, y))$. The graph of $f(x, y)$ is thus a surface in three-dimensional space. (See Fig. 1.) We will

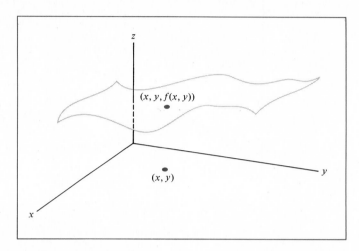

FIGURE 1   Graph of $f(x, y)$.

not employ graphical analysis to study $f(x, y)$ for two reasons. First, it is fairly difficult to draw good graphs in three-dimensional space. This fact makes graphical analysis inconvenient for those not artistically inclined. (See Fig. 2.*) Second, and more significantly, however, we will not use graphical analysis because it is a tool that exists only for functions of one or two variables. It is not possible to graph a function of three or more variables.

Throughout this book we have shown how functions of one variable arise in applications. Functions of several variables are equally important. In Example 2 we present an application to architectural design and in Example 3 an application to economics. These two examples will recur throughout the chapter to illustrate the various techniques that will be developed.

In designing a building, it is necessary to know, at least approximately, how much heat the building loses per day. The heat loss affects many aspects of the design, such as the size of the heating plant, the size and location of duct work, and so on. A building loses heat through its sides, roof, and floor. How much heat is lost will generally differ for each face of the building and will depend on such factors as insulation, materials used in construction, exposure (north, south, east, or west), and climate. It is possible to estimate how much heat is lost per square foot of each face. Using this data, one can construct a heat-loss function as in the following example.

_____

* These graphs were drawn by Norton Starr at the University of Waterloo Computing Centre.

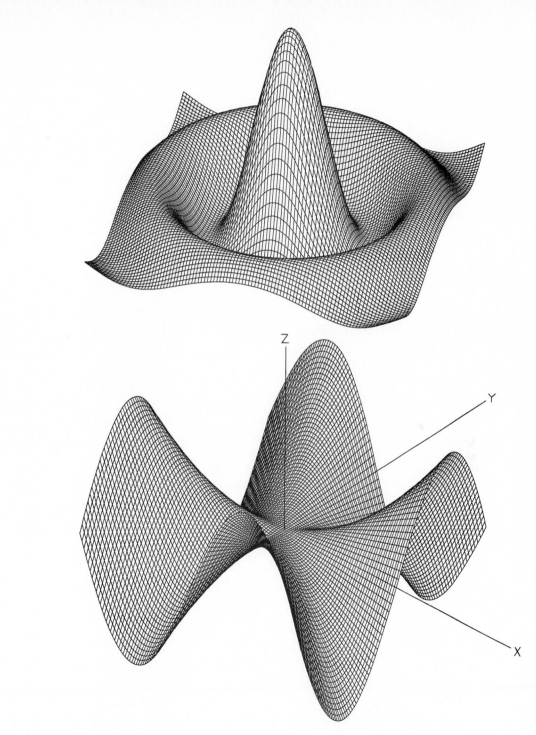

FIGURE 2

EXAMPLE 2   A rectangular industrial building of dimensions x, y, and z is shown in Fig. 3(a). In Fig. 3(b) we give the amount of heat lost per day by each side of the building, measured in suitable units of heat per square foot. Let $f(x, y, z)$ be the total daily heat loss for such a building.

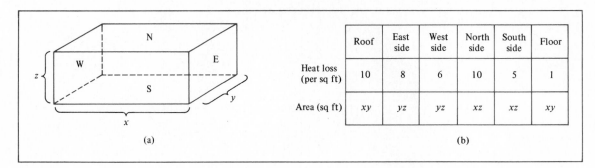

	Roof	East side	West side	North side	South side	Floor
Heat loss (per sq ft)	10	8	6	10	5	1
Area (sq ft)	xy	yz	yz	xz	xz	xy

(a)   (b)

FIGURE 3

(a) Find a formula for $f(x, y, z)$.

(b) Find the total daily heat loss if the building has length 100 feet, width 70 feet, and height 50 feet.

Solution   (a) The total heat loss is the sum of the amount of heat loss through each side of the building. The heat loss through the roof is

[heat loss per square foot of roof] · [area of roof in square feet] $= 10xy.$

Similarly, the heat loss through the east side is $8yz$. Continuing in this way, we see that the total daily heat loss is

$$f(x, y, z) = 10xy + 8yz + 6yz + 10xz + 5xz + 1 \cdot xy.$$

We collect terms to obtain

$$f(x, y, z) = 11xy + 14yz + 15xz.$$

(b) The amount of heat loss when $x = 100$, $y = 70$, and $z = 50$ is given by $f(100, 70, 50)$, which equals

$$f(100, 70, 50) = 11(100)(70) + 14(70)(50) + 15(100)(50)$$

$$= 77{,}000 + 49{,}000 + 75{,}000 = 201{,}000.$$

In Section 3 we will determine the dimensions x, y, z that minimize the heat loss for a building of specified volume.

The costs of a manufacturing process can generally be classified as one of two types: cost of labor and cost of capital. The meaning of the cost of labor is clear. By the cost of capital, we mean the cost of buildings, tools, machines, and similar items used in the production process. A manufacturer usually has some control over the relative portions of labor and capital utilized in his production process. He can completely automate production so that labor is at a minimum, or he can utilize mostly labor and little capital. Suppose that x units of labor and y units of capital are used.* Let $f(x, y)$ denote the number of units of finished product that are manufactured. Economists have found that $f(x, y)$ is often a function of the form

$$f(x, y) = Cx^A y^{1-A},$$

where A and C are constants, $0 < A < 1$. Such a function is called a *Cobb-Douglas production function*.

EXAMPLE 3   (*Production in a Firm*)   Suppose that during a certain time period the number of units of goods produced when utilizing x units of labor and y units of capital is $f(x, y) = 60x^{3/4}y^{1/4}$.

(a) How many units of goods will be produced by using 81 units of labor and 16 units of capital?

(b) Show that whenever the amounts of labor and capital being used are doubled, so is the production. (Economists say that the production function has "constant returns to scale.")

Solution   (a) $f(81, 16) = 60(81)^{3/4} \cdot (16)^{1/4} = 60 \cdot 27 \cdot 2 = 3240$. There will be 3240 units of goods produced.

(b) Utilization of a units of labor and b units of capital results in the production of $f(a, b) = 60a^{3/4}b^{1/4}$ units of goods. Utilizing $2a$ and $2b$ units of labor and capital, respectively, results in $f(2a, 2b)$ units produced. Set $x = 2a$ and $y = 2b$. Then we see that

$$\begin{aligned}
f(2a, 2b) &= 60(2a)^{3/4}(2b)^{1/4} \\
&= 60 \cdot 2^{3/4} \cdot a^{3/4} \cdot 2^{1/4} \cdot b^{1/4} \\
&= 60 \cdot 2^{(3/4 + 1/4)} \cdot a^{3/4}b^{1/4} \\
&= 2^1 \cdot 60a^{3/4}b^{1/4} \\
&= 2f(a, b).
\end{aligned}$$

Level Curves   It is possible graphically to depict a function $f(x, y)$ of two variables using a family of curves called level curves. Let c be any number. Then the graph of the equation $f(x, y) = c$ is a curve in the xy-plane called the *level curve of height c*. This curve describes all points of height c on the graph of the

---

     * Economists normally use L and K, respectively, for labor and capital. However, for simplicity, we use x and y.

function $f(x, y)$. As c varies, we have a family of level curves indicating the sets of points on which $f(x, y)$ assumes various values c. In Fig. 4, we have drawn the graph and various level curves for the function $f(x, y) = x^2 + y^2$.

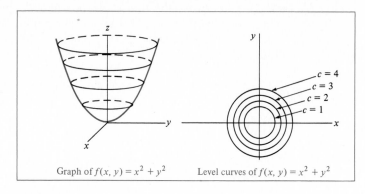

Graph of $f(x, y) = x^2 + y^2$ Level curves of $f(x, y) = x^2 + y^2$

FIGURE 4

Level curves often have interesting physical interpretations. For example, let (x, y) specify the coordinates of a point on the earth (x = latitude, y = longitude) and $f(x, y)$ the current temperature at location (x, y). Then the level curves of the function $f(x, y)$ indicate the locations having equal temperatures. Such curves are called *isotherms*. In Fig. 5 we have drawn several typical isotherms.

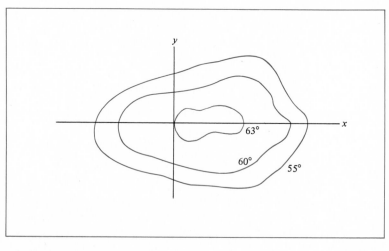

FIGURE 5

EXAMPLE 4   Determine the level curve at height 600 for the production function $f(x, y) = 60x^{3/4}y^{1/4}$ of Example 3.

Solution   The level curve is the graph of $f(x, y) = 600$, or

$$60x^{3/4}y^{1/4} = 600$$

$$y^{1/4} = \frac{10}{x^{3/4}}$$

$$y = \frac{10,000}{x^3}.$$

Of course, since x and y represent quantities of labor and capital, they must both be positive. We have sketched the graph of the level curve in Fig. 6. The points

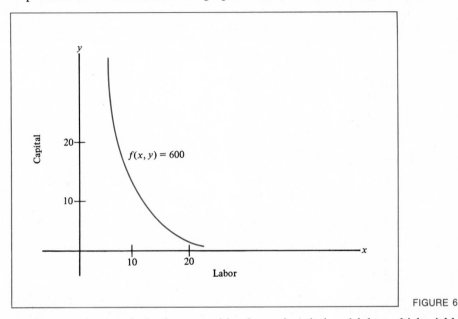

$f(x, y) = 600$

Capital

Labor

FIGURE 6

on the curve are precisely those combinations of capital and labor which yield 600 units of production. Since all points on the curve yield the same amount of production, the level curve is also called an *isoquant* ("iso" means equal, "quant" means "amount").

PRACTICE PROBLEMS 1

1. Let $f(x, y, z) = x^2 + y/(x - z) - 4$. Compute $f(3, 5, 2)$.

2. Suppose that in a certain country the daily demand for coffee is given by $f(p_1, p_2) = 16p_1/p_2$ thousand pounds, where p_1 and p_2 are the respective prices of tea and coffee per pound. Compute and interpret $f(3, 4)$.

EXERCISES 1

1. Let $f(x, y) = x^2 + 8y$. Compute $f(1, 0)$, $f(0, 1)$, and $f(3, 2)$.

2. Let $g(x, y) = 3xe^y$. Compute $g(2, 1)$, $g(1, 0)$, and $g(0, 0)$.

3. Let $f(L, K) = 3\sqrt{LK}$. Compute $f(0, 1)$, $f(3, 12)$, and $f(a, b)$.

4. Let $f(p, q) = pe^{q/p}$. Compute $f(1, 0)$, $f(3, 12)$, and $f(a, b)$.

5. Let $f(x, y, z) = x/(y - z)$. Compute $f(2, 3, 4)$ and $f(7, 46, 44)$.

6. Let $f(x, y, z) = x^2 e^{\sqrt{y^2 + z^2}}$. Compute $f(1, 0, 1)$ and $f(5, 2, 3)$.

7. Let $f(x, y) = xy$. Show that $f(2 + h, 3) - f(2, 3) = 3h$.

8. Let $f(x, y) = xy$. Show that $f(2, 3 + K) - f(2, 3) = 2K$.

9. Let $f(x, y) = \dfrac{x^2 + 3xy + 3y^2}{x + y}$. Show that $f(2a, 2b) = 2f(a, b)$.

10. Let $f(x, y) = 75x^A y^{1-A}$, where $0 < A < 1$. Show that $f(2a, 2b) = 2f(a, b)$.

11. The present value of A dollars to be paid t years in the future (assuming a 5% continuous interest rate) is $P(A, t) = Ae^{-.05t}$. Find and interpret $P(100, 13.8)$.

12. Refer to Example 3. Suppose that labor costs $100 per unit and capital costs $200 per unit. Express as a function of two variables, $C(x, y)$, the cost of utilizing x units of labor and y units of capital.

Draw the level curves of heights 0, 1, and 2 for each of the following functions.

13. $f(x, y) = 2x + y$ 14. $f(x, y) = -x^2 + y$

15. Suppose that a topographic map is viewed as the graph of a certain function $f(x, y)$. What are the level curves?

16. A certain production process uses labor and capital. If the quantities of these commodities are x and y, respectively, then the total cost is $100x + 200y$ dollars. Draw the level curves of height 600, 800, and 1000 for this function. Explain the significance of these curves. (Economists frequently refer to these lines as *budget lines* or *isocost lines*.)

SOLUTIONS TO PRACTICE PROBLEMS 1

1. Substitute 3 for x, 5 for y, and 2 for z.

$$f(3, 5, 2) = 3^2 + \frac{5}{3 - 2} - 4 = 10.$$

2. To compute $f(3, 4)$, substitute 3 for p_1 and 4 for p_2 into $f(p_1, p_2) = 16p_1/p_2$. Thus

$$f(3, 4) = 16 \cdot \tfrac{3}{4} = 12.$$

Therefore, if the price of tea is $3 per pound and the price of coffee is $4 per pound, then 12,000 pounds of coffee will be sold each day. (Notice that as the price of coffee increases, the demand decreases.)

7.2. Partial Derivatives

In Chapter 1 we introduced the notion of a derivative to measure the rate at which a function $f(x)$ is changing with respect to changes in the variable x. Let us now study the analog of the derivative for functions of two (or more) variables.

Let $f(x, y)$ be a function of the two variables x and y. Since we want to know how $f(x, y)$ changes both with respect to changes in the variable x and changes in the variable y, we shall define two derivatives of $f(x, y)$ (to be called "partial derivatives"), one with respect to each of the variables. The *partial derivative of* $f(x, y)$ *with respect to* x, written $\dfrac{\partial f}{\partial x}$, is the derivative of $f(x, y)$, where y is treated as a constant and $f(x, y)$ is considered as a function of x alone. The *partial derivative of* $f(x, y)$ *with respect to* y, written $\dfrac{\partial f}{\partial y}$, is the derivative of $f(x, y)$, where x is treated as a constant.

EXAMPLE 1 Let $f(x, y) = 5x^3y^2$. Compute $\dfrac{\partial f}{\partial x}$ and $\dfrac{\partial f}{\partial y}$.

Solution To compute $\dfrac{\partial f}{\partial x}$, we think of $f(x, y)$ written as

$$f(x, y) = [5y^2]x^3,$$

where the brackets emphasize that $5y^2$ is to be treated as a constant. Therefore, when differentiating with respect to x, $f(x, y)$ is just a constant times x^3. Recall that if k is any constant, then

$$\frac{d}{dx}(kx^3) = 3 \cdot k \cdot x^2.$$

Thus

$$\frac{\partial f}{\partial x} = 3 \cdot [5y^2] \cdot x^2 = 15x^2y^2.$$

After some practice, it is unnecessary to place the y^2 in front of the x^3 before differentiating.

Now, in order to compute $\dfrac{\partial f}{\partial y}$, we think of

$$f(x, y) = [5x^3]y^2.$$

When differentiating with respect to y, $f(x, y)$ is simply a constant (namely $5x^3$) times y^2. Hence

$$\frac{\partial f}{\partial y} = 2 \cdot [5x^3] \cdot y = 10x^3y.$$

EXAMPLE 2 Let $f(x, y) = 3x^2 + 2xy + 5y$. Compute $\dfrac{\partial f}{\partial x}$ and $\dfrac{\partial f}{\partial y}$.

Solution To compute $\dfrac{\partial f}{\partial x}$, we think of

$$f(x, y) = 3x^2 + [2y]x + [5y].$$

Now we differentiate $f(x, y)$ as if it were a quadratic polynomial in x:

$$\frac{\partial f}{\partial x} = 6x + [2y] + 0 = 6x + 2y.$$

Note that $5y$ is treated as a constant when differentiating with respect to x, so the partial derivative of $5y$ with respect to x is zero.

To compute $\frac{\partial f}{\partial y}$, we think of

$$f(x, y) = [3x^2] + [2x]y + 5y.$$

Then

$$\frac{\partial f}{\partial y} = 0 + [2x] + 5 = 2x + 5.$$

Note that $3x^2$ is treated as a constant when differentiating with respect to y, so the partial derivative of $3x^2$ with respect to y is zero.

EXAMPLE 3 Compute $\frac{\partial f}{\partial x}$ and $\frac{\partial f}{\partial y}$ for each of the following.

(a) $f(x, y) = (4x + 3y - 5)^8$

(b) $f(x, y) = e^{xy^2}$

(c) $f(x, y) = y/(x + 3y)$

Solution (a) To compute $\frac{\partial f}{\partial x}$, we think of

$$f(x, y) = (4x + [3y - 5])^8.$$

By the general power rule,

$$\frac{\partial f}{\partial x} = 8 \cdot (4x + [3y - 5])^7 \cdot 4 = 32(4x + 3y - 5)^7.$$

Here we used the fact that the derivative of $4x + 3y - 5$ with respect to x is just 4.

To compute $\frac{\partial f}{\partial y}$, we think of

$$f(x, y) = ([4x] + 3y - 5)^8.$$

Then

$$\frac{\partial f}{\partial y} = 8 \cdot ([4x] + 3y - 5)^7 \cdot 3 = 24(4x + 3y - 5)^7.$$

(b) To compute $\frac{\partial f}{\partial x}$, we observe that

$$f(x, y) = e^{x[y^2]},$$

so that

$$\frac{\partial f}{\partial x} = [y^2]e^{x[y^2]} = y^2 e^{xy^2}.$$

To compute $\dfrac{\partial f}{\partial y}$, we think of

$$f(x, y) = e^{[x]y^2}.$$

Thus

$$\frac{\partial f}{\partial y} = e^{[x]y^2} \cdot 2[x]y = 2xye^{xy^2}.$$

(c) To compute $\dfrac{\partial f}{\partial x}$, we use the general power rule to differentiate $[y](x + [3y])^{-1}$
with respect to x:

$$\frac{\partial f}{\partial x} = (-1) \cdot [y](x + [3y])^{-2} \cdot 1 = -\frac{y}{(x + 3y)^2}.$$

To compute $\dfrac{\partial f}{\partial y}$, we use the quotient rule to differentiate

$$f(x, y) = \frac{y}{[x] + 3y}$$

with respect to y. We find that

$$\frac{\partial f}{\partial y} = \frac{([x] + 3y) \cdot 1 - y \cdot 3}{([x] + 3y)^2} = \frac{x}{(x + 3y)^2}.$$

The use of brackets to highlight constants is helpful initially in order to compute partial derivatives. From now on we shall merely form a mental picture of those terms to be treated as constants and dispense with brackets.

A partial derivative of a function of several variables is also a function of several variables and hence can be evaluated at specific values of the variables. We write

$$\frac{\partial f}{\partial x}(a, b)$$

for $\dfrac{\partial f}{\partial x}$ evaluated at $x = a$, $y = b$. Similarly,

$$\frac{\partial f}{\partial y}(a, b)$$

denotes the function $\dfrac{\partial f}{\partial y}$ evaluated at $x = a$, $y = b$.

EXAMPLE 4 Let $f(x, y) = 3x^2 + 2xy + 5y$.

(a) Calculate $\dfrac{\partial f}{\partial x}(1, 4)$.

(b) Evaluate $\dfrac{\partial f}{\partial y}$ at $(x, y) = (1, 4)$.

Solution (a) $\dfrac{\partial f}{\partial x} = 6x + 2y, \dfrac{\partial f}{\partial x}(1, 4) = 6 \cdot 1 + 2 \cdot 4 = 14$.

(b) $\dfrac{\partial f}{\partial y} = 2x + 5, \dfrac{\partial f}{\partial y}(1, 4) = 2 \cdot 1 + 5 = 7$.

Since $\dfrac{\partial f}{\partial x}$ is simply the ordinary derivative with y held constant, $\dfrac{\partial f}{\partial x}$ gives the rate of change of $f(x, y)$ with respect to x for y held constant. In other words, keeping y constant and increasing x by one (small) unit produces a change in $f(x, y)$ that is approximately given by $\dfrac{\partial f}{\partial x}$. An analogous interpretation holds for $\dfrac{\partial f}{\partial y}$.

EXAMPLE 5 Interpret the partial derivatives of $f(x, y) = 3x^2 + 2xy + 5y$ calculated in Example 4.

Solution We showed in Example 4 that

$$\frac{\partial f}{\partial x}(1, 4) = 14, \qquad \frac{\partial f}{\partial y}(1, 4) = 7.$$

The fact that $\dfrac{\partial f}{\partial x}(1, 4) = 14$ means that if y is kept constant at 4 and x is allowed to vary near 1, then $f(x, y)$ changes at a rate equal to 14 times the change in x. That is, if x increases by one small unit, then $f(x, y)$ increases by approximately 14 units. If x increases by h units (h small), then $f(x, y)$ increases approximately $14 \cdot h$ units. That is, we have

$$f(1 + h, 4) - f(1, 4) \approx 14 \cdot h.$$

Similarly, the fact that $\dfrac{\partial f}{\partial y}(1, 4) = 7$ means that if we keep x constant at 1 and let y vary near 4, then $f(x, y)$ changes at a rate equal to seven times the change in y. So for a small value of k, we have

$$f(1, 4 + k) - f(1, 4) \approx 7 \cdot k.$$

We can generalize the interpretations of $\dfrac{\partial f}{\partial x}$ and $\dfrac{\partial f}{\partial y}$ given in Example 5 to yield the following general fact.

Let $f(x, y)$ be a function of two variables. Then if h and k are small, we have

$$f(a + h, b) - f(a, b) \approx \frac{\partial f}{\partial x}(a, b) \cdot h$$

$$f(a, b + k) - f(a, b) \approx \frac{\partial f}{\partial y}(a, b) \cdot k.$$

Partial derivatives can be computed for functions of any number of variables. When taking the partial with respect to one variable, we treat the other variables as constants.

EXAMPLE 6 Let $f(x, y, z) = x^2 yz - 3z$.

(a) Compute $\dfrac{\partial f}{\partial x}, \dfrac{\partial f}{\partial y}$, and $\dfrac{\partial f}{\partial z}$.   (b) Calculate $\dfrac{\partial f}{\partial z}(2, 3, 1)$.

Solution (a) $\dfrac{\partial f}{\partial x} = 2xyz, \qquad \dfrac{\partial f}{\partial y} = x^2 z, \qquad \dfrac{\partial f}{\partial z} = x^2 y - 3.$

(b) $\dfrac{\partial f}{\partial z}(2, 3, 1) = 2^2 \cdot 3 - 3 = 12 - 3 = 9.$

EXAMPLE 7 Let $f(x, y, z)$ be the heat-loss function computed in Example 2 of Section 1. That is, $f(x, y, z) = 11xy + 14yz + 15xz$. Calculate and interpret $\dfrac{\partial f}{\partial x}(10, 7, 5)$.

Solution We have

$$\frac{\partial f}{\partial x} = 11y + 15z$$

$$\frac{\partial f}{\partial x}(10, 7, 5) = 11 \cdot 7 + 15 \cdot 5 = 77 + 75 = 152.$$

The quantity $\dfrac{\partial f}{\partial x}$ is commonly referred to as the *marginal heat loss with respect to change in x*. Specifically, if x is changed from 10 by h units (where h is small) and the values of y and z remain fixed at 7 and 5, then the amount of heat loss will change by approximately $152 \cdot h$ units.

EXAMPLE 8 (*Production*)   Consider the production function $f(x, y) = 60x^{3/4}y^{1/4}$, which gives the number of units of goods produced when utilizing x units of labor and y units of capital.

(a) Find $\dfrac{\partial f}{\partial x}$ and $\dfrac{\partial f}{\partial y}$.

(b) Evaluate $\dfrac{\partial f}{\partial x}$ and $\dfrac{\partial f}{\partial y}$ at $x = 81$, $y = 16$.

(c) Interpret the numbers computed in part (b).

Solution (a) $\dfrac{\partial f}{\partial x} = 60 \cdot \dfrac{3}{4} x^{-1/4} \cdot y^{1/4} = 45x^{-1/4}y^{1/4} = 45\dfrac{y^{1/4}}{x^{1/4}}$

$\dfrac{\partial f}{\partial y} = 60 \cdot \dfrac{1}{4} x^{3/4} y^{-3/4} = 15x^{3/4}y^{-3/4} = 15\dfrac{x^{3/4}}{y^{3/4}}.$

(b) $\dfrac{\partial f}{\partial x}(81, 16) = 45 \cdot \dfrac{16^{1/4}}{81^{1/4}} = 45 \cdot \dfrac{2}{3} = 30$

$\dfrac{\partial f}{\partial y}(81, 16) = 15 \cdot \dfrac{(81)^{3/4}}{(16)^{3/4}} = 15 \cdot \dfrac{27}{8} = \dfrac{405}{8} = 50\tfrac{5}{8}.$

(c) The quantities $\dfrac{\partial f}{\partial x}$ and $\dfrac{\partial f}{\partial y}$ are referred to as the *marginal productivity of labor* and the *marginal productivity of capital*. If the amount of capital is held fixed at $y = 16$ and the amount of labor increases by 1 unit, then the quantity of goods produced will increase by approximately 30 units. Similarly, an increase in capital of 1 unit (with labor fixed at 81) results in an increase in production of approximately $50\tfrac{5}{8}$ units of goods.

Just as we formed second derivatives in the case of one variable, we can form second partial derivatives of a function $f(x, y)$ of two variables. Since $\dfrac{\partial f}{\partial x}$ is a function of x and y, we can differentiate it with respect to x or y. The partial derivative of $\dfrac{\partial f}{\partial x}$ with respect to x is denoted by $\dfrac{\partial^2 f}{\partial x^2}$. The partial derivative of $\dfrac{\partial f}{\partial x}$ with respect to y is denoted by $\dfrac{\partial^2 f}{\partial y\,\partial x}$. Similarly, the partial derivative of the function $\dfrac{\partial f}{\partial y}$ with respect to x is denoted by $\dfrac{\partial^2 f}{\partial x\,\partial y}$, and the partial derivative of $\dfrac{\partial f}{\partial y}$ with respect to y is denoted by $\dfrac{\partial^2 f}{\partial y^2}$. Almost all functions $f(x, y)$ encountered in applications [and all functions $f(x, y)$ in this text] have the property that

$$\frac{\partial^2 f}{\partial y\,\partial x} = \frac{\partial^2 f}{\partial x\,\partial y}.$$

EXAMPLE 9 Let $f(x, y) = x^2 + 3xy + 2y^2$. Calculate $\dfrac{\partial^2 f}{\partial x^2}$, $\dfrac{\partial^2 f}{\partial y^2}$, $\dfrac{\partial^2 f}{\partial x\,\partial y}$, and $\dfrac{\partial^2 f}{\partial y\,\partial x}$.

Solution First we compute $\dfrac{\partial f}{\partial x}$ and $\dfrac{\partial f}{\partial y}$.

$$\frac{\partial f}{\partial x} = 2x + 3y, \qquad \frac{\partial f}{\partial y} = 3x + 4y.$$

To compute $\dfrac{\partial^2 f}{\partial x^2}$, we differentiate $\dfrac{\partial f}{\partial x}$ with respect to x:

$$\frac{\partial^2 f}{\partial x^2} = 2.$$

Similarly, to compute $\dfrac{\partial^2 f}{\partial y^2}$, we differentiate $\dfrac{\partial f}{\partial y}$ with respect to y:

$$\frac{\partial^2 f}{\partial y^2} = 4.$$

To compute $\dfrac{\partial^2 f}{\partial x \, \partial y}$, we differentiate $\dfrac{\partial f}{\partial y}$ with respect to x:

$$\frac{\partial^2 f}{\partial x \, \partial y} = 3.$$

Finally, to compute $\dfrac{\partial^2 f}{\partial y \, \partial x}$, we differentiate $\dfrac{\partial f}{\partial x}$ with respect to y:

$$\frac{\partial^2 f}{\partial y \, \partial x} = 3.$$

PRACTICE PROBLEMS 2

1. The number of TV sets sold per week by an appliance store is given by a function of two variables, $f(x, y)$, where x is the price per TV set and y is the amount of money spent weekly on advertising. Suppose that the current price is $400 per set and that currently $2000 per week is being spent for advertising.

   (a) Would you expect $\dfrac{\partial f}{\partial x}(400, 2000)$ to be positive or negative?

   (b) Would you expect $\dfrac{\partial f}{\partial y}(400, 2000)$ to be positive or negative?

2. The monthly mortgage payment for a house is a function of two variables, $f(A, r)$, where A is the amount of the mortgage and the interest rate is $r\%$. For a 30-year mortgage, $f(92{,}000, 14) = 1090.08$ and $\dfrac{\partial f}{\partial x}(92{,}000, 14) = 72.82$. What is the significance of the number 72.82?

EXERCISES 2

Find $\dfrac{\partial f}{\partial x}$ and $\dfrac{\partial f}{\partial y}$ for each of the following functions.

1. $f(x, y) = 5xy$

2. $f(x, y) = 3x^2 + 2y + 1$

3. $f(x, y) = 2x^2 e^y$

4. $f(x, y) = x + e^{xy}$

5. $f(x, y) = \dfrac{y^2}{x}$

6. $f(x, y) = \dfrac{x}{1 + e^y}$

7. $f(x, y) = (2x - y + 5)^2$

8. $f(x, y) = (9x^2 y + 3x)^{12}$

9. $f(x, y) = x^2 e^{3x} \ln y$

10. $f(x, y) = (x - \ln y)e^{xy}$

11. $f(x, y) = \dfrac{x - y}{x + y}$

12. $f(x, y) = \dfrac{2xy}{e^x}$

13. Let $f(L, K) = 3\sqrt{LK}$. Compute $\dfrac{\partial f}{\partial L}$.

14. Let $f(p, q) = e^{q/p}$. Compute $\dfrac{\partial f}{\partial q}$ and $\dfrac{\partial f}{\partial p}$.

15. Let $f(x, y, z) = (1 + x^2 y)/z$. Compute $\dfrac{\partial f}{\partial x}, \dfrac{\partial f}{\partial y}$, and $\dfrac{\partial f}{\partial z}$.

16. Let $f(x, y, z) = x^2 y + 3yz - z^2$. Compute $\dfrac{\partial f}{\partial x}, \dfrac{\partial f}{\partial y}$, and $\dfrac{\partial f}{\partial z}$.

17. Let $f(x, y, z) = xze^{yz}$. Find $\dfrac{\partial f}{\partial x}, \dfrac{\partial f}{\partial y}$, and $\dfrac{\partial f}{\partial z}$.

18. Let $f(x, y, z) = ze^{z/xy}$. Find $\dfrac{\partial f}{\partial x}, \dfrac{\partial f}{\partial y}$, and $\dfrac{\partial f}{\partial z}$.

19. Let $f(x, y) = x^2 + 2xy + y^2 + 3x + 5y$. Compute $\dfrac{\partial f}{\partial x}(2, -3)$ and $\dfrac{\partial f}{\partial y}(2, -3)$.

20. Let $f(x, y) = xye^{2x-y}$. Evaluate $\dfrac{\partial f}{\partial x}$ and $\dfrac{\partial f}{\partial y}$ at $(x, y) = (1, 2)$.

21. Let $f(x, y, z) = xy^2 z + 5$. Evaluate $\dfrac{\partial f}{\partial y}$ at $(x, y, z) = (2, -1, 3)$.

22. Let $f(x, y, z) = \dfrac{x}{y - z}$. Compute $\dfrac{\partial f}{\partial y}(2, -1, 3)$.

23. Let $f(x, y) = x^3 y + 2xy^2$. Find $\dfrac{\partial^2 f}{\partial x^2}, \dfrac{\partial^2 f}{\partial y^2}, \dfrac{\partial^2 f}{\partial x \, \partial y}$, and $\dfrac{\partial^2 f}{\partial y \, \partial x}$.

24. Let $f(x, y) = xe^y + x^4 y + y^3$. Find $\dfrac{\partial^2 f}{\partial x^2}, \dfrac{\partial^2 f}{\partial y^2}, \dfrac{\partial^2 f}{\partial x \, \partial y}$, and $\dfrac{\partial^2 f}{\partial y \, \partial x}$.

25. A farmer can produce $f(x, y) = 200\sqrt{6x^2 + y^2}$ units of produce by utilizing x units of labor and y units of capital. (The capital is used to rent or purchase land, materials, and equipment.)

   (a) Calculate the marginal productivities of labor and capital when $x = 10$ and $y = 5$.

   (b) Use the result of part (a) to determine the approximate effect on production of utilizing 5 units of capital but cutting back to $9\frac{1}{2}$ units of labor.

26. The productivity of a country is given by $f(x, y) = 300x^{2/3}y^{1/3}$, where x and y are the amounts of labor and capital.

   (a) Compute the marginal productivities of labor and capital when $x = 125$ and $y = 64$.

   (b) What would be the approximate effect of utilizing 125 units of labor but cutting back to 62 units of capital?

27. In a certain suburban community commuters have the choice of getting into the city by bus or train. The demand for these modes of transportation varies with their cost. Let $f(p_1, p_2)$ be the number of people who will take the bus when p_1 is the price of the bus ride and p_2 is the price of the train ride. Explain why $\dfrac{\partial f}{\partial p_1} < 0$ and $\dfrac{\partial f}{\partial p_2} > 0$.

28. The demand for a certain gas-guzzling car is given by $f(p_1, p_2)$, where p_1 is the price of the car and p_2 is the price of gasoline. Explain why $\dfrac{\partial f}{\partial p_1} < 0$ and $\dfrac{\partial f}{\partial p_2} < 0$.

29. Using data collected from 1929–1941, Richard Stone* determined that the yearly quantity Q of beer consumed in the United Kingdom was approximately given by the formula $Q = f(m, p, r, s)$, where

$$f(m, p, r, s) = (1.058)m^{.136}p^{-.727}r^{.914}s^{.816}$$

and m is the aggregate real income (personal income after direct taxes, adjusted for retail price changes), p is the average retail price of the commodity (in this case, beer), r is the average retail price level of all other consumer goods and services, and s is a measure of the strength of the beer. Determine which partial derivatives are positive and which are negative and give interpretations. (For example, since $\dfrac{\partial f}{\partial r} > 0$, people buy more beer when the prices of other goods increase and the other factors remain constant.)

30. Richard Stone (see Exercise 29) determined that the yearly consumption of food in the United States was given by

$$f(m, p, r) = (2.186)m^{.595}p^{-.543}r^{.922}.$$

Determine which partial derivatives are positive and which are negative and give interpretations of these facts.

31. The volume (V) of a certain amount of a gas is determined by the temperature (T) and the pressure (P) by the formula, $V = .08(T/P)$. Calculate and interpret $\dfrac{\partial V}{\partial P}$ and $\dfrac{\partial V}{\partial T}$ when $P = 20$, $T = 300$.

---

* Richard Stone, "The Analysis of Market Demand," *Journal of the Royal Statistical Society*, CVIII (1945), 286–391.

32. For the production function, $f(x, y) = 60x^{3/4}y^{1/4}$, considered in Example 8, think of $f(x, y)$ as the revenue when utilizing x units of labor and y units of capital. Under actual operating conditions, say $x = a$ and $y = b$, $\dfrac{\partial f}{\partial x}(a, b)$ is referred to as the *wage per unit of labor* and $\dfrac{\partial f}{\partial y}(a, b)$ is referred to as the *wage per unit of capital*. Show that

$$f(a, b) = a \cdot \left[\frac{\partial f}{\partial x}(a, b) \right] + b \cdot \left[\frac{\partial f}{\partial y}(a, b) \right].$$

(This equation shows how the revenue is distributed between labor and capital.)

33. Compute $\dfrac{\partial^2 f}{\partial x^2}$ where $f(x, y) = 60x^{3/4}y^{1/4}$, a production function (where x is units of labor). Explain why $\dfrac{\partial^2 f}{\partial x^2}$ is always negative.

34. Compute $\dfrac{\partial^2 f}{\partial y^2}$ where $f(x, y) = 60x^{3/4}y^{1/4}$, a production function (where y is units of capital). Explain why $\dfrac{\partial^2 f}{\partial y^2}$ is always negative.

35. Let $f(x, y) = 3x^2 + 2xy + 5y$, as in Example 5. Show that

$$f(1 + h, 4) - f(1, 4) = 14h + 3h^2.$$

Thus the error in approximating $f(1 + h, 4) - f(1, 4)$ by $14h$ is $3h^2$. (If $h = .01$, for instance, the error is only .0003.)

36. Physicians, particularly pediatricians, sometimes need to know the body surface area of a patient. For instance, the surface area is used to adjust the results of certain tests of kidney performance. Tables are available that give the approximate body surface A in square meters of a person who weighs W kilograms and is H centimeters tall. The following empirical formula* is also used:

$$A = .007W^{.425}H^{.725}.$$

Evaluate $\dfrac{\partial A}{\partial W}$ and $\dfrac{\partial A}{\partial H}$ when $W = 54$, $H = 165$, and give a physical interpretation of your answers. You may use the approximations: $(54)^{.425} \approx 5.4$, $(54)^{-.575} \approx .10$, $(165)^{.725} \approx 40.5$, $(165)^{-.275} \approx .25$.

SOLUTIONS TO PRACTICE PROBLEMS 2

1. (a) Negative. $\dfrac{\partial f}{\partial x}(400, 2000)$ is approximately the change in sales due to a \$1 increase in x (price). Since raising prices lowers sales, we would expect $\dfrac{\partial f}{\partial x}(400, 2000)$ to be negative.

---

* See J. Routh, *Mathematical Preparation for Laboratory Technicians* (Philadelphia: W. B. Saunders Co., 1971), p. 92.

(b) Positive. $\dfrac{\partial f}{\partial y}$ (400, 2000) is approximately the change in sales due to a $1 increase in advertising. Since spending more money on advertising brings in more customers, we would expect sales to increase; that is, $\dfrac{\partial f}{\partial y}$ (400, 2000) is most likely positive.

2. If the interest rate is raised from 14% to 15%, then the monthly payment will increase by about $72.82. [An increase to $14\frac{1}{2}$% causes an increase in the monthly payment of $\frac{1}{2} \cdot (72.82)$ or $36.41, and so on.]

7.3. Maxima and Minima of Functions of Several Variables

Previously, we studied how to determine the maxima and minima of functions of a single variable. Let us extend that discussion to functions of several variables.

If $f(x, y)$ is a function of two variables, then we say that $f(x, y)$ has a *maximum* when $x = a$, $y = b$ if $f(x, y)$ is at most equal to $f(a, b)$ whenever x is near a and y is near b. Geometrically, the graph of $f(x, y)$ has a peak at the point (a, b). [See Fig. 1(a).] Similarly, we say that $f(x, y)$ has a *minimum* when $x = a$, $y = b$

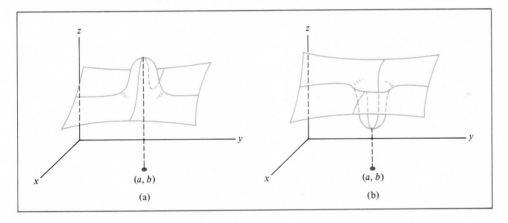

FIGURE 1   Maximum and minimum points.

if $f(x, y)$ is at least equal to $f(a, b)$ whenever x is near a and y is near b. Geometrically, the graph of $f(x, y)$ has a pit with bottom at the point (a, b). [See Fig. 1(b).]

Suppose that $f(x, y)$ has a maximum at $(x, y) = (a, b)$. Then, keeping y constant at b, $f(x, y)$ becomes a function of the variable x with a maximum at $x = a$. Therefore, its derivative with respect to x is zero at $x = a$. That is,

$$\frac{\partial f}{\partial x}(a, b) = 0.$$

Similarly, keeping x constant at a, $f(x, y)$ becomes a function of the variable y with a maximum at $y = b$. Therefore, its derivative with respect to y is zero at

$y = b$. That is,

$$\frac{\partial f}{\partial y}(a, b) = 0.$$

Similar considerations apply if $f(x, y)$ has a minimum at $(x, y) = (a, b)$. Thus we have the following test for extrema in two variables.

First-Derivative Test for Extrema If $f(x, y)$ has either a maximum or a minimum at $(x, y) = (a, b)$, then

$$\frac{\partial f}{\partial x}(a, b) = 0$$

and

$$\frac{\partial f}{\partial y}(a, b) = 0.$$

This result is quite helpful in finding maximum and minimum points, especially when additional information about the function is known.

EXAMPLE 1 The function $f(x, y) = 3x^2 - 4xy + 3y^2 + 8x - 17y + 5$ has the graph pictured in Fig. 2. Find the point (x, y) at which $f(x, y)$ attains its minimum.

Solution We look for those values of x and y at which both partial derivatives are zero. The partial derivatives are

$$\frac{\partial f}{\partial x} = 6x - 4y + 8$$

$$\frac{\partial f}{\partial y} = -4x + 6y - 17.$$

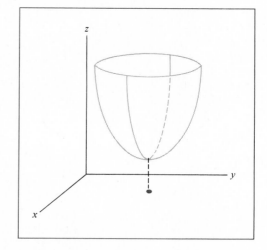

FIGURE 2   Graph of $f(x, y) = 3x^2 - 4xy + 3y^2 + 8x - 17y + 5$.

Setting $\dfrac{\partial f}{\partial x} = 0$ and $\dfrac{\partial f}{\partial y} = 0$, we obtain

$$6x - 4y + 8 = 0 \qquad \text{or} \qquad y = \frac{6x + 8}{4}$$

$$-4x + 6y - 17 = 0 \qquad \text{or} \qquad y = \frac{4x + 17}{6}.$$

By equating these two expressions for y, we have

$$\frac{6x + 8}{4} = \frac{4x + 17}{6}.$$

Cross-multiplying, we see that

$$36x + 48 = 16x + 68$$

$$20x = 20$$

$$x = 1.$$

When we substitute this value for x into our first equation for y in terms of x, we obtain

$$y = \frac{6x + 8}{4} = \frac{6 \cdot 1 + 8}{4} = \frac{7}{2}.$$

If $f(x, y)$ has a minimum, it must occur where $\frac{\partial f}{\partial x} = 0$ and $\frac{\partial f}{\partial y} = 0$. We have determined that the partial derivatives are zero only when $x = 1$, $y = \frac{7}{2}$. From Fig. 2 we know that $f(x, y)$ has a minimum, so it must be at $(x, y) = (1, \frac{7}{2})$.

EXAMPLE 2 (*Price Discrimination*) A monopolist markets his product in two countries and can charge different amounts in each country. Let x be the number of units to be sold in the first country and y the number of units to be sold in the second country. Due to the laws of demand, the monopolist must set the price at $97 - (x/10)$ dollars in the first country and $83 - (y/20)$ dollars in the second country in order to sell all the units. The cost of producing these units is $20,000 + 3(x + y)$. Find the values of x and y that maximize the profit.

Solution Let $f(x, y)$ be the profit derived from selling x units in the first country and y in the second. Then

$$f(x, y) = [\text{revenue from first country}] + [\text{revenue from second country}] - [\text{cost}]$$

$$= \left(97 - \frac{x}{10}\right)x + \left(83 - \frac{y}{20}\right)y - [20,000 + 3(x + y)]$$

$$= 97x - \frac{x^2}{10} + 83y - \frac{y^2}{20} - 20,000 - 3x - 3y$$

$$= 94x - \frac{x^2}{10} + 80y - \frac{y^2}{20} - 20,000.$$

To find where $f(x, y)$ has its maximum value, we look for those values of x and y at which both partial derivatives are zero.

$$\frac{\partial f}{\partial x} = 94 - \frac{x}{5}$$

$$\frac{\partial f}{\partial y} = 80 - \frac{y}{10}.$$

We set $\dfrac{\partial f}{\partial x} = 0$ and $\dfrac{\partial f}{\partial y} = 0$ to obtain

$$94 - \frac{x}{5} = 0 \qquad \text{or} \qquad x = 470$$

$$80 - \frac{y}{10} = 0 \qquad \text{or} \qquad y = 800.$$

Therefore, the firm should adjust its prices to levels where it will sell 470 units in the first country and 800 units in the second country.

EXAMPLE 3 Suppose that we want to design a rectangular building having volume 147,840 cubic feet. Assuming that the daily loss of heat is given by

$$11xy + 14yz + 15xz,$$

where x, y, and z are, respectively, the length, width, and height of the building, find the dimensions of the building for which the daily heat loss is minimal.

Solution We must minimize the function

$$11xy + 14yz + 15xz, \tag{1}$$

where x, y, z satisfy the constraint equation

$$xyz = 147,840.$$

For simplicity, let us denote 147,840 by V. Then $xyz = V$, so that $z = V/xy$. We substitute this expression for z into the objective function (1) to obtain a heat-loss function $g(x, y)$ of two variables—namely

$$g(x, y) = 11xy + 14y\frac{V}{xy} + 15x\frac{V}{xy}$$

$$= 11xy + \frac{14V}{x} + \frac{15V}{y}.$$

To minimize this function, we first compute the partial derivatives with respect to x and y; then we equate them to zero.

$$\frac{\partial g}{\partial x} = 11y - \frac{14V}{x^2} = 0$$

$$\frac{\partial g}{\partial y} = 11x - \frac{15V}{y^2} = 0.$$

These two equations yield

$$y = \frac{14V}{11x^2} \tag{2}$$

$$11xy^2 = 15V. \tag{3}$$

If we substitute the value of y from (2) into (3), we see that

$$11x\left(\frac{14V}{11x^2}\right)^2 = 15V$$

$$\frac{14^2V^2}{11x^3} = 15V$$

$$x^3 = \frac{14^2 \cdot V^2}{11 \cdot 15 \cdot V} = \frac{14^2 \cdot V}{11 \cdot 15}$$

$$= \frac{14^2 \cdot 147{,}840}{11 \cdot 15}$$

$$= 175{,}616.$$

Therefore, we see that (using a calculator or a table of cube roots)

$$x = 56.$$

From equation (2) we find that

$$y = \frac{14 \cdot V}{11x^2} = \frac{14 \cdot 147{,}840}{11 \cdot 56^2} = 60.$$

Finally,

$$z = \frac{V}{xy} = \frac{147{,}840}{56 \cdot 60} = 44.$$

Thus the building should be 56 feet long, 60 feet wide, and 44 feet high in order to minimize the heat loss.*

When considering a function of two variables, we find points (x, y) at which $f(x, y)$ has a potential maximum or minimum by setting $\frac{\partial f}{\partial x}$ and $\frac{\partial f}{\partial y}$ equal to zero and solving for x and y. However, if we are given no additional information about $f(x, y)$, it may be difficult to determine whether we have found a maximum or a minimum (or neither). In the case of functions of one variable, we studied concavity and deduced the second-derivative test. There is an analog of the second-derivative test for functions of two variables, but it is much more complicated than the one-variable test. We state it without proof.

---

* For further discussion of this heat-loss problem, as well as other examples of optimization in architectural design, see L. March, "Elementary Models of Built Forms," Chapter 3 in *Urban Space and Structures*, L. Martin and L. March, eds. (Cambridge: Cambridge University Press, 1972).

Second-Derivative Test for Maxima and Minima Suppose that $f(x, y)$ is a function and (a, b) is a point at which

$$\frac{\partial f}{\partial x}(a, b) = 0 \qquad \text{and} \qquad \frac{\partial f}{\partial y}(a, b) = 0,$$

and let

$$D(x, y) = \frac{\partial^2 f}{\partial x^2} \cdot \frac{\partial^2 f}{\partial y^2} - \left(\frac{\partial^2 f}{\partial x \, \partial y}\right)^2.$$

1. If

$$D(a, b) > 0 \qquad \text{and} \qquad \frac{\partial^2 f}{\partial x^2}(a, b) > 0,$$

then $f(x, y)$ has a minimum at (a, b).

2. If

$$D(a, b) > 0 \qquad \text{and} \qquad \frac{\partial^2 f}{\partial x^2}(a, b) < 0,$$

then $f(x, y)$ has a maximum at (a, b).

3. If

$$D(a, b) < 0,$$

then $f(x, y)$ has neither a maximum nor a minimum at (a, b).

4. If $D(a, b) = 0$, then no conclusion can be drawn from this test.

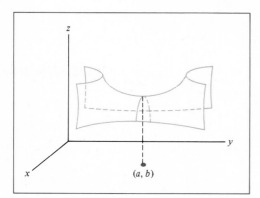

FIGURE 3

The saddle-shaped graph in Fig. 3 illustrates a function $f(x, y)$ for which $D(a, b) < 0$. Both partial derivatives are zero at $(x, y) = (a, b)$ and yet the function has neither a maximum nor a minimum there. (Observe that the function has a maximum with respect to x when y is held constant and a minimum with respect to y when x is held constant.)

EXAMPLE 4 Let $f(x, y) = y^3 - x^2 + 6x - 12y + 5$. Find all possible maximum and minimum points of $f(x, y)$. Use the second-derivative test to determine the nature of each such point.

Solution Since

$$\frac{\partial f}{\partial x} = -2x + 6, \qquad \frac{\partial f}{\partial y} = 3y^2 - 12,$$

we find that $f(x, y)$ has a potential extreme point when

$$-2x + 6 = 0$$
$$3y^2 - 12 = 0.$$

The first equation implies that $x = 3$. From the second equation we have

$$3y^2 = 12$$
$$y^2 = 4$$
$$y = \pm 2.$$

Thus $\dfrac{\partial f}{\partial x}$ and $\dfrac{\partial f}{\partial y}$ are both zero when $(x, y) = (3, 2)$ and when $(x, y) = (3, -2)$. To apply the second-derivative test, we compute

$$\frac{\partial^2 f}{\partial x^2} = -2, \qquad \frac{\partial^2 f}{\partial y^2} = 6y, \qquad \frac{\partial^2 f}{\partial x\, \partial y} = 0,$$

and

$$D(x, y) = \frac{\partial^2 f}{\partial x^2} \cdot \frac{\partial^2 f}{\partial y^2} - \left(\frac{\partial^2 f}{\partial x\, \partial y}\right)^2 = (-2)(6y) - 0 = -12y. \qquad (4)$$

Since $D(3, 2) = -24$ is negative, case 3 of the second derivative test says that $f(x, y)$ has neither a maximum nor a minimum at $(3, 2)$. Next, note that

$$D(3, -2) = 24 > 0, \qquad \frac{\partial^2 f}{\partial x^2}(3, -2) = -2 < 0.$$

Thus, by case 2 of the second-derivative test, the function $f(x, y)$ has a maximum at $(3, -2)$.

In this section we have restricted ourselves to functions of two variables, but the case of three or more variables is handled in a similar fashion. For instance, here is the first-derivative test for a function of three variables.

If $f(x, y, z)$ has a maximum or a minimum at $(x, y, z) = (a, b, c)$, then

$$\frac{\partial f}{\partial x}(a, b, c) = 0$$

$$\frac{\partial f}{\partial y}(a, b, c) = 0$$

$$\frac{\partial f}{\partial z}(a, b, c) = 0.$$

1. Find all points (x, y) where $f(x, y) = x^3 - 3xy + \frac{1}{2}y^2 + 8$ has a possible maximum or minimum.

2. Apply the second-derivative test to the function $g(x, y)$ of Example 3 to confirm that a minimum point actually occurs when $x = 56$ and $y = 60$.

EXERCISES 3

Find all points (x, y) where $f(x, y)$ has a possible maximum or minimum.

1. $f(x, y) = x^2 - 3y^2 + 4x + 6y + 8$

2. $f(x, y) = \frac{1}{2}x^2 + y^2 - 3x + 2y - 5$

3. $f(x, y) = x^2 - 5xy + 6y^2 + 3x - 2y + 4$

4. $f(x, y) = -3x^2 + 7xy - 4y^2 + x + y$

5. $f(x, y) = x^3 + y^2 - 3x + 6y$

6. $f(x, y) = x^2 - y^3 + 5x + 12y + 1$

7. $f(x, y) = \frac{1}{3}x^3 - 2y^3 - 5x + 6y - 5$

8. $f(x, y) = x^4 - 8xy + 2y^2 - 3$

9. The function $f(x, y) = 2x + 3y + 9 - x^2 - xy - y^2$ has a maximum at some point (x, y). Find the values of x and y where this maximum occurs.

10. The function $f(x, y) = \frac{1}{2}x^2 + 2xy + 3y^2 - x + 2y$ has a minimum at some point (x, y). Find the values of x and y where this minimum occurs.

Find all points (x, y) where $f(x, y)$ has a possible maximum or minimum. Then use the second-derivative test to determine, if possible, the nature of $f(x, y)$ at each of these points. If the second-derivative test is inconclusive, so state.

11. $f(x, y) = x^2 - 2xy + 4y^2$

12. $f(x, y) = 2x^2 + 3xy + 5y^2$

13. $f(x, y) = -2x^2 + 2xy - y^2 + 4x - 6y + 5$

14. $f(x, y) = -x^2 - 8xy - y^2$

15. $f(x, y) = x^2 + 2xy + 5y^2 + 2x + 10y - 3$

16. $f(x, y) = x^2 - 2xy + 3y^2 + 4x - 16y + 22$

17. $f(x, y) = x^3 - y^2 - 3x + 4y$

18. $f(x, y) = x^3 - 2xy + 4y$

19. $f(x, y) = 2x^2 + y^3 - x - 12y + 7$

20. $f(x, y) = x^2 + 4xy + 2y^4$

21. Find the possible values of x, y, z at which

$$f(x, y, z) = 2x^2 + 3y^2 + z^2 - 2x - y - z$$

assumes its minimum value.

22. Find the possible values of x, y, z at which

$$f(x, y, z) = 5 + 8x - 4y + x^2 + y^2 + z^2$$

assumes its minimum value.

23. U.S. postal rules require that the length plus the girth of a package cannot exceed 84 inches in order to be mailed. Find the dimensions of the rectangular package of greatest volume that can be mailed. [*Note:* From Fig. 4 we see that $84 = $ (length) $+$ (girth) $= l + (2x + 2y)$ and volume $= xyl$.]

FIGURE 4

24. Find the dimensions of the rectangular box of least surface area that has a volume of 100 cubic inches. [*Note:* From Fig. 4 we see that the surface area $= 2xy + 2xl + 2yl$ and volume $= 100 = xyl$.]

25. A company manufactures and sells two products, call them I and II, that sell for $10 and $9 per unit, respectively. The cost of producing x units of product I and y units of product II is

$$400 + 2x + 3y + .01(3x^2 + xy + 3y^2).$$

Find the values of x and y that maximize the company's profit. [*Note:* Profit $=$ (revenue) $-$ (cost).]

26. A monopolist manufactures and sells two competing products, call them I and II, that cost $30 and $20 per unit, respectively, to produce. The revenue from marketing x units of product I and y units of product II is $98x + 112y - .04xy - .1x^2 - .2y^2$. Find the values of x and y that maximize the monopolist's profits.

27. A company manufactures and sells two products, call them I and II, that sell for $\$p_1$ and $\$p_{II}$ per unit, respectively. Let $C(x, y)$ be the cost of producing x units of product I and y units of product II. Show that if the company's profit is maximized when $x = a$, $y = b$, then

$$\frac{\partial C}{\partial x}(a, b) = p_I \quad \text{and} \quad \frac{\partial C}{\partial y}(a, b) = p_{II}.$$

28. A monopolist manufactures and sells two competing products, call them I and II, that cost $\$p_I$ and $\$p_{II}$ per unit, respectively, to produce. Let $R(x, y)$ be the revenue from marketing x units of product I and y units of product II. Show that if the monopolist's profit is maximized when $x = a$, $y = b$, then

$$\frac{\partial R}{\partial x}(a, b) = p_I \quad \text{and} \quad \frac{\partial R}{\partial y}(a, b) = p_{II}.$$

SOLUTIONS TO PRACTICE PROBLEMS 3

1. Compute the first partial derivatives of $f(x, y)$ and solve the system of equations that results from setting the partials equal to zero.

$$\frac{\partial f}{\partial x} = 3x^2 - 3y = 0$$

$$\frac{\partial f}{\partial y} = -3x + y = 0.$$

Solve each equation for y in terms of x.

$$\begin{cases} y = x^2 \\ y = 3x. \end{cases}$$

Equate expressions for y and solve for x.

$$x^2 = 3x$$

$$x^2 - 3x = 0$$

$$x(x - 3) = 0$$

$$x = 0 \quad \text{or} \quad x = 3,$$

When $x = 0$, $y = 0^2 = 0$. When $x = 3$, $y = 3^2 = 9$. Therefore, the possible maximum or minimum points are $(0, 0)$ and $(3, 9)$.

2. We have

$$g(x, y) = 11xy + \frac{14V}{x} + \frac{15V}{y},$$

$$\frac{\partial g}{\partial x} = 11y - \frac{14V}{x^2}, \quad \text{and} \quad \frac{\partial g}{\partial y} = 11x - \frac{15V}{y^2}.$$

Now,

$$\frac{\partial^2 g}{\partial x^2} = \frac{28V}{x^3}, \quad \frac{\partial^2 g}{\partial y^2} = \frac{30V}{y^3}, \quad \text{and} \quad \frac{\partial^2 g}{\partial x \, \partial y} = 11.$$

Therefore,

$$D(x, y) = \frac{28V}{x^3} \cdot \frac{30V}{y^3} - (11)^2$$

$$D(56, 60) = \frac{28(147,840)}{(56)^3} \cdot \frac{30(147,840)}{(60)^3} - 121$$

$$= 484 - 121 = 363 > 0,$$

and

$$\frac{\partial^2 g}{\partial x^2}(56, 60) = \frac{28(147,840)}{(56)^3} > 0.$$

It follows that $g(x, y)$ has a minimum at $x = 56$, $y = 60$.

7.4. Lagrange Multipliers and Constrained Optimization

We have seen a number of optimization problems in which we were required to minimize (or maximize) an objective function where the variables were subject to a constraint equation. For instance, in Example 4 of Section 2.5, we minimized the cost of a rectangular enclosure by minimizing the objective function $21x + 14y$, where x and y were subject to the constraint equation $600 - xy = 0$. In the preceding section (Example 3) we minimized the daily heat loss from a building by minimizing the objective function $11xy + 14yz + 15xz$, subject to the constraint equation $147,840 - xyz = 0$.

In this section we introduce a powerful technique for solving problems of this type. Let us begin with the following general problem, which involves two variables.

> **PROBLEM** Let $f(x, y)$ and $g(x, y)$ be functions of two variables. Find values of x and y that maximize (or minimize) the objective function $f(x, y)$ and that also satisfy the constraint equation $g(x, y) = 0$.

Of course, if we can solve the equation $g(x, y) = 0$ for one variable in terms of the other and substitute the resulting expression into $f(x, y)$, we arrive at a function of a single variable that can be maximized (or minimized) by using the methods of Chapter 2. However, this technique can be unsatisfactory for two reasons. First, it may be difficult to solve the equation $g(x, y) = 0$ for x or for y. For example, if $g(x, y) = x^4 + 5x^3y + 7x^2y^3 + y^5 - 17 = 0$, then it is difficult to write y as a function of x or x as a function of y. Second, even if $g(x, y) = 0$ can be solved for one variable in terms of the other, substitution of the result into $f(x, y)$ may yield a complicated function.

One clever idea for handling the preceding problem was discovered by the eighteenth-century mathematician Lagrange, and the technique that he pioneered today bears his name—the method of *Lagrange multipliers*. The basic idea of this method is to replace $f(x, y)$ by an auxiliary function of three variables $F(x, y, \lambda)$, defined as

$$F(x, y, \lambda) = f(x, y) + \lambda g(x, y).$$

The new variable λ (lambda) is called a *Lagrange multiplier* and always multiplies the constraint function $g(x, y)$. The following theorem is stated without proof.

> **Theorem** Suppose that, subject to the constraint $g(x, y) = 0$, the function $f(x, y)$ has a maximum or a minimum at $(x, y) = (a, b)$. Then there is a value of λ, say $\lambda = c$, such that the partial derivatives of $F(x, y, \lambda)$ all equal zero at $(x, y, \lambda) = (a, b, c)$.

The theorem implies that if we locate all points (x, y, λ) where the partial derivatives of $F(x, y, \lambda)$ are all zero, then among the corresponding points (x, y) we will find all possible places where $f(x, y)$ may have a constrained maximum or minimum. Thus the first step in the method of Lagrange multipliers is to set the partial derivatives of $F(x, y, \lambda)$ equal to zero and solve for x, y, and λ:

$$\frac{\partial F}{\partial x} = 0 \tag{L-1}$$

$$\frac{\partial F}{\partial y} = 0 \tag{L-2}$$

$$\frac{\partial F}{\partial \lambda} = 0. \tag{L-3}$$

From the definition of $F(x, y, \lambda)$, we see that $\dfrac{\partial F}{\partial \lambda} = g(x, y)$. Thus the third equation (L-3) is just the original constraint equation $g(x, y) = 0$. So when we find a point (x, y, λ) that satisfies (L-1), (L-2), and (L-3), the coordinates x and y will automatically satisfy the constraint equation.

Let us see how this method works in practice.

EXAMPLE 1   Minimize $x^2 + y^2$, subject to the constraint $2x + 3y - 4 = 0$.

Solution   Here we have $f(x, y) = x^2 + y^2$, $g(x, y) = 2x + 3y - 4$, and

$$F(x, y, \lambda) = x^2 + y^2 + \lambda(2x + 3y - 4).$$

Equations (L-1) to (L-3) read

$$\frac{\partial F}{\partial x} = 2x + 2\lambda = 0 \tag{1}$$

$$\frac{\partial F}{\partial y} = 2y + 3\lambda = 0 \tag{2}$$

$$\frac{\partial F}{\partial \lambda} = 2x + 3y - 4 = 0. \tag{3}$$

Let us solve the first two equations for λ:

$$\lambda = -x \tag{4}$$

$$\lambda = -\tfrac{2}{3}y.$$

If we equate these two expressions for λ, we obtain

$$-x = -\tfrac{2}{3}y$$

$$x = \tfrac{2}{3}y. \tag{5}$$

Then, substituting this expression for x into the equation (3), we have

$$2(\tfrac{2}{3}y) + 3y - 4 = 0$$

$$\tfrac{13}{3}y = 4$$

$$y = \tfrac{12}{13}.$$

Using this value for y, equations (4) and (5) give us the value of x and λ:

$$x = \tfrac{2}{3}y = \tfrac{2}{3}(\tfrac{12}{13}) = \tfrac{8}{13}$$

$$\lambda = -x = -\tfrac{8}{13}.$$

Therefore, the partial derivatives of $F(x, y, \lambda)$ are zero when $x = \tfrac{8}{13}$, $y = \tfrac{12}{13}$, and $\lambda = -\tfrac{8}{13}$. So the minimum value of $x^2 + y^2$, subject to the constraint $2x + 3y - 4 = 0$, is

$$(\tfrac{8}{13})^2 + (\tfrac{12}{13})^2 = \tfrac{16}{13}.$$

The preceding technique for solving three equations in the three variables x, y, λ can usually be applied to solve Lagrange multiplier problems. Here is the basic procedure.

1. Solve (L-1) and (L-2) for λ in terms of x and y; then equate the resulting expressions for λ.
2. Solve the resulting equation for one of the variables.
3. Substitute the expression so derived into the equation (L-3) and solve the resulting equation of one variable.
4. Use the one known variable and the equations of steps 1 and 2 to determine the other two variables.

EXAMPLE 2 Using Lagrange multipliers, minimize $21x + 14y$, subject to the constraint $600 - xy = 0$, where x and y are restricted to positive values. (This problem arose in Example 4 of Section 2.5, where $21x + 14y$ was the cost of building a 600 square foot enclosure having dimensions x and y.)

Solution We have $f(x, y) = 21x + 14y$, $g(x, y) = 600 - xy$, and

$$F(x, y, \lambda) = 21x + 14y + \lambda(600 - xy).$$

The equations (L-1) to (L-3), in this case, are

$$\frac{\partial F}{\partial x} = 21 - \lambda y = 0$$

$$\frac{\partial F}{\partial y} = 14 - \lambda x = 0$$

$$\frac{\partial F}{\partial \lambda} = 600 - xy = 0.$$

From the first two equations we see that

$$\lambda = \frac{21}{y} = \frac{14}{x} \quad \text{(step 1).}$$

Therefore,

$$21x = 14y$$

and

$$x = \tfrac{2}{3}y \quad \text{(step 2).}$$

Substituting this expression for x into the third equation, we derive

$$600 - (\tfrac{2}{3}y)y = 0$$

$$y^2 = \tfrac{3}{2} \cdot 600 = 900$$

$$y = \pm 30 \quad \text{(step 3).}$$

We discard the case $y = -30$ because we are interested only in positive values of x and y. Using $y = 30$, we find that

$$\left.\begin{array}{l} x = \tfrac{2}{3}(30) = 20 \\ \lambda = \tfrac{14}{20} = \tfrac{7}{10}. \end{array}\right\} \quad \text{(step 4).}$$

So the minimum value of $21x + 14y$ with x and y subject to the constraint occurs when $x = 20$, $y = 30$, and $\lambda = \tfrac{7}{10}$. And the minimum value of $21x + 14y$, subject to the constraint $600 - xy = 0$, is

$$21 \cdot (20) + 14 \cdot (30) = 840.$$

EXAMPLE 3   *(Production)*   Suppose that x units of labor and y units of capital can produce $f(x, y) = 60x^{3/4}y^{1/4}$ units of a certain product. Also suppose that each unit of labor costs \$100, whereas each unit of capital costs \$200. Assume that \$30,000 is available to spend on production. How many units of labor and how many of capital should be utilized in order to maximize production?

Solution   The cost of x units of labor and y units of capital equals $100x + 200y$. Therefore, since we want to use all the available money (\$30,000), we must satisfy the constraint equation

$$100x + 200y = 30{,}000$$

or

$$g(x, y) = 30{,}000 - 100x - 200y = 0.$$

Our objective function is $f(x, y) = 60x^{3/4}y^{1/4}$. In this case, we have

$$F(x, y, \lambda) = 60x^{3/4}y^{1/4} + \lambda(30{,}000 - 100x - 200y).$$

The equations (L-1) to (L-3) read

$$\frac{\partial F}{\partial x} = 45x^{-1/4}y^{1/4} - 100\lambda = 0 \tag{L-1}$$

$$\frac{\partial F}{\partial y} = 15x^{3/4}y^{-3/4} - 200\lambda = 0 \tag{L-2}$$

$$\frac{\partial F}{\partial \lambda} = 30{,}000 - 100x - 200y = 0. \tag{L-3}$$

By solving the first two equations for λ, we see that

$$\lambda = \tfrac{45}{100}x^{-1/4}y^{1/4} = \tfrac{9}{20}x^{-1/4}y^{1/4}$$
$$\lambda = \tfrac{15}{200}x^{3/4}y^{-3/4} = \tfrac{3}{40}x^{3/4}y^{-3/4}.$$

Therefore, we must have

$$\tfrac{9}{20}x^{-1/4}y^{1/4} = \tfrac{3}{40}x^{3/4}y^{-3/4}.$$

To solve for y in terms of x, let us multiply both sides of this equation by $x^{1/4}y^{3/4}$:

$$\tfrac{9}{20}y = \tfrac{3}{40}x$$

or

$$y = \tfrac{1}{6}x.$$

Inserting this result in (L-3), we find that

$$100x + 200(\tfrac{1}{6}x) = 30{,}000$$

$$\frac{400x}{3} = 30{,}000$$

$$x = 225.$$

Hence

$$y = \tfrac{225}{6} = 37.5.$$

So maximum production is achieved by using 225 units of labor and 37.5 units of capital.

In Example 3 it turns out that, at the optimum value of x and y,

$$\lambda = \tfrac{9}{20}x^{-1/4}y^{1/4} = \tfrac{9}{20}(225)^{-1/4}(37.5)^{1/4} \approx .2875$$

$$\frac{\partial f}{\partial x} = 45x^{-1/4}y^{1/4} = 45(225)^{-1/4}(37.5)^{1/4} \tag{6}$$

$$\frac{\partial f}{\partial y} = 15x^{3/4}y^{-3/4} = 15(225)^{3/4}(37.5)^{-3/4}. \tag{7}$$

It can be shown that the Lagrange multiplier λ can be interpreted as the marginal productivity of money. That is, if one additional dollar is available, then approximately .2875 additional units of the product can be produced.

Recall that the partial derivatives $\dfrac{\partial f}{\partial x}$ and $\dfrac{\partial f}{\partial y}$ are called the marginal productivity of labor and capital, respectively. From (6) and (7) we have

$$\frac{[\text{marginal productivity of labor}]}{[\text{marginal productivity of capital}]} = \frac{45(225)^{-1/4}(37.5)^{1/4}}{15(225)^{3/4}(37.5)^{-3/4}}$$

$$= \frac{45}{15}(225)^{-1}(37.5)^{1}$$

$$= \frac{3(37.5)}{225} = \frac{37.5}{75} = \frac{1}{2}.$$

On the other hand,

$$\frac{[\text{cost per unit of labor}]}{[\text{cost per unit of capital}]} = \frac{100}{200} = \frac{1}{2}.$$

This result illustrates the following law of economics. *If labor and capital are at their optimal levels, then the ratio of their marginal productivities equals the ratio of their unit costs.*

The method of Lagrange multipliers generalizes to functions of any number of variables. For instance, we can maximize $f(x, y, z)$, subject to the constraint equation $g(x, y, z) = 0$, by forming the Lagrange function

$$F(x, y, z, \lambda) = f(x, y, z) + \lambda g(x, y, z).$$

The analogs of equations (L-1) to (L-3) are

$$\frac{\partial F}{\partial x} = 0$$

$$\frac{\partial F}{\partial y} = 0$$

$$\frac{\partial F}{\partial z} = 0$$

$$\frac{\partial F}{\partial \lambda} = 0.$$

Let us now show how we can solve the heat-loss problem of Section 3 by using this method.

EXAMPLE 4 Use Lagrange multipliers to find the values of x, y, z that minimize the objective function

$$f(x, y, z) = 11xy + 14yz + 15xz,$$

subject to the constraint

$$xyz = 147,840.$$

Solution   The Lagrange function is

$$F(x, y, z, \lambda) = 11xy + 14yz + 15xz + \lambda(147,840 - xyz).$$

The conditions for a minimum are

$$\frac{\partial F}{\partial x} = 11y + 15z - \lambda yz = 0$$

$$\frac{\partial F}{\partial y} = 11x + 14z - \lambda xz = 0$$

$$\frac{\partial F}{\partial z} = 14y + 15x - \lambda xy = 0$$

$$\frac{\partial F}{\partial \lambda} = 147,840 - xyz = 0. \tag{8}$$

From the first three equations we have

$$\left. \begin{aligned} \lambda &= \frac{11y + 15z}{yz} = \frac{11}{z} + \frac{15}{y} \\[2mm] \lambda &= \frac{11x + 14z}{xz} = \frac{11}{z} + \frac{14}{x} \\[2mm] \lambda &= \frac{14y + 15x}{xy} = \frac{14}{x} + \frac{15}{y} \end{aligned} \right\}. \tag{9}$$

Let us equate the first two expressions for λ:

$$\frac{11}{z} + \frac{15}{y} = \frac{11}{z} + \frac{14}{x}$$

$$\frac{15}{y} = \frac{14}{x}$$

$$x = \frac{14}{15} y.$$

Next, we equate the second and third expressions for λ in (9):

$$\frac{11}{z} + \frac{14}{x} = \frac{14}{x} + \frac{15}{y}$$

$$\frac{11}{z} = \frac{15}{y}$$

$$z = \frac{11}{15} y.$$

We now substitute the expressions for x and z into the constraint equation (8) and obtain

$$\tfrac{14}{15}y \cdot y \cdot \tfrac{11}{15}y = 147{,}840$$

$$y^3 = \frac{(147{,}840)(15)^2}{(14)(11)} = 216{,}000$$

$$y = 60.$$

From this, we find that

$$x = \tfrac{14}{15}(60) = 56 \qquad \text{and} \qquad z = \tfrac{11}{15}(60) = 44.$$

We conclude that the heat loss is minimized when $x = 56$, $y = 60$, and $z = 44$.

In the solution of Example 4, we found that at the optimal values of x, y, and z,

$$\frac{14}{x} = \frac{15}{y} = \frac{11}{z}.$$

Referring to Example 2 of Section 1, we see that 14 is the combined heat loss through the east and west sides of the building, 15 is the heat loss through the north and south sides of the building, and 11 is the heat loss through the floor and roof. Thus we have that under optimal conditions

$$\frac{[\text{heat loss through east and west sides}]}{[\text{distance between east and west sides}]} = \frac{[\text{heat loss through north and south sides}]}{[\text{distance between north and south sides}]}$$

$$= \frac{[\text{heat loss through floor and roof}]}{[\text{distance between floor and roof}]}.$$

This is a principle of optimal design: minimal heat loss occurs when the distance between each pair of opposite sides is some fixed constant times the heat loss from the pair of sides.

The value of λ in Example 4 corresponding to the optimal values of x, y, and z is

$$\lambda = \frac{11}{z} + \frac{15}{y} = \frac{11}{44} + \frac{15}{60} = \frac{1}{2}.$$

One can show that the Lagrange multiplier λ is the marginal heat loss with respect to volume. That is, if a building of volume slightly more than 147,840 cubic feet is optimally designed, then $\frac{1}{2}$ unit of additional heat will be lost for each additional cubic foot of volume.

PRACTICE PROBLEMS 4

1. Let $F(x, y, \lambda) = 2x + 3y + \lambda(90 - 6x^{1/3}y^{2/3})$. Find $\dfrac{\partial F}{\partial x}$.

2. Refer to Exercise 23 of Section 3. What is the function $F(x, y, \lambda)$ when the exercise is solved using the method of Lagrange multipliers?

EXERCISES 4

Solve the following exercises by the method of Lagrange multipliers.

1. Minimize the function $x^2 + 3y^2 + 10$, subject to the constraint $8 - x - y = 0$.

2. Maximize the function $x^2 - y^2$, subject to the constraint $2x + y - 3 = 0$.

3. Maximize $x^2 + xy - 3y^2$, subject to the constraint $2 - x - 2y = 0$.

4. Minimize $\frac{1}{2}x^2 - 3xy + y^2 + \frac{1}{2}$, subject to the constraint $3x - y - 1 = 0$.

5. Find the values of x, y that maximize the function
$$-2x^2 - 2xy - \frac{3}{2}y^2 + x + 2y,$$
subject to the constraint $x + y - \frac{5}{2} = 0$.

6. Find the values of x, y that minimize the function
$$x^2 + xy + y^2 - 2x - 5y,$$
subject to the constraint $1 - x + y = 0$.

7. Find the values of x, y, z that maximize the function
$$3x + 5y + z - x^2 - y^2 - z^2,$$
subject to the constraint $6 - x - y - z = 0$.

8. Find the values of x, y, z that minimize the function
$$x^2 + y^2 + z^2 - 3x - 5y - z,$$
subject to the constraint $20 - 2x - y - z = 0$.

9. The material for a rectangular box costs $2 per square foot for the top and $1 per square foot for the sides and bottom. Using Lagrange multipliers, find the dimensions for which the volume of the box is 12 cubic feet and the cost of the materials is minimized. [Referring to Fig. 2(a), the cost will be $3xy + 2xz + 2yz$.]

10. Use Lagrange multipliers to find the three positive numbers whose sum is 15 and whose product is as large as possible.

11. Find the dimensions of the rectangle of maximum area that can be inscribed in the unit circle. [See Fig. 1(a).] That is, find the values of x, y that maximize $4xy$, subject to the constraint $x^2 + y^2 = 1$.

FIGURE 1

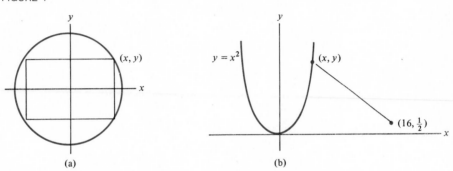

(a)

(b)

12. Find the point on the parabola $y = x^2$ that has minimal distance from the point $(16, \frac{1}{2})$. [See Fig. 1(b).] [*Suggestion:* If d denotes the distance from (x, y) to $(16, \frac{1}{2})$, then $d^2 = (x - 16)^2 + (y - \frac{1}{2})^2$. If d^2 is minimized, then d will be minimized. Thus it suffices to minimize $(x - 16)^2 + (y - \frac{1}{2})^2$, subject to the constraint $y - x^2 = 0$.]

13. Find the dimensions of an open rectangular glass tank of volume 32 cubic feet for which the amount of material needed to construct the tank is minimized. [See Fig. 2(a).] That is, find the values of x, y, z that minimize $xy + 2xz + 2yz$, subject to the constraint $32 - xyz = 0$.

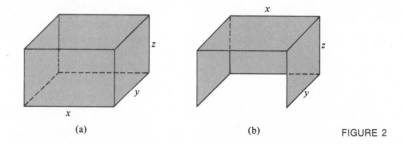

(a)   (b)   FIGURE 2

14. A shelter for use at the beach has a back, two sides, and a top made of canvas. [See Fig. 2(b).] Find the dimensions that maximize the volume and require 96 square feet of canvas. That is, find the values of x, y, z that maximize xyz, subject to the constraint $xz + xy + 2yz - 96 = 0$.

15. The production function for a firm is $f(x, y) = 64x^{3/4}y^{1/4}$, where x and y are the number of units of labor and capital utilized. Suppose that labor costs \$96 per unit and capital costs \$162 per unit and that the firm decides to produce 3456 units of goods.

   (a) Determine the amounts of labor and capital that should be utilized in order to minimize the cost. That is, find the values of x, y that minimize $96x + 162y$, subject to the constraint $3456 - 64x^{3/4}y^{1/4} = 0$.

   (b) Find the value of λ at the optimal level of production.

   (c) Show that, at the optimal level of production, we have

   $$\frac{[\text{marginal productivity of labor}]}{[\text{marginal productivity of capital}]} = \frac{[\text{unit price of labor}]}{[\text{unit price of capital}]}.$$

16. The production function for a firm is $f(x, y) = 1000\sqrt{6x^2 + y^2}$, where x and y are, respectively, the number of units of labor and capital utilized. Suppose that labor costs \$480 per unit and capital costs \$40 per unit and that the firm has \$5000 to spend.

   (a) Determine the amounts of labor and capital that should be utilized in order to maximize production.

   (b) Find the value of λ.

   (c) Repeat part (c) of Exercise 15 for $f(x, y) = 1000\sqrt{6x^2 + y^2}$.

17. Suppose that a firm makes two products A and B that use the same raw materials. Given a fixed amount of raw materials and a fixed amount of manpower, the firm must decide how much of its resources should be allocated to the production of A and how much to

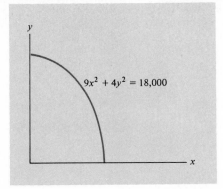

FIGURE 3

B. If x units of A and y units of B are produced, suppose that x and y must satisfy

$$9x^2 + 4y^2 = 18,\!000.$$

The graph of this equation (for $x \geqslant 0$, $y \geqslant 0$) is called a *production possibilities curve* (Fig. 3). A point (x, y) on this curve represents a *production schedule* for the firm, committing it to produce x units of A and y units of B. The reason for the relationship between x and y involves the limitations on personnel and raw materials available to the firm. Suppose that each unit of A yields a \$3 profit, whereas each unit of B yields a \$4 profit. Then the profit of the firm is

$$P(x, y) = 3x + 4y.$$

Find the production schedule that maximizes the profit function $P(x, y)$.

18. Consider the monopolist of Example 2, Section 3, who sells his goods in two countries. Suppose that he must set the same price in each country. That is, $97 - (x/10) = 83 - (y/20)$. Find the values of x and y that maximize profits under this new restriction.

19. Let $f(x, y)$ be any production function where x represents labor (costing \$$a$ per unit) and y represents capital (costing \$$b$ per unit). Assuming that \$$c$ is available, show that, at the values of x, y that maximize production,

$$\frac{\dfrac{\partial f}{\partial x}}{\dfrac{\partial f}{\partial y}} = \frac{a}{b}.$$

[*Note:* Let $F(x, y, \lambda) = f(x, y) + \lambda(c - ax - by)$. The result follows from (L-1) and (L-2).]

20. By applying the result in Exercise 19 to the production function $f(x, y) = kx^{\alpha}y^{\beta}$, show that, for the values of x, y that maximize production, we have

$$\frac{y}{x} = \frac{a\beta}{b\alpha}.$$

(This tells us that the ratio of capital to labor does not depend on the amount of money available nor on the level of production but only on the numbers a, b, α, and β.)

SOLUTIONS TO PRACTICE PROBLEMS 4

1. The function can be written as

$$F(x, y, \lambda) = 2x + 3y + \lambda \cdot 90 - \lambda \cdot 6x^{1/3}y^{2/3}.$$

When differentiating with respect to x, both y and λ should be treated as constants (so $\lambda \cdot 90$ and $\lambda \cdot 6$ are also regarded as constants).

$$\frac{\partial F}{\partial x} = 2 - \lambda \cdot 6 \cdot \tfrac{1}{3}x^{-2/3} \cdot y^{2/3}$$
$$= 2 - \lambda \cdot 2 \cdot x^{-2/3}y^{2/3}.$$

[*Note:* It is not necessary to write out the multiplication by λ as we did. Most people just do this mentally and then differentiate.]

2. The quantity to be maximized is the volume xyl. The constraint is that length plus girth is 84. This translates to $84 = l + 2x + 2y$ or $84 - l - 2x - 2y = 0$. Therefore,

$$F(x, y, l, \lambda) = xyl + \lambda(84 - l - 2x - 2y).$$

7.5. Total Differentials and Their Applications

In Section 2 we saw that the partial derivatives of $f(x, y)$ indicate how much the function changes with respect to small changes in its variables. In particular, if h and k are small, then, at the point (a, b), a change in x of h units produces a change in $f(x, y)$ of approximately $\left[\dfrac{\partial f}{\partial x}(a, b)\right] \cdot h$ units, and a change in y of k units produces a change in $f(x, y)$ of approximately $\left[\dfrac{\partial f}{\partial y}(a, b)\right] \cdot k$ units. Therefore, it is not surprising that when both coordinates are changed, the change in $f(x, y)$ is approximated by the sum of these terms. Precisely, we have

$$f(a + h, b + k) - f(a, b) \approx \left[\frac{\partial f}{\partial x}(a, b)\right] \cdot h + \left[\frac{\partial f}{\partial y}(a, b)\right] \cdot k, \qquad (1)$$

where the approximation improves as h, k approach zero.

The expression on the right side of (1) is usually called a *total differential*. Its value depends on h and k, as well as the partial derivatives at $x = a$, $y = b$.

Formula (1) generalizes to functions of any number of variables. For example, for a function of three variables, $f(x, y, z)$, we have

$$f(a + h, b + k, c + l) - f(a, b, c) \approx \frac{\partial f}{\partial x} \cdot h + \frac{\partial f}{\partial y} \cdot k + \frac{\partial f}{\partial z} \cdot l, \qquad (2)$$

where each of the partial derivatives is evaluated at $x = a$, $y = b$, $z = c$.

We shall use total differentials to approximate function values, estimate errors, and interpret Lagrange multipliers.

EXAMPLE 1   Approximate $60 \cdot (80)^{3/4} \cdot (17)^{1/4}$.

Solution   Let $f(x, y) = 60x^{3/4}y^{1/4}$, $a = 81$, $b = 16$, $h = -1$, and $k = 1$. Then, by formula (1), we have

$$f(81 - 1, 16 + 1) - f(81, 16) \approx \left[\frac{\partial f}{\partial x}(81, 16)\right] \cdot (-1) + \left[\frac{\partial f}{\partial y}(81, 16)\right] \cdot 1.$$

Using the values computed in Example 3 of Section 1 and Example 8 of Section 2, we have

$$f(80, 17) - 3240 \approx 30(-1) + 50\tfrac{5}{8}(1) = 20\tfrac{5}{8}$$

$$f(80, 17) \approx 20\tfrac{5}{8} + 3240 = 3260\tfrac{5}{8}.$$

Therefore $60 \cdot (80)^{3/4} \cdot (17)^{1/4} \approx 3260\tfrac{5}{8}$. [The actual value of $60 \cdot (80)^{3/4} \cdot (17)^{1/4}$ is $3258.97 \ldots$, so that our approximation yields an error of about 1.65.]

EXAMPLE 2 A rectangle is measured and found to have dimensions of 3.3 and 4.6 millimeters, with an error of at most .1 millimeter in each measurement. Approximate the maximum error that might occur if these numbers are used to compute the area of the rectangle.

Solution The true dimensions are $3.3 + h$ and $4.6 + k$, where $|h|$ and $|k|$ are at most .1 millimeter. Let $f(x, y) = x \cdot y$. Then

$$[\text{error}] = [\text{true area}] - [\text{computed area}]$$

$$= f(3.3 + h, 4.6 + k) - f(3.3, 4.6)$$

$$\approx \left[\frac{\partial f}{\partial x}(3.3, 4.6) \right] \cdot h + \left[\frac{\partial f}{\partial y}(3.3, 4.6) \right] \cdot k.$$

Since $\dfrac{\partial f}{\partial x} = y$ and $\dfrac{\partial f}{\partial y} = x$,

$$[\text{error}] \approx (4.6) \cdot h + (3.3) \cdot k.$$

In addition, since h and k might be positive or negative, the safest estimate is

$$|\text{error}| \leq |(4.6)h| + |(3.3)k|$$

$$\leq (4.6)(.1) + (3.3)(.1)$$

$$\leq .46 + .33 = .79.$$

Consequently, the maximum error in the area of the rectangle is approximately .79 square millimeter.

EXAMPLE 3 *(Cardiac Output)* The quantity of blood pumped through the lungs per minute can be computed by measuring the amount of oxygen (O_2) used by the body per minute and the concentration of oxygen in well-chosen blood samples. Let

$C =$ number of liters of blood pumped into lungs per minute,

$x =$ number of milliliters of O_2 utilized by the body per minute (computed by analyzing expelled air),

$y =$ number of milliliters of O_2 per liter of arterial blood (blood sampled from an artery after absorbing O_2 from the lungs),

$z =$ number of milliliters of O_2 per liter of mixed venous blood (blood sampled from veins before entering the lungs).

Then, for each minute,

[amount of O_2 utilized] = [amount of O_2 gained per liter of blood]
·[number of liters of blood pumped];

that is,

$$x = (y - z) \cdot C.$$

Equivalently,

$$C = \frac{x}{y - z}.$$

Suppose that, for a particular person at rest, the measurements $x = 250$, $y = 180$, $z = 140$ are taken and used to compute $C = 6.25$. Approximate the maximum error in C if each measurement might be in error by at most 1%.

Solution Let $f(x, y, z) = x/(y - z)$. The partial derivatives of $f(x, y, z)$ are

$$\frac{\partial f}{\partial x} = \frac{1}{y - z},$$

$$\frac{\partial f}{\partial y} = \frac{-x}{(y - z)^2},$$

$$\frac{\partial f}{\partial z} = \frac{x}{(y - z)^2}.$$

The values of the partial derivatives when $x = 250$, $y = 180$, and $z = 140$ are

$$\frac{\partial f}{\partial x} = \frac{1}{180 - 140} = \frac{1}{40},$$

$$\frac{\partial f}{\partial y} = \frac{-250}{(40)^2} = -\frac{5}{32},$$

$$\frac{\partial f}{\partial z} = \frac{250}{(40)^2} = \frac{5}{32}.$$

By formula (2), the error in C is approximately

$$\frac{1}{40} h - \frac{5}{32} k + \frac{5}{32} l,$$

where h, k, and l are the errors in the measurements of x, y, and z, respectively. Since each error is at most 1%,

$$|h| \leq 2.5$$
$$|k| \leq 1.8$$
$$|l| \leq 1.4.$$

Moreover, since h, k, and l could each be positive or negative, the safest estimate is

$$|\text{error}| \leq \left| \frac{1}{40} h \right| + \left| \frac{-5}{32} k \right| + \left| \frac{5}{32} l \right|$$

$$\leq \frac{1}{40}(2.5) + \frac{5}{32}(1.8) + \frac{5}{32}(1.4) = .5625.$$

Since .5625 is 9% of 6.25, we see that small errors in the measurements of the oxygen can result in large errors in the computation of cardiac output.

EXAMPLE 4   (*Production*)   In Example 3 of Section 4 we found that, with $30,000 to spend, the production function $f(x, y) = 60x^{3/4}y^{1/4}$ was maximized when $x = 225$ units of labor (costing $100 per unit) and $y = 37.5$ units of capital (costing $200 per unit) were employed. We also found that $\lambda \approx .2875$. From (6) and (7) of Section 4 it follows that

$$\frac{\partial f}{\partial x}(225, 37.5) = 100\lambda, \qquad \frac{\partial f}{\partial y}(225, 37.5) = 200\lambda.$$

Show that λ is the marginal productivity of money. That is, if one extra dollar is spent, then approximately λ additional units can be produced.

Solution   With one additional dollar, we can purchase h units of labor and k units of capital, where

$$100 \cdot h + 200 \cdot k = 1.$$

By (1), the additional number of units of goods produced is

$$f(225 + h, 37.5 + k) - f(225, 37.5) \approx \left[\frac{\partial f}{\partial x}(225, 37.5) \right] \cdot h + \left[\frac{\partial f}{\partial y}(225, 37.5) \right] \cdot k$$

$$\approx (100\lambda)h + (200\lambda)k$$

$$= \lambda(100h + 200k) = \lambda \cdot 1 = \lambda.$$

Therefore, approximately an additional .2875 unit of goods can be produced.

PRACTICE PROBLEMS 5

1. Let $f(x, y) = x^4 - (y/x^2)$. Use the total differential to approximate $f(1.02, 1.99)$.

2. Consider the production function $f(x, y) = 60x^{3/4}y^{1/4}$ of Example 8, Section 2. Suppose that currently 81 units of labor and 16 units of capital are being employed. Use the total differential to estimate the increase in production from employing two more units of labor and one more unit of capital.

EXERCISES 5

1. Let $f(x, y) = 2x^3 + 3xy$. Approximate $f(1.001, 1)$.

2. Let $f(x, y) = x^{100}y^3$. Approximate $f(.99, 2.02)$.

3. Let $f(x, y) = ye^x$. Approximate $f(.01, .98)$.

4. Let $f(x, y) = xe^{-y}$. Approximate $f(7.06, .03)$.

5. Let $f(x, y) = y/(1 - x)$. Approximate $f(-.01, 5)$.

6. Let $f(x, y) = \sqrt{x}/y$. Approximate $f(3.98, .99)$.

7. Let $f(x, y) = \sqrt{xy}$. Approximate $f(11.9, 3.2)$.

8. Let $f(x, y) = x^{2/3}y^3$. Approximate $f(7.9, -1.2)$.

9. Let $f(x, y, z) = x^{10}y^2z^3$. Approximate $f(1.01, 2.97, 1.98)$.

10. Let $f(x, y, z) = \sqrt{z}e^{x-y^2}$. Approximate $f(4.1, 1.9, 9.1)$.

11. Suppose that the profit for a firm selling two products at prices p and q, respectively, is given by

$$P(p, q) = -100{,}000 + 5000p + 10{,}000q - 50p^2 - 100q^2 + 10pq.$$

Thus when $p = 100$ and $q = 50$, the total profit of the firm is \$200,000. Use the total differential to approximate the effect of a \$1 increase in the price p and a \$.20 increase in the price q.

12. Let S denote the sales of a certain product, p the unit price, and a the number of dollars spent per unit in advertising. Studies have shown that

$$S = 100{,}000 + 20{,}000\left(1 - \frac{1}{a + 1}\right)e^{-p}.$$

(a) Calculate the total sales when $a = \$1$, $p = \$3$.

(b) Approximate the effect of reducing a to \$.95 and increasing p to \$3.10.

13. The volume of a cylinder is $\pi x^2 y$, where x is the radius of the base and y is the height. A cylindrical container is measured and found to have radius 20 millimeters and height 100 millimeters. Find the approximate error in calculating the volume of the container if each measurement is in error by at most 1 millimeter.

14. The volume of a certain gas is related to the temperature and pressure of the gas by the formula $V = .08(T/P)$, where V is the volume in liters, P is the pressure in atmospheres, and T is the temperature in degrees Kelvin ($^\circ\text{K} = {}^\circ\text{C} + 273$). A container of the gas is found to have temperature 293°K and pressure 20 atmospheres. Find the approximate maximum error in computing the volume if the temperature and pressure readings might be in error by at most 1 unit.

15. The solution to Exercise 15 of Section 4 is $x = 81$, $y = 16$, $\lambda = 3$. Show that λ is the marginal cost of production. That is, the cost of producing one additional unit of goods is approximately \$3.

16. The solution to Exercise 16 of Section 4 is $x = 10$, $y = 5$, $\lambda = 5$. Show that λ is the marginal productivity of money. That is, if $1 additional is available, then approximately five more units of goods can be produced.

17. *(Renal Clearance)* An important function of a person's kidneys is to remove urea (a nitrogenous end product of protein decomposition) from the blood. A standard clinical measure of the health of the kidneys is the so-called maximum clearance rate at which the blood is cleared of urea:

$$C = \frac{UV}{S},$$

where C is expressed in milliliters of blood cleared per minute, U is the concentration of urea in the urine (in mg/100 ml), S is the concentration of urea in the blood (in mg/100 ml), and V is the volume of urine excreted per minute (ml/min).

Suppose that laboratory tests for a patient indicate that $U = 200$, $V = 3.00$, and $S = 12.0$, so that $C = 50$ ml/min. Estimate the maximum error in C if the error in each measurement does not exceed 1%.

18. *(Renal Clearance, continued)* For certain physiological reasons, physicians use the formula

$$f(U, V, S) = \frac{U\sqrt{V}}{S}$$

to calculate the "standard urea clearance" whenever the value of V (see Exercise 17) drops below 2 ml/min. Suppose that laboratory tests for a patient indicate that $U = 500$, $V = 1.44$, $S = 20$. Estimate the maximum error in $f(U, V, S)$ if the error in each measurement does not exceed 1%.

SOLUTIONS TO PRACTICE PROBLEMS 5

1. Evaluating the function at $(1.02, 1.99)$ is not as easy as evaluating it at $(1, 2)$, a nearby point.

$$f(1, 2) = 1^4 - \frac{2}{1^2} = 1 - 2 = -1$$

$$\frac{\partial f}{\partial x} = 4x^3 + 2\frac{y}{x^3}; \qquad \frac{\partial f}{\partial y} = -\frac{1}{x^2}$$

$$\frac{\partial f}{\partial x}(1, 2) = 4 + 4 = 8; \qquad \frac{\partial f}{\partial y}(1, 2) = -1.$$

$(1.02, 1.99) = (1 + h, 2 + k)$ where $h = .02$, $k = -.01$. By (1),

$$f(1.02, 1.99) - f(1, 2) \approx \left[\frac{\partial f}{\partial x}(1, 2)\right] \cdot h + \left[\frac{\partial f}{\partial y}(1, 2)\right] \cdot k$$

$$f(1.02, 1.99) - (-1) \approx (8) \cdot (.02) + (-1)(-.01)$$

$$f(1.02, 1.99) + 1 \approx .16 + .01 = .17$$

$$f(1.02, 1.99) \approx .17 - 1 = -.83.$$

2. We have shown in Example 8 of Section 2 that

$$\frac{\partial f}{\partial x}(81, 16) = 30 \qquad \text{and} \qquad \frac{\partial f}{\partial y}(81, 16) = 50\tfrac{5}{8}.$$

By (1),

$$f(83, 17) - f(81, 16) \approx \left[\frac{\partial f}{\partial x}(81, 16)\right] \cdot 2 + \left[\frac{\partial f}{\partial y}(81, 16)\right] \cdot 1$$

$$\approx 30 \cdot 2 + 50\tfrac{5}{8} \cdot 1$$

$$= 110\tfrac{5}{8}.$$

That is, approximately $110\tfrac{5}{8}$ additional units can be produced.

7.6. The Method of Least Squares

Modern man compiles graphs of literally thousands of different quantities: the purchasing value of the dollar as a function of time, the pressure of a fixed volume of air as a function of temperature, the average income of people as a function of their years of formal education, or the incidence of strokes as a function of blood pressure. The observed points on such graphs tend to be irregularly distributed due both to the complicated nature of the phenomena underlying them as well as to errors made in observation (for example, a given procedure for measuring average income may not count certain groups). In spite of the imperfect nature of the data, we are often faced with the problem of making assessments and predictions based on them. Roughly speaking, this problem amounts to filtering the sources of errors in the data and isolating the basic underlying trend. Frequently, on the basis of a suspicion or a working hypothesis, we may suspect that the underlying trend is linear—that is, the data should lie on a straight line. But which straight line? This is the problem that the *method of least squares* attempts to answer. To be more specific, let us consider the following problem:

> **PROBLEM** Given observed data points $(x_1, y_1), (x_2, y_2), \ldots, (x_N, y_N)$ on a graph, find the straight line that "best" fits these points.

In order to completely understand the statement of the problem being considered, we must define what it means for a line to "best" fit a set of points. If (x_i, y_i) is one of our observed points, then we will measure how far it is from a given line $y = Ax + B$ by the vertical distance from the point to the line. Since the point on the line with x-coordinate x_i is $(x_i, Ax_i + B)$, this vertical distance is the distance between the y-coordinates $Ax_i + B$ and y_i (see Fig. 1). If $E_i = (Ax_i + B) - y_i$, then either E_i or $-E_i$ is the vertical distance from (x_i, y_i) to the line. It turns out to be convenient to work with the square of this vertical distance, namely

$$E_i^2 = (Ax_i + B - y_i)^2.$$

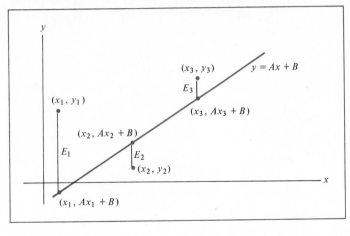

FIGURE 1

The total error in approximating the data points $(x_1, y_1), \ldots, (x_N, y_N)$ by the line $y = Ax + B$ is usually measured by the sum E of the squares of the vertical distances from the points to the line,

$$E = E_1^2 + E_2^2 + \cdots + E_N^2.$$

E is called the *least-squares error* of the observed points with respect to the line. If all the observed points lie on the line $y = Ax + B$, then all E_i are zero and the error E is zero. If a given observed point is far away from the line, the corresponding E_i^2 is large and hence makes a large contribution to the error E.

In general, we cannot expect to find a line $y = Ax + B$ that fits the observed points so well that the error E is zero. Actually, this situation will occur only if the observed points lie on a straight line. However, we can rephrase our original problem as follows:

PROBLEM Given observed data points $(x_1, y_1), (x_2, y_2), \ldots, (x_N, y_N)$, find a straight line $y = Ax + B$ for which the error E is as small as possible.

It turns out that this problem is a minimization problem in the two variables A and B and so can be solved by using the methods of Section 3. Let us consider an example.

EXAMPLE 1 Find the straight line that minimizes the least-squares error for the points $(1, 4)$, $(2, 5)$, $(3, 8)$.

Solution Let the straight line be $y = Ax + B$. When $x = 1, 2, 3$, the y-coordinate of the corresponding point of the line is $A + B, 2A + B, 3A + B$, respectively. Therefore, the squares of the vertical distances from the points $(1, 4), (2, 5), (3, 8)$ are,

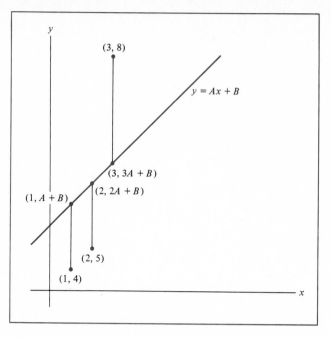

FIGURE 2

respectively,

$$E_1^2 = (A + B - 4)^2$$
$$E_2^2 = (2A + B - 5)^2$$
$$E_3^2 = (3A + B - 8)^2.$$

(See Fig. 2.) Thus the least-squares error is

$$E_1^2 + E_2^2 + E_3^2 = (A + B - 4)^2 + (2A + B - 5)^2 + (3A + B - 8)^2.$$

This error obviously depends on the choice of A and B. Let $f(A, B)$ denote this least-squares error. We want to find values of A and B that minimize $f(A, B)$. To do so, we take partial derivatives with respect to A and B and set the partial derivatives equal to zero:

$$\frac{\partial f}{\partial A} = 2(A + B - 4) + 2(2A + B - 5) \cdot 2 + 2(3A + B - 8) \cdot 3$$

$$= 28A + 12B - 76 = 0,$$

$$\frac{\partial f}{\partial B} = 2(A + B - 4) + 2(2A + B - 5) + 2(3A + B - 8)$$

$$= 12A + 6B - 34 = 0.$$

In order to find A and B, we must solve the system of simultaneous linear equations

$$28A + 12B = 76$$
$$12A + 6B = 34.$$

Multiplying the second equation by 2 and subtracting from the first equation, we have $4A = 8$ or $A = 2$. Therefore, $B = \frac{5}{3}$, and the straight line that minimizes the least-squares error is $y = 2x + \frac{5}{3}$.

We may follow a similar procedure for finding the line $y = Ax + B$ when more observed points are given.

EXAMPLE 2 The following table* gives the crude male death rate for lung cancer in 1950 and the per capita consumption of cigarettes in 1930 in various countries.

	Cigarette consumption (per capita)	Lung cancer deaths (per million males)
Norway	250	95
Sweden	300	120
Denmark	350	165
Australia	470	170

(a) Use the method of least squares to obtain the straight line that best fits these data.

(b) In 1930 the per capita cigarette consumption in Finland was 1100. Use the straight line found in part (a) to estimate the male lung cancer death rate in Finland in 1950.

Solution (a) The points are plotted in Fig. 3. Let the line that best fits these points have equation $y = Ax + B$. The least-squares error of the given data points is

$$E = (250A + B - 95)^2 + (300A + B - 120)^2$$
$$+ (350A + B - 165)^2 + (470A + B - 170)^2.$$

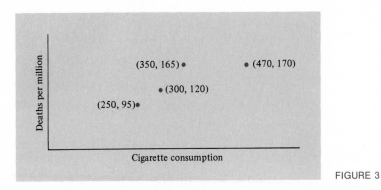

FIGURE 3

* These data were obtained from *Smoking and Health*, Report of the Advisory Committee to the Surgeon General of the Public Health Service, U.S. Department of Health, Education, and Welfare, Washington, D.C., Public Health Service Publication No. 1103, p. 176.

We determine the values of A and B for which E is minimized by taking the partial derivatives with respect to A and B and setting them equal to zero. Computing the partial derivatives, we get

$$\frac{\partial E}{\partial A} = 2(250A + B - 95) \cdot 250 + 2(300A + B - 120) \cdot 300$$

$$+ 2(350A + B - 165) \cdot 350 + 2(470A + B - 170) \cdot 470$$

$$\frac{\partial E}{\partial B} = 2(250A + B - 95) + 2(300A + B - 120)$$

$$+ 2(350A + B - 165) + 2(470A + B - 170).$$

If we set $\dfrac{\partial E}{\partial A} = 0$ and $\dfrac{\partial E}{\partial B} = 0$, the following system of equations results:

$$495,900A + 1370B = 197,400 \qquad (1)$$

$$1370A + 4B = 550. \qquad (2)$$

There are several ways to solve this system of equations. We will solve equation (2) for B and then insert this expression for B into equation (1) so as to obtain an equation in A alone. Solving (2) for B gives

$$B = \frac{550 - 1370A}{4} = 137.5 - 342.5A. \qquad (3)$$

Now substitute (3) into (1) to obtain

$$495,900A + 1370(137.5 - 342.5A) = 197,400$$

$$[495,900 - (1370)(342.5)]A = 197,400 - (1370)(137.5)$$

$$26,675A = 9025$$

$$A = \frac{9025}{26,675} \approx .338.$$

Then substitute this value for A into (3) to find

$$B = 137.5 - 342.5(.338) = 21.735.$$

Therefore, the straight line that best fits these points is

$$y = .338x + 21.735.$$

(b) We use the straight line to estimate the lung cancer death rate in Finland by setting $x = 1100$. Then we get

$$y = .338(1100) + 21.735 = 393.535 \approx 394.$$

Therefore, we estimate the lung cancer death rate in Finland to be 394. (*Note:* The actual rate was 350.)

PRACTICE PROBLEMS 6

1. Let $E = (A + B + 2)^2 + (3A + B)^2 + (6A + B - 8)^2$. What is $\dfrac{\partial E}{\partial A}$?

2. Find the formula (of the type in Problem 1) that gives the least-squares error for the points $(1, 10), (5, 8), (7, 0)$.

EXERCISES 6

Find the straight line that best fits the following data points, where "best" is meant in the sense of least squares.

1. $(1, 0), (2, -1), (3, -4)$ 2. $(2, 2), (3, 0), (7, -1)$

3. $(1, 6), (3, 0), (6, 10)$ 4. $(1, 1), (3, 2), (5, 4)$

5. $(0, 1), (1, 1), (2, 2), (3, 2)$ 6. $(0, 2), (1, 2), (2, 2), (3, 3)$

7. $(1, 0), (2, 1), (4, 2), (5, 3)$ 8. $(2, 3), (3, 2), (5, 1), (6, 0)$

9. $(1, 0), (2, 1), (3, 5), (15, 11), (-1, 4)$ 10. $(0, 7), (-1, 5), (1, 9), (3, 14), (-2, 0)$

11. An ecologist wished to know whether certain species of aquatic insects have their ecological range limited by temperature. He collected the following data, relating the average daily temperature at different portions of a creek with the elevation of that portion of the creek (above sea level).*

Elevation (kilometers)	Average temperature (degrees Celsius)
2.7	11.2
2.8	10
3.0	8.5
3.5	7.5

(a) Find the straight line that provides the best least-squares fit to these data.

(b) Use the linear function to estimate the average daily temperature for this creek at altitude 3.2 kilometers.

12. A soap manufacturer hopes to determine the demand curve for his soap by setting different prices in each of three cities of the same size and observing the volume of sales in each city. The results are given on page 392.

---

    * The authors express their thanks to Dr. J. David Allen, Department of Zoology, University of Maryland, for suggesting this exercise.

Price (cents)	Volume of sales (thousands of cases)
30	120
35	100
40	90

(a) Let x be the price and y the volume. Find the straight line that best fits these data.

(b) Use the demand curve of part (a) to estimate the volume of sales if the price is 50 cents.

13. The trend of sales of a certain new car dealer is as shown in the table.

Week number	Number of cars sold
1	38
2	40
3	41
4	39
5	45

Make a prediction of the number of cars sold during the seventh week.

14. A company analyzes production and profit figures and determines that in recent years productivity and profits have been related as shown in the table below.

Productivity (thousands of units)	Profits
78	$52,000
83	55,000
100	68,000
110	77,000
129	97,000

For the current year, their plant has been increased to allow production of 150,000 units. Estimate the profits.

SOLUTIONS TO PRACTICE PROBLEMS 6

1. $\dfrac{\partial E}{\partial A} = 2(A + B + 2) \cdot 1 + 2(3A + B) \cdot 3 + 2(6A + B - 8) \cdot 6$

$= (2A + 2B + 4) + (18A + 6B) + (72A + 12B - 96)$

$= 92A + 20B - 92.$

(Notice that we used the general power rule when differentiating and so had to always multiply by the derivative of the quantity inside the parentheses. Also, you might be tempted to first square the terms in the expression for E and then differentiate. We recommend that you resist that temptation.)

2. $E = (A + B - 10)^2 + (5A + B - 8)^2 + (7A + B)^2$. In general, E is a sum of squares, one for each point being fitted. The point (a, b) gives rise to the term $(aA + B - b)^2$.

7.7. Double Integrals

Up to this point, our discussion of calculus of several variables has been confined to the study of differentiation. Let us now take up the topic of integration of functions of several variables. As has been the case throughout most of this chapter, we restrict our discussion to functions $f(x, y)$ of two variables.

Let us begin with some motivation. Before we define the concept of integral for functions of several variables, let us review the essential features of the integral in one variable.

Consider the definite integral $\int_a^b f(x)\, dx$. To write down this integral, it takes two pieces of information. The first is the function $f(x)$. The second is the interval over which the integration is to be performed. In this case, the interval is the portion of the x-axis from $x = a$ to $x = b$. The value of the definite integral is a number. In case the function $f(x)$ is nonnegative throughout the interval from $x = a$ to $x = b$, this number equals the area under the graph of $f(x)$ from $x = a$ to $x = b$. (See Fig. 1.) If $f(x)$ is negative for some values of x in the interval, then the integral still equals the area under the graph, but areas below the x-axis are counted as negative.

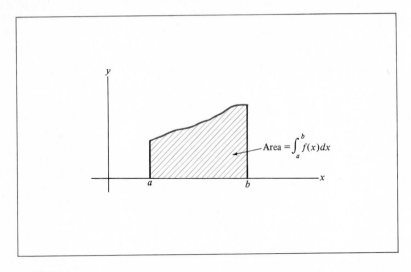

FIGURE 1

Let us generalize the above ingredients to a function $f(x, y)$ of two variables. First, we must provide a two-dimensional analog of the interval from $x = a$ to $x = b$. This is easy. We take a two-dimensional region R of the plane, such as the region shown in Fig. 2. As our generalization of $f(x)$, we take a function $f(x, y)$ of two variables. Our generalization of the definite integral is denoted

$$\iint\limits_R f(x, y)\, dx\, dy$$

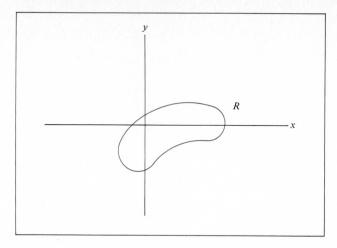

FIGURE 2   A region in the xy-plane.

and is called the *double integral of f(x, y) over the region R*. The value of the double integral is a number defined as follows. For the sake of simplicity, let us begin by assuming that $f(x, y) \geq 0$ for all points (x, y) in the region R. (This is the analog of the assumption that $f(x) \geq 0$ for all x in the interval from $x = a$ to $x = b$.) This means that the graph of f lies above the region R in three-dimensional space. (See Fig. 3.) The portion of the graph over R determines a solid figure. (See Fig. 4.)

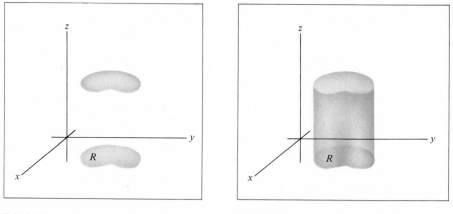

FIGURE 3   FIGURE 4

This figure is called the *solid bounded by f(x, y) over the region R*. We define the double integral $\iint\limits_{R} f(x, y)\, dx\, dy$ to be the volume of this solid. In case the graph of $f(x, y)$ lies partially above the region R and partially below, we define the double integral to be the volume of the solid above the region minus the volume of the solid below the region. That is, we count volumes below the xy-plane as negative.

Now that we have defined the notion of a double integral, we must learn how to calculate its value. To do so, let us introduce the notion of an iterated integral. Let $f(x, y)$ be a function of two variables, let $g(x)$ and $h(x)$ be two functions of x alone, and let a and b be numbers. Then an *iterated integral* is an expression of the form

$$\int_a^b \left(\int_{g(x)}^{h(x)} f(x, y)\, dy \right) dx.$$

To explain the meaning of this collection of symbols, we proceed from the inside out. We evaluate the integral

$$\int_{g(x)}^{h(x)} f(x, y)\, dy$$

by considering $f(x, y)$ as a function of y alone. We treat x as a constant in this integration. So we evaluate the integral by first finding an antiderivative $F(x, y)$ with respect to y. The integral above is then evaluated as

$$F(x, h(x)) - F(x, g(x)).$$

That is, we evaluate the antiderivative between the limits $y = g(x)$ and $y = h(x)$. This gives us a function of x alone. To complete the evaluation of the iterated integral, we integrate this function from $x = a$ to $x = b$. The next two examples illustrate the procedure for evaluating iterated integrals.

EXAMPLE 1 Evaluate the iterated integral $\int_1^2 \left(\int_3^4 (y - x)\, dy \right) dx.$

Solution Here $g(x)$ and $h(x)$ are constant functions: $g(x) = 3$ and $h(x) = 4$. We evaluate the inner integral first. The variable in this integral is y, so we treat x as a constant.

$$\int_3^4 (y - x)\, dy = \left. (\tfrac{1}{2}y^2 - xy) \right|_3^4$$

$$= (\tfrac{1}{2} \cdot 16 - x \cdot 4) - (\tfrac{1}{2} \cdot 9 - x \cdot 3)$$

$$= 8 - 4x - \tfrac{9}{2} + 3x$$

$$= \tfrac{7}{2} - x.$$

Now we carry out the integration with respect to x:

$$\int_1^2 (\tfrac{7}{2} - x)\, dx = \left. \tfrac{7}{2}x - \tfrac{1}{2}x^2 \right|_1^2$$

$$= (\tfrac{7}{2} \cdot 2 - \tfrac{1}{2} \cdot 4) - (\tfrac{7}{2} - \tfrac{1}{2} \cdot 1)$$

$$= (7 - 2) - (3) = 2.$$

So the value of the iterated integral is 2.

EXAMPLE 2 Evaluate the iterated integral

$$\int_0^1 \left(\int_{\sqrt{x}}^{x+1} 2xy\, dy \right) dx.$$

Solution We evaluate the inner integral first.

$$\int_{\sqrt{x}}^{x+1} 2xy\,dy = xy^2\Big|_{\sqrt{x}}^{x+1}$$

$$= x(x+1)^2 - x(\sqrt{x})^2$$

$$= x(x^2 + 2x + 1) - x \cdot x$$

$$= x^3 + 2x^2 + x - x^2$$

$$= x^3 + x^2 + x.$$

Now we evaluate the outer integral.

$$\int_0^1 (x^3 + x^2 + x)\,dx = \tfrac{1}{4}x^4 + \tfrac{1}{3}x^3 + \tfrac{1}{2}x^2\Big|_0^1$$

$$= \tfrac{1}{4} + \tfrac{1}{3} + \tfrac{1}{2} = \tfrac{13}{12}.$$

So the value of the iterated integral is $\tfrac{13}{12}$.

Let us now return to the discussion of the double integral $\iint\limits_R f(x,y)\,dx\,dy$.

When the region R has a special form, the double integral may be expressed as an iterated integral. Namely let us suppose that R is bounded by the graphs of $y = g(x)$, $y = h(x)$ and by the vertical lines $x = a$ and $x = b$. (See Fig. 5.) In this case, we have the following fundamental result, which we cite without proof.

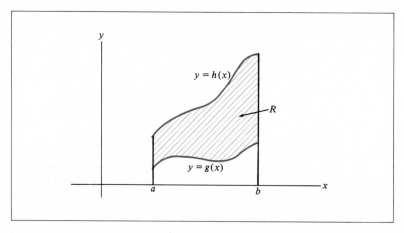

FIGURE 5

Let R be the region in the xy-plane bounded by the graphs of $y = g(x)$, $y = h(x)$, and the vertical lines $x = a$, $x = b$. Then

$$\iint\limits_R f(x,y)\,dx\,dy = \int_a^b\left(\int_{g(x)}^{h(x)} f(x,y)\,dy\right)dx.$$

Since the value of the double integral gives the volume of the solid bounded by the graph $f(x, y)$ over the region R, the result above may be used to calculate volumes, as the next two examples show.

EXAMPLE 3 Calculate the volume of the solid bounded above by the function $y - x$ and lying over the rectangular region $R: 1 \le x \le 2, 3 \le y \le 4$. (See Fig. 6.)

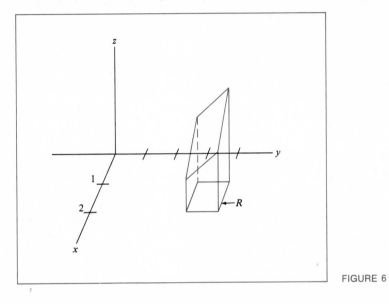

FIGURE 6

Solution The desired volume is given by the double integral $\iint\limits_{R} (y - x) \, dx \, dy$. By the result cited above, this double integral is equal to the iterated integral

$$\int_1^2 \left(\int_3^4 (y - x) \, dy \right) dx.$$

The value of this iterated integral was shown in Example 1 to be 2, so the volume of the solid shown in Fig. 6 is 2.

EXAMPLE 4 Calculate $\iint\limits_{R} 2xy \, dx \, dy$, where R is the region shown in Fig. 7.

Solution The region R is bounded below by $y = \sqrt{x}$, above by $y = x + 1$, on the left by $x = 0$ and on the right by $x = 1$. Therefore,

$$\iint\limits_{R} 2xy \, dx \, dy = \int_0^1 \left(\int_{\sqrt{x}}^{x+1} 2xy \, dy \right) dx$$

$$= \tfrac{13}{12} \quad \text{(by Example 2)}.$$

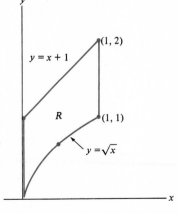

FIGURE 7

In our discussion, we have confined ourselves to iterated integrals in which the inner integral was with respect to y. In a completely analogous manner, we may treat iterated integrals in which the inner integral is with respect to x. Such iterated integrals may be used to evaluate double integrals over regions R bounded by curves of the form $x = g(y)$, $x = h(y)$, and horizontal lines $y = a$, $y = b$. The computations are analogous to those given above.

PRACTICE PROBLEMS 7

1. Calculate the iterated integral

$$\int_0^2 \left(\int_0^{x/2} e^{2y-x} \, dy \right) dx.$$

2. Calculate

$$\iint_R e^{2y-x} \, dx \, dy$$

where R is the region in Fig. 8.

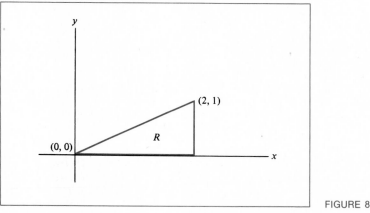

FIGURE 8

EXERCISES 7

Calculate the following iterated integrals.

1. $\int_0^1 \left(\int_0^1 e^{x+y} \, dy \right) dx$

2. $\int_{-1}^1 \left(\int_0^2 (3x^3 + y^2) \, dy \right) dx.$

3. $\int_{-2}^0 \left(\int_{-1}^1 xe^{xy} \, dy \right) dx$

4. $\int_1^3 \left(\int_2^5 (2y - 3x) \, dy \right) dx$

5. $\int_1^4 \left(\int_x^{x^2} xy \, dy \right) dx$

6. $\int_0^2 \left(\int_x^{5x} y \, dy \right) dx$

7. $\int_{-1}^{1} \left(\int_{x}^{2x} (x + y) \, dy \right) dx$

8. $\int_{0}^{1} \left(\int_{0}^{x} (x + e^y) \, dy \right) dx$

Let R be the rectangle consisting of all points (x, y) such that $0 \le x \le 2$, $2 \le y \le 3$. Calculate the following double integrals. Interpret each as a volume.

9. $\iint_{R} xy^2 \, dx \, dy$

10. $\iint_{R} (xy + y^2) \, dx \, dy$

11. $\iint_{R} e^{-x-y} \, dx \, dy$

12. $\iint_{R} e^{y-x} \, dx \, dy$

Calculate the volumes over the following regions R and bounded above by the graph of $f(x, y) = x^2 + y^2$.

13. R is the rectangle bounded by the lines $x = 1$, $x = 3$, $y = 0$, $y = 1$.

14. R is the region bounded by the lines $x = 0$, $x = 1$ and the curves $y = 0$ and $y = \sqrt[3]{x}$.

SOLUTIONS TO PRACTICE PROBLEMS 7

1.
$$\int_{0}^{2} \left(\int_{0}^{x/2} e^{2y-x} \, dy \right) dx = \int_{0}^{2} \left(\tfrac{1}{2} e^{2y-x} \Big|_{0}^{x/2} \right) dx$$

$$= \int_{0}^{2} \left(\tfrac{1}{2} e^{2(x/2)-x} - \tfrac{1}{2} e^{2 \cdot 0 - x} \right) dx$$

$$= \int_{0}^{2} \left(\tfrac{1}{2} - \tfrac{1}{2} e^{-x} \right) dx$$

$$= \tfrac{1}{2} x + \tfrac{1}{2} e^{-x} \Big|_{0}^{2}$$

$$= \tfrac{1}{2} \cdot 2 + \tfrac{1}{2} e^{-2} - \tfrac{1}{2} \cdot 0 - \tfrac{1}{2} e^{-0}$$

$$= 1 + \tfrac{1}{2} e^{-2} - 0 - \tfrac{1}{2}$$

$$= \tfrac{1}{2} + \tfrac{1}{2} e^{-2}.$$

2. The line passing through the points $(0, 0)$ and $(2, 1)$ has equation $y = x/2$. Hence the region R is bounded below by $y = 0$, above by $y = x/2$, on the left by $x = 0$ and on the right by $x = 2$. Therefore,

$$\iint_{R} e^{2y-x} \, dx \, dy = \int_{0}^{2} \left(\int_{0}^{x/2} e^{2y-x} \, dy \right) dx = \tfrac{1}{2} + \tfrac{1}{2} e^{-2}$$

by Problem 1.

Chapter 7: CHECKLIST

☐ Level curve
☐ Partial derivative
☐ $\dfrac{\partial f}{\partial x}(a, b)$, $\dfrac{\partial f}{\partial y}(a, b)$
☐ Second partial derivatives

- ☐ Maxima and minima in several variables
- ☐ Second-derivative test in two variables
- ☐ Method of Lagrange multipliers:
     objective function
     constraint equation
- ☐ Total differential
- ☐ Method of least squares
- ☐ Double integral
- ☐ Iterated integral

Chapter 7: SUPPLEMENTARY EXERCISES

1. Let $f(x, y) = x\sqrt{y}/(1 + x)$. Compute $f(2, 9)$, $f(5, 1)$, and $f(0, 0)$.

2. Let $f(x, y, z) = x^2 e^{y/z}$. Compute $f(-1, 0, 1)$, $f(1, 3, 3)$, and $f(5, -2, 2)$.

3. If A dollars is deposited in a bank (assuming a 6% continuous interest rate), the amount in the account after t years is $f(A, t) = Ae^{.06t}$. Find and interpret $f(10, 11.5)$.

4. Let $f(x, y, \lambda) = xy + \lambda(5 - x - y)$. Find $f(1, 2, 3)$.

5. Let $f(x, y) = 3x^2 + xy + 5y^2$. Find $\dfrac{\partial f}{\partial x}$ and $\dfrac{\partial f}{\partial y}$.

6. Let $f(x, y) = 3x - \frac{1}{2}y^4 + 1$. Find $\dfrac{\partial f}{\partial x}$ and $\dfrac{\partial f}{\partial y}$.

7. Let $f(x, y) = e^{x/y}$. Find $\dfrac{\partial f}{\partial x}$ and $\dfrac{\partial f}{\partial y}$.

8. Let $f(x, y) = x/(x - 2y)$. Find $\dfrac{\partial f}{\partial x}$ and $\dfrac{\partial f}{\partial y}$.

9. Let $f(x, y, z) = x^3 - yz^2$. Find $\dfrac{\partial f}{\partial x}, \dfrac{\partial f}{\partial y}$ and $\dfrac{\partial f}{\partial z}$.

10. Let $f(x, y, \lambda) = xy + \lambda(5 - x - y)$. Find $\dfrac{\partial f}{\partial x}, \dfrac{\partial f}{\partial y}$, and $\dfrac{\partial f}{\partial \lambda}$.

11. Let $f(x, y) = x^3 y + 8$. Compute $\dfrac{\partial f}{\partial x}(1, 2)$ and $\dfrac{\partial f}{\partial y}(1, 2)$.

12. Let $f(x, y, z) = (x + y)z$. Evaluate $\dfrac{\partial f}{\partial y}$ at $(x, y, z) = (2, 3, 4)$.

13. Let $f(x, y) = x^5 - 2x^3 y + \dfrac{1}{2}y^4$. Find $\dfrac{\partial^2 f}{\partial x^2}, \dfrac{\partial^2 f}{\partial y^2}, \dfrac{\partial^2 f}{\partial x \, \partial y}$, and $\dfrac{\partial^2 f}{\partial y \, \partial x}$.

14. Let $f(x, y) = 2x^3 + x^2 y - y^2$. Compute $\dfrac{\partial^2 f}{\partial x^2}, \dfrac{\partial^2 f}{\partial y^2}$, and $\dfrac{\partial^2 f}{\partial x \, \partial y}$ at $(x, y) = (1, 2)$.

15. A dealer in a certain brand of electronic calculator finds that (within certain limits) the number of calculators he can sell is given by $f(p, t) = -p + 6t - .02pt$, where p is the

price of the calculator and t is the number of dollars spent on advertising. Compute $\dfrac{\partial f}{\partial p}$ (25, 10,000) and $\dfrac{\partial f}{\partial t}$ (25, 10,000) and interpret these numbers.

16. Suppose that the crime rate in a certain city can be approximated by a function $f(x, y, z)$, where x is the unemployment rate, y is the amount of social services available, and z is the size of the police force. Explain why $\dfrac{\partial f}{\partial x} > 0$, $\dfrac{\partial f}{\partial y} < 0$, and $\dfrac{\partial f}{\partial z} < 0$.

In Exercises 17–20, find all points (x, y) where $f(x, y)$ has a possible maximum or minimum.

17. $f(x, y) = -x^2 + 2y^2 + 6x - 8y + 5$.

18. $f(x, y) = x^2 + 3xy - y^2 - x - 8y + 4$.

19. $f(x, y) = x^3 + 3x^2 + 3y^2 - 6y + 7$.

20. $f(x, y) = \frac{1}{2}x^2 + 4xy + y^3 + 8y^2 + 3x + 2$.

In Exercises 21–23, find all points (x, y) where $f(x, y)$ has a possible maximum or minimum. Then use the second-derivative test to determine, if possible, the nature of $f(x, y)$ at each of these points. If the second-derivative test is inconclusive, so state.

21. $f(x, y) = x^2 + 3xy + 4y^2 - 13x - 30y + 12$.

22. $f(x, y) = 7x^2 - 5xy + y^2 + x - y + 6$.

23. $f(x, y) = x^3 + y^2 - 3x - 8y + 12$.

24. Find the values of x, y, z at which
$$f(x, y, z) = x^2 + 4y^2 + 5z^2 - 6x + 8y + 3$$
assumes its minimum value.

25. Using the method of Lagrange multipliers, maximize the function $3x^2 + 2xy - y^2$, subject to the constraint $5 - 2x - y = 0$.

26. Using the method of Lagrange multipliers, find the values of x, y that minimize the function $-x^2 - 3xy - \frac{1}{2}y^2 + y + 10$, subject to the constraint $10 - x - y = 0$.

27. Using the method of Lagrange multipliers, find the values of x, y, z that minimize the function $3x^2 + 2y^2 + z^2 + 4x + y + 3z$, subject to the constraint $4 - x - y - z = 0$.

28. Using the method of Lagrange multipliers, find the dimensions of the rectangular box of volume 1000 cubic inches for which the sum of the dimensions is minimized.

29. A person wants to plant a rectangular garden along one side of a house and put a fence on the other three sides of the garden. (See Fig. 1.) Using the method of Lagrange

FIGURE 1

multipliers, find the dimensions of the garden of greatest area that can be enclosed by using 40 feet of fencing.

30. The solution to Exercise 29 is $x = 10$, $y = 20$, $\lambda = 10$. Suppose that one additional foot of fencing becomes available and that now the optimal dimensions for the garden are $10 + h$ and $20 + k$, where $2h + k = 1$. Use the total differential to show that the increased area is approximately $\lambda = 10$ square feet. In other words, λ is the marginal change in area with respect to change in the length of fencing.

31. Let $f(x, y) = x\sqrt{y}$. Use the total differential to approximate $f(12.1, 3.8)$.

32. Let $f(x, y) = x^5 + (y/x)$. Use the total differential to approximate $f(.98, 2.99)$.

33. Let $f(x, y, z) = (x^2 + y)/(z - 1)$. Use the total differential to approximate $f(1.98, 1.99, 2.01)$.

34. The surface area of a square-ended rectangular box is $2x^2 + 4xy$, where x is the dimension of the square side. Such a box is measured and found to have $x = 30$ millimeters and $y = 50$ millimeters. Estimate the maximum error in calculating the surface area of the box if each measurement is in error by at most 1 millimeter.

In Exercises 35–37, find the straight line that best fits the following data points, where "best" is meant in the sense of least squares.

35. (1, 1), (2, 3), (3, 6)   36. (1,1), (3, 4), (5, 7)

37. (0, 1), (1, −1), (2, −3), (3, −5),

In Exercises 38 and 39, calculate the iterated integral.

38. $\int_0^4 \left(\int_0^1 x\sqrt{y} + y\, dx \right) dy$ 39. $\int_1^4 \left(\int_0^5 (2xy^4 + 3)\, dx \right) dy$

In Exercises 40 and 41, let R be the rectangle consisting of all points (x, y) such that $0 \le x \le 4$, $1 \le y \le 3$, and calculate the double integral.

40. $\iint_R (2x + 3y)\, dx\, dy$ 41. $\iint_R 5\, dx\, dy$

42. The present value of y dollars after x years at 15% continuous interest is $f(x, y) = ye^{-.15x}$. Sketch some sample level curves. (Economists call this collection of level curves a "discount system.")

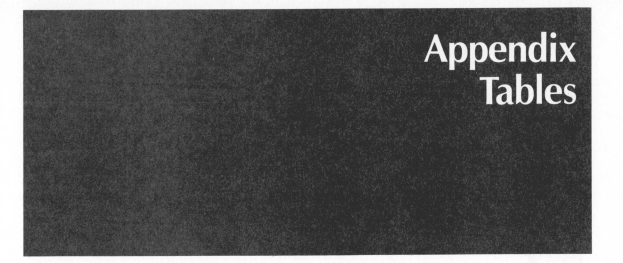

Appendix
Tables

TABLE 1   The Exponential Function

x	e^x	e^{-x}	x	e^x	e^{-x}	x	e^x	e^{-x}
.00	1.00000	1.00000	**.30**	1.34986	.74082	**.60**	1.82212	.54881
.01	1.01005	.99005	.31	1.36343	.73345	.61	1.84043	.54335
.02	1.02020	.98020	.32	1.37713	.72615	.62	1.85893	.53794
.03	1.03045	.97045	.33	1.39097	.71892	.63	1.87761	.53259
.04	1.04081	.96079	.34	1.40495	.71177	.64	1.89648	.52729
.05	1.05127	.95123	.35	1.41907	.70469	.65	1.91554	.52205
.06	1.06184	.94176	.36	1.43333	.69768	.66	1.93479	.51685
.07	1.07251	.93239	.37	1.44773	.69073	.67	1.95424	.51171
.08	1.08329	.92312	.38	1.46228	.68386	.68	1.97388	.50662
.09	1.09417	.91393	.39	1.47698	.67706	.69	1.99372	.50158
.10	1.10517	.90484	**.40**	1.49182	.67032	**.70**	2.01375	.49659
.11	1.11628	.89583	.41	1.50682	.66365	.71	2.03399	.49164
.12	1.12750	.88692	.42	1.52196	.65705	.72	2.05443	.48675
.13	1.13883	.87810	.43	1.53726	.65051	.73	2.07508	.48191
.14	1.15027	.86936	.44	1.55271	.64404	.74	2.09594	.47711
.15	1.16183	.86071	.45	1.56831	.63763	.75	2.11700	.47237
.16	1.17351	.85214	.46	1.58407	.63128	.76	2.13828	.46767
.17	1.18530	.84366	.47	1.59999	.62500	.77	2.15977	.46301
.18	1.19722	.83527	.48	1.61607	.61878	.78	2.18147	.45841
.19	1.20925	.82696	.49	1.63232	.61263	.79	2.20340	.45384
.20	1.22140	.81873	**.50**	1.64872	.60653	**.80**	2.22554	.44933
.21	1.23368	.81058	.51	1.66529	.60050	.81	2.24791	.44486
.22	1.24608	.80252	.52	1.68203	.59452	.82	2.27050	.44043
.23	1.25860	.79453	.53	1.69893	.58860	.83	2.29332	.43605
.24	1.27125	.78663	.54	1.71601	.58275	.84	2.31637	.43171
.25	1.28403	.77880	.55	1.73325	.57695	.85	2.33965	.42741
.26	1.29693	.77105	.56	1.75067	.57121	.86	2.36316	.42316
.27	1.30996	.76338	.57	1.76827	.56553	.87	2.38691	.41895
.28	1.32313	.75578	.58	1.78604	.55990	.88	2.41090	.41478
.29	1.33643	.74826	.59	1.80399	.55433	.89	2.43513	.41066

TABLE 1   The Exponential Function (*continued*)

x	e^x	e^{-x}	x	e^x	e^{-x}	x	e^x	e^{-x}
.90	2.45960	.40657	**1.40**	4.05520	.24660	**3.0**	20.086	.04979
.91	2.48432	.40252	1.41	4.09596	.24414	3.1	22.198	.04505
.92	2.50929	.39852	1.42	4.13712	.24171	3.2	24.533	.04076
.93	2.53451	.39455	1.43	4.17870	.23931	3.3	27.113	.03688
.94	2.55998	.39063	1.44	4.22070	.23693	3.4	29.964	.03337
.95	2.58571	.38674	1.45	4.26311	.23457	3.5	33.115	.03020
.96	2.61170	.38289	1.46	4.30596	.23224	3.6	36.598	.02732
.97	2.63794	.37908	1.47	4.34924	.22993	3.7	40.447	.02472
.98	2.66446	.37531	1.48	4.39295	.22764	3.8	44.701	.02237
.99	2.69123	.37158	1.49	4.43710	.22537	3.9	49.402	.02024
1.00	2.71828	.36788	**1.50**	4.48169	.22313	**4.0**	54.598	.01832
1.01	2.74560	.36422	1.51	4.52673	.22091	4.1	60.340	.01657
1.02	2.77319	.36059	1.52	4.57223	.21871	4.2	66.686	.01500
1.03	2.80107	.35701	1.53	4.61818	.21654	4.3	73.700	.01357
1.04	2.82922	.35345	1.54	4.66459	.21438	4.4	81.451	.01228
1.05	2.85765	.34994	1.55	4.71147	.21225	4.5	90.017	.01111
1.06	2.88637	.34646	1.56	4.75882	.21014	4.6	99.484	.01005
1.07	2.91538	.34301	1.57	4.80665	.20805	4.7	109.947	.00910
1.08	2.94468	.33960	1.58	4.85496	.20598	4.8	121.510	.00823
1.09	2.97427	.33622	1.59	4.90375	.20393	4.9	134.290	.00745
1.10	3.00417	.33287	**1.60**	4.95303	.20190	**5.0**	148.41	.00674
1.11	3.03436	.32956	1.61	5.00281	.19989	5.1	164.02	.00610
1.12	3.06485	.32628	1.62	5.05309	.19790	5.2	181.27	.00552
1.13	3.09566	.32303	1.63	5.10387	.19593	5.3	200.34	.00499
1.14	3.12677	.31982	1.64	5.15517	.19398	5.4	221.41	.00452
1.15	3.15819	.31664	1.65	5.20698	.19205	5.5	244.69	.00409
1.16	3.18993	.31349	1.66	5.25931	.19014	5.6	270.43	.00370
1.17	3.22199	.31037	1.67	5.31217	.18825	5.7	298.87	.00335
1.18	3.25437	.30728	1.68	5.36556	.18637	5.8	330.30	.00303
1.19	3.28708	.30422	1.69	5.41948	.18452	5.9	365.04	.00274
1.20	3.32012	.30119	**1.70**	5.47395	.18268	**6.0**	403.43	.00248
1.21	3.35348	.29820	1.71	5.52896	.18087	6.1	445.86	.00224
1.22	3.38719	.29523	1.72	5.58453	.17907	6.2	492.75	.00203
1.23	3.42123	.29229	1.73	5.64065	.17728	6.3	544.57	.00184
1.24	3.45561	.28938	1.74	5.69734	.17552	6.4	601.85	.00166
1.25	3.49034	.28650	1.75	5.75460	.17377	6.5	665.14	.00150
1.26	3.52542	.28365	1.80	6.04965	.16530	6.6	735.10	.00136
1.27	3.56085	.28083	1.85	6.35982	.15724	6.7	812.41	.00123
1.28	3.59664	.27804	1.90	6.68589	.14957	6.8	897.85	.00111
1.29	3.63279	.27527	1.95	7.02869	.14227	6.9	992.27	.00101
1.30	3.66930	.27253	**2.0**	7.3891	.13534	**7.0**	1096.6	.00091
1.31	3.70617	.26982	2.1	8.1662	.12246	7.5	1808.0	.00055
1.32	3.74342	.26714	2.2	9.0250	.11080	8.0	2981.0	.00034
1.33	3.78104	.26448	2.3	9.9742	.10026	8.5	4914.8	.00020
1.34	3.81904	.26185	2.4	11.0232	.09072	9.0	8103.1	.00012
1.35	3.85743	.25924	2.5	12.1825	.08208	9.5	13360	.00007
1.36	3.89619	.25666	2.6	13.4637	.07427	10.0	22026	.00005
1.37	3.93535	.25411	2.7	14.8797	.06721	10.5	36316	.00003
1.38	3.97490	.25158	2.8	16.4446	.06081	11.0	59874	.00002
1.39	4.01485	.24908	2.9	18.1741	.05502	11.5	98716	.00001

TABLE 2   The Natural Logarithm Function

x	$\ln x$	x	$\ln x$	x	$\ln x$	x	$\ln x$
		0.40	−0.91629	0.80	−0.22314	1.20	0.18232
0.01	−4.60517	0.41	−0.89160	0.81	−0.21072	1.21	0.19062
0.02	−3.91202	0.42	−0.86750	0.82	−0.19845	1.22	0.19885
0.03	−3.50656	0.43	−0.84397	0.83	−0.18633	1.23	0.20701
0.04	−3.21888	0.44	−0.82098	0.84	−0.17435	1.24	0.21511
0.05	−2.99573	0.45	−0.79851	0.85	−0.16252	1.25	0.22314
0.06	−2.81341	0.46	−0.77653	0.86	−0.15082	1.26	0.23111
0.07	−2.65926	0.47	−0.75502	0.87	−0.13926	1.27	0.23902
0.08	−2.52573	0.48	−0.73397	0.88	−0.12783	1.28	0.24686
0.09	−2.40795	0.49	−0.71335	0.89	−0.11653	1.29	0.25464
0.10	−2.30259	0.50	−0.69315	0.90	−0.10536	1.30	0.26236
0.11	−2.20727	0.51	−0.67334	0.91	−0.09431	1.31	0.27003
0.12	−2.12026	0.52	−0.65393	0.92	−0.08338	1.32	0.27763
0.13	−2.04022	0.53	−0.63488	0.93	−0.07257	1.33	0.28518
0.14	−1.96611	0.54	−0.61619	0.94	−0.06188	1.34	0.29267
0.15	−1.89712	0.55	−0.59784	0.95	−0.05129	1.35	0.30010
0.16	−1.83258	0.56	−0.57982	0.96	−0.04082	1.36	0.30748
0.17	−1.77196	0.57	−0.56212	0.97	−0.03046	1.37	0.31481
0.18	−1.71480	0.58	−0.54473	0.98	−0.02020	1.38	0.32208
0.19	−1.66073	0.59	−0.52763	0.99	−0.01005	1.39	0.32930
0.20	−1.60944	0.60	−0.51083	1.00	0.00000	1.40	0.33647
0.21	−1.56065	0.61	−0.49430	1.01	0.00995	1.41	0.34359
0.22	−1.51413	0.62	−0.47804	1.02	0.01980	1.42	0.35066
0.23	−1.46968	0.63	−0.46204	1.03	0.02956	1.43	0.35767
0.24	−1.42712	0.64	−0.44629	1.04	0.03922	1.44	0.36464
0.25	−1.38629	0.65	−0.43078	1.05	0.04879	1.45	0.37156
0.26	−1.34707	0.66	−0.41552	1.06	0.05827	1.46	0.37844
0.27	−1.30933	0.67	−0.40048	1.07	0.06766	1.47	0.38526
0.28	−1.27297	0.68	−0.38566	1.08	0.07696	1.48	0.39204
0.29	−1.23787	0.69	−0.37106	1.09	0.08618	1.49	0.39878
0.30	−1.20397	0.70	−0.35667	1.10	0.09531	1.50	0.40547
0.31	−1.17118	0.71	−0.34249	1.11	0.10436	1.51	0.41211
0.32	−1.13943	0.72	−0.32850	1.12	0.11333	1.52	0.41871
0.33	−1.10866	0.73	−0.31471	1.13	0.12222	1.53	0.42527
0.34	−1.07881	0.74	−0.30111	1.14	0.13103	1.54	0.43178
0.35	−1.04982	0.75	−0.28768	1.15	0.13976	1.55	0.43825
0.36	−1.02165	0.76	−0.27444	1.16	0.14842	1.56	0.44469
0.37	−0.99425	0.77	−0.26136	1.17	0.15700	1.57	0.45108
0.38	−0.96758	0.78	−0.24846	1.18	0.16551	1.58	0.45742
0.39	−0.94161	0.79	−0.23572	1.19	0.17395	1.59	0.46373

TABLE 2   The Natural Logarithm Function (*continued*)

x	$\ln x$	x	$\ln x$	x	$\ln x$	x	$\ln x$
1.60	0.47000	2.00	0.69315	6.00	1.79176	10.0	2.30259
1.61	0.47623	2.10	0.74194	6.10	1.80829	11.0	2.39790
1.62	0.48243	2.20	0.78846	6.20	1.82455	12.0	2.48491
1.63	0.48858	2.30	0.83291	6.30	1.84055	13.0	2.56495
1.64	0.49470	2.40	0.87547	6.40	1.85630	14.0	2.63906
1.65	0.50078	2.50	0.91629	6.50	1.87180	15.0	2.70805
1.66	0.50682	2.60	0.95551	6.60	1.88707	16.0	2.77259
1.67	0.51282	2.70	0.99325	6.70	1.90211	17.0	2.83321
1.68	0.51879	2.80	1.02962	6.80	1.91692	18.0	2.89037
1.69	0.52473	2.90	1.06471	6.90	1.93152	19.0	2.94444
1.70	0.53063	3.00	1.09861	7.00	1.94591	20.0	2.99573
1.71	0.53649	3.10	1.13140	7.10	1.96009	21.0	3.04452
1.72	0.54232	3.20	1.16315	7.20	1.97408	22.0	3.09104
1.73	0.54812	3.30	1.19392	7.30	1.98787	23.0	3.13549
1.74	0.55389	3.40	1.22378	7.40	2.00148	24.0	3.17805
1.75	0.55962	3.50	1.25276	7.50	2.01490	25.0	3.21888
1.76	0.56531	3.60	1.28093	7.60	2.02815	26.0	3.25810
1.77	0.57098	3.70	1.30833	7.70	2.04122	27.0	3.29584
1.78	0.57661	3.80	1.33500	7.80	2.05412	28.0	3.33220
1.79	0.58222	3.90	1.36098	7.90	2.06686	29.0	3.36730
1.80	0.58779	4.00	1.38629	8.00	2.07944	30.0	3.40120
1.81	0.59333	4.10	1.41099	8.10	2.09186	31.0	3.43399
1.82	0.59884	4.20	1.43508	8.20	2.10413	32.0	3.46574
1.83	0.60432	4.30	1.45862	8.30	2.11626	33.0	3.49651
1.84	0.60977	4.40	1.48160	8.40	2.12823	34.0	3.52636
1.85	0.61519	4.50	1.50408	8.50	2.14007	35.0	3.55535
1.86	0.62058	4.60	1.52606	8.60	2.15176	36.0	3.58352
1.87	0.62594	4.70	1.54756	8.70	2.16332	37.0	3.61092
1.88	0.63127	4.80	1.56862	8.80	2.17475	38.0	3.63759
1.89	0.63658	4.90	1.58924	8.90	2.18605	39.0	3.66356
1.90	0.64185	5.00	1.60944	9.00	2.19722	40.0	3.68888
1.91	0.64710	5.10	1.62924	9.10	2.20827	41.0	3.71357
1.92	0.65233	5.20	1.64866	9.20	2.21920	42.0	3.73767
1.93	0.65752	5.30	1.66771	9.30	2.23001	43.0	3.76120
1.94	0.66269	5.40	1.68640	9.40	2.24071	44.0	3.78419
1.95	0.66783	5.50	1.70475	9.50	2.25129	45.0	3.80666
1.96	0.67294	5.60	1.72277	9.60	2.26176	46.0	3.82864
1.97	0.67803	5.70	1.74047	9.70	2.27213	47.0	3.85015
1.98	0.68310	5.80	1.75786	9.80	2.28238	48.0	3.87120
1.99	0.68813	5.90	1.77495	9.90	2.29253	49.0	3.89182

TABLE 2   The Natural Logarithm Function (*continued*)

x	$\ln x$	x	$\ln x$	x	$\ln x$	x	$\ln x$
50.0	3.91202	90.0	4.49981	400.	5.99146	800.	6.68461
51.0	3.93183	91.0	4.51086	410.	6.01616	810.	6.69703
52.0	3.95124	92.0	4.52179	420.	6.04025	820.	6.70930
53.0	3.97029	93.0	4.53260	430.	6.06379	830.	6.72143
54.0	3.98898	94.0	4.54329	440.	6.08677	840.	6.73340
55.0	4.00733	95.0	4.55388	450.	6.10925	850.	6.74524
56.0	4.02535	96.0	4.56435	460.	6.13123	860.	6.75693
57.0	4.04305	97.0	4.57471	470.	6.15273	870.	6.76849
58.0	4.06044	98.0	4.58497	480.	6.17379	880.	6.77992
59.0	4.07754	99.0	4.59512	490.	6.19441	890.	6.79122
60.0	4.09434	100.	4.60517	500.	6.21461	900.	6.80239
61.0	4.11087	110.	4.70048	510.	6.23441	910.	6.81344
62.0	4.12713	120.	4.78749	520.	6.25383	920.	6.82437
63.0	4.14313	130.	4.86753	530.	6.27288	930.	6.83518
64.0	4.15888	140.	4.94164	540.	6.29157	940.	6.84588
65.0	4.17439	150.	5.01064	550.	6.30992	950.	6.85646
66.0	4.18965	160.	5.07517	560.	6.32794	960.	6.86693
67.0	4.20469	170.	5.13580	570.	6.34564	970.	6.87730
68.0	4.21951	180.	5.19296	580.	6.36303	980.	6.88755
69.0	4.23411	190.	5.24702	590.	6.38012	990.	6.89770
70.0	4.24850	200.	5.29832	600.	6.39693	1000.	6.90776
71.0	4.26268	210.	5.34711	610.	6.41346	—	—
72.0	4.27667	220.	5.39363	620.	6.42972	—	—
73.0	4.29046	230.	5.43808	630.	6.44572	—	—
74.0	4.30407	240.	5.48064	640.	6.46147	—	—
75.0	4.31749	250.	5.52146	650.	6.47697	—	—
76.0	4.33073	260.	5.56068	660.	6.49224	—	—
77.0	4.34381	270.	5.59842	670.	6.50728	—	—
78.0	4.35671	280.	5.63479	680.	6.52209	—	—
79.0	4.36945	290.	5.66988	690.	6.53669	—	—
80.0	4.38203	300.	5.70378	700.	6.55108	—	—
81.0	4.39445	310.	5.73657	710.	6.56526	—	—
82.0	4.40672	320.	5.76832	720.	6.57925	—	—
83.0	4.41884	330.	5.79909	730.	6.59304	—	—
84.0	4.43082	340.	5.82895	740.	6.60665	—	—
85.0	4.44265	350.	5.85793	750.	6.62007	—	—
86.0	4.45435	360.	5.88610	760.	6.63332	—	—
87.0	4.46591	370.	5.91350	770.	6.64639	—	—
88.0	4.47734	380.	5.94017	780.	6.65929	—	—
89.0	4.48864	390.	5.96615	790.	6.67203	—	—

TABLE 3   Table of Integrals

Forms Involving $ax + b$

1. $\displaystyle \int \frac{1}{x(ax + b)}\, dx = \frac{1}{b} \ln\left|\frac{x}{ax + b}\right| + C$

2. $\displaystyle \int \frac{x}{ax + b}\, dx = \frac{x}{a} - \frac{b}{a^2} \ln|ax + b| + C$

Forms Involving $(ax + b)(cx + d)$

3. $\displaystyle \int \frac{1}{(ax + b)(cx + d)}\, dx = \frac{1}{ad - bc} \ln\left|\frac{ax + b}{cx + d}\right| + C$

4. $\displaystyle \int \frac{x}{\cdot(ax + b)(cx + d)}\, dx = \frac{1}{ad - bc}\left[\frac{d}{c}\ln|cx + d| - \frac{b}{a}\ln|ax + b|\right] + C$

Forms Involving $\sqrt{ax + b}$

5. $\displaystyle \int \frac{x}{\sqrt{ax + b}}\, dx = \frac{2ax - 4b}{3a^2}\sqrt{ax + b} + C$

6. $\displaystyle \int \frac{1}{x\sqrt{ax + b}}\, dx = \frac{1}{\sqrt{b}}\ln\left|\frac{\sqrt{ax + b} - \sqrt{b}}{\sqrt{ax + b} + \sqrt{b}}\right| + C, \qquad (b > 0)$

Forms Involving $\sqrt{x^2 \pm a^2}$

7. $\displaystyle \int \sqrt{x^2 \pm a^2}\, dx = \frac{x}{2}\sqrt{x^2 \pm a^2} \pm \frac{a^2}{2}\ln\left|x + \sqrt{x^2 \pm a^2}\right| + C$

8. $\displaystyle \int \frac{1}{\sqrt{x^2 \pm a^2}}\, dx = \ln\left|x + \sqrt{x^2 \pm a^2}\right| + C$

Forms Involving $\sqrt{a^2 - x^2}$

9. $\displaystyle \int \frac{1}{x\sqrt{a^2 - x^2}}\, dx = -\frac{1}{a}\ln\left|\frac{a + \sqrt{a^2 - x^2}}{x}\right| + C$

10. $\displaystyle \int \frac{\sqrt{a^2 - x^2}}{x}\, dx = \sqrt{a^2 - x^2} - a\ln\left|\frac{a + \sqrt{a^2 - x^2}}{x}\right| + C$

Forms Involving e^x, $\ln x$

11. $\displaystyle \int x^m e^{kx}\, dx = \frac{x^m e^{kx}}{k} - \frac{m}{k}\int x^{m-1} e^{kx}\, dx$

12. $\displaystyle \int \frac{1}{a + be^{kx}}\, dx = \frac{x}{a} - \frac{1}{ak}\ln|a + be^{kx}| + C$

13. $\displaystyle \int x^m \ln ax\, dx = \frac{x^{m+1}}{m + 1}\ln ax - \frac{x^{m+1}}{(m + 1)^2} + C \qquad (m \neq -1)$

Answers to Odd-Numbered Exercises

CHAPTER 0

EXERCISES 0.1, page 11

1. $0, 10, 0, 70$ **3.** $0, 0, -\frac{9}{8}, a^3 + a^2 - a - 1$ **5.** $\frac{1}{3}, 3, \dfrac{a + 1}{a + 2}$ **7.** $a^2 - 1, a^2 + 2a$ **9.** (a) 1980 sales, (b) 60

11. $x \neq 1, 2$ **13.** $x < 3$ **15.** Function **17.** Not a function **19.** Not a function **21.** 1 **23.** 3 **25.** Positive
27. Positive **29.** $-1, 5, 9$ **31.** $.03$ **33.** $.04$ **35.** No **37.** Yes **39.** $(a + 1)^3$ **41.** $1, 3, 4$ **43.** $\pi, 3, 12$

45. $f(x) = \begin{cases} .06x & \text{for } 50 \leq x \leq 300 \\ .02x + 12 & \text{for } 300 < x \leq 600 \\ .015x + 15 & \text{for } 600 < x \end{cases}$

EXERCISES 0.2, page 20

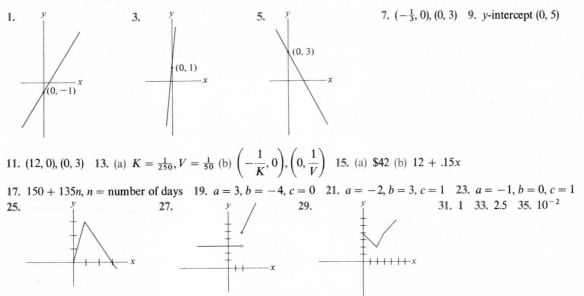

1. **3.** **5.** **7.** $\left(-\frac{1}{3}, 0\right), (0, 3)$ **9.** y-intercept $(0, 5)$

11. $(12, 0), (0, 3)$ **13.** (a) $K = \frac{1}{250}, V = \frac{1}{50}$ (b) $\left(-\dfrac{1}{K}, 0\right), \left(0, \dfrac{1}{V}\right)$ **15.** (a) \$42 (b) $12 + .15x$

17. $150 + 135n, n = $ number of days **19.** $a = 3, b = -4, c = 0$ **21.** $a = -2, b = 3, c = 1$ **23.** $a = -1, b = 0, c = 1$
25. **27.** **29.** **31.** 1 **33.** 2.5 **35.** 10^{-2}

EXERCISES 0.3, page 25

1. $x^2 + 9x + 1$ 3. $9x^3 + 9x$ 5. $\dfrac{t}{9} + \dfrac{1}{9t}$ 7. $\dfrac{3x + 1}{x^2 - x - 6}$ 9. $\dfrac{4x}{x^2 - 12x + 32}$ 11. $\dfrac{2x^2 + 5x + 50}{x^2 - 100}$

13. $\dfrac{2x^2 - 2x + 10}{x^2 + 3x - 10}$ 15. $\dfrac{-x^2 + 5x}{x^2 + 3x - 10}$ 17. $\dfrac{x^2 + 5x}{-x^2 + 7x - 10}$ 19. $\dfrac{-x^2 + 3x + 4}{x^2 + 5x - 6}$ 21. $\dfrac{-x^2 - 3x}{x^2 + 15x + 50}$

23. $\dfrac{5u - 1}{5u + 1}$ 25. $\left(\dfrac{x}{1-x}\right)^6$ 27. $\left(\dfrac{x}{1-x}\right)^3 - 5\left(\dfrac{x}{1-x}\right)^2 + 1$ 29. $\dfrac{t^3 - 5t^2 + 1}{-t^3 + 5t^2}$ 31. $2xh + h^2$ 33. $4 - 2t - h$

35. (a) $C(A(t)) = 3000 + 1600t - 40t^2$ (b) \$6040

EXERCISES 0.4, page 32

1. $2, \frac{3}{2}$ 3. $\frac{3}{2}$ 5. No zeros   7. $1, -\frac{1}{5}$ 9. $5, 4$ 11. $2 + \sqrt{6}/3, 2 - \sqrt{6}/3$ 13. $(x + 5)(x + 3)$ 15. $(x - 4)(x + 4)$
17. $3(x + 2)^2$ 19. $-2(x - 3)(x + 5)$ 21. $x(3 - x)$ 23. $-2x(x - \sqrt{3})(x + \sqrt{3})$ 25. $(-1, 1), (5, 19)$
27. $(-1, 9), (4, 4)$ 29. $(0, 8), (2, 6)$ 31. $(0, 5), (2 - \sqrt{3}, 25 - 23\sqrt{3}/2), (2 + \sqrt{3}, 25 + 23\sqrt{3}/2)$ 33. $-7, 3$
35. $-2, 3$ 37. -7

EXERCISES 0.5, page 36

1. 27 3. 1 5. $.0001$ 7. -16 9. 4 11. $.01$ 13. $\frac{1}{6}$ 15. 100 17. 16 19. 125 21. 1 23. 4 25. $\frac{1}{2}$

27. 1000 29. 10 31. 6 33. 16 35. 18 37. $\frac{4}{9}$ 39. 7 41. x^6y^6 43. x^3y^3 45. $\dfrac{1}{\sqrt{x}}$ 47. $\dfrac{x^{12}}{y^6}$ 49. $x^{12}y^{20}$

51. x^2y^6 53. $16x^4$ 55. x^2 57. $\dfrac{1}{x^7}$ 59. x 61. $\dfrac{27x^6}{8y^3}$ 63. $2\sqrt{x}$ 65. $\dfrac{1}{8x^6}$ 67. $\dfrac{1}{32x^2}$ 69. $9x^3$ 71. $x - 1$

73. $1 + 6\sqrt{x}$ 77. 16 79. $\frac{1}{4}$ 81. 8 83. $\frac{1}{32}$

CHAPTER 0: SUPPLEMENTARY EXERCISES, page 39

1. $2, 27\frac{1}{3}, -2, -2\frac{1}{8}, \dfrac{5\sqrt{2}}{2}$ 3. $a^2 - 4a + 2$ 5. $x \neq 0, -3$ 7. All x 9. Yes   11. $5x(x - 1)(x + 4)$

13. $(-1)(x - 6)(x + 3)$ 15. $-\frac{2}{5}, 1$ 17. $\left(\dfrac{5 + \sqrt{45}}{10}, \dfrac{\sqrt{45}}{5}\right), \left(\dfrac{5 - \sqrt{45}}{10}, -\dfrac{\sqrt{45}}{5}\right)$ 19. $x^2 + x - 1$ 21. $x^{5/2} - 2x^{3/2}$

23. $x^{3/2} - 2x^{1/2}$ 25. $\dfrac{x^2 - x + 1}{x^2 - 1}$ 27. $-\dfrac{3x^2 + 1}{3x^2 + 4x + 1}$ 29. $\dfrac{-3x^2 + 9x - 10}{3x^2 - 5x - 8}$ 31. $\dfrac{1}{x^4} - \dfrac{2}{x^2} + 4$ 33. $(\sqrt{x} - 1)^2$

35. $\dfrac{1}{(\sqrt{x} - 1)^2} - \dfrac{2}{\sqrt{x} - 1} + 4$ 37. $27, 32, 4$ 39. $301 + 10t + .04t^2$ 41. $x^2 + 2x + 1$ 43. x

CHAPTER 1

EXERCISES 1.1, page 50

1. -5 3. 0 5. $\frac{2}{7}$ 7. $-\frac{2}{3}$ 9. $y = 3x - 1$ 11. $y = x + 1$ 13. $y = 35 - 7x$ 15. $y = 4$ 17. $y = \dfrac{x}{2}$

19. $y = -2x$ 21. $y = 6 - 2x$ 23. $y = \frac{1}{2}x - \frac{1}{2}$ 25. $(2, 5); (0, 1); (-1, -1)$ 27. $(0, -\frac{5}{4}); (1, -\frac{3}{2}); (-2, -\frac{3}{4})$

29. l_1 **31.**

33.

35. $y - 1 = 2(x - 1)$ **37.** $y - \frac{1}{4} = -(x + \frac{1}{2})$
39. If the monopolist wants to sell one more unit of goods, then the price per unit must be lowered by 2 cents. No one is willing to pay \$7 or more for a unit of goods.

EXERCISES 1.2, page 56

1.   **3.**   **5.**

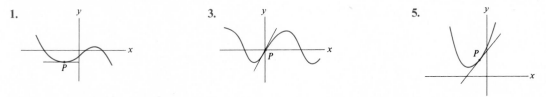

7. 1   **9.** -3 **11.** $\frac{1}{3}$ **13.** -4, $y - 4 = -4(x + 2)$ **15.** $\frac{8}{3}$, $y - \frac{16}{9} = \frac{8}{3}(x - \frac{4}{3})$ **17.** $y - 2.25 = 3(x - 1.5)$
19. $(\frac{5}{6}, \frac{25}{36})$ **21.** $(-\frac{1}{4}, \frac{1}{16})$ **23.** 12   **25.** $\frac{3}{4}$ **27.** $y + 1 = 3(x + 1)$

EXERCISES 1.3, page 65

1. 2   **3.** $8x^7$ **5.** $\frac{5}{2}x^{3/2}$ **7.** $\frac{1}{3}x^{-2/3}$ **9.** $-2x^{-3}$ **11.** $-\frac{1}{4}x^{-5/4}$ **13.** 0   **15.** $-3x^{-4}$ **17.** -192 **19.** $-\frac{1}{9}$ **21.** -1
23. $\frac{9}{2}$ **25.** 108   **27.** $\frac{1}{6}$ **29.** 25, -10 **31.** $\frac{1}{32}, -\frac{5}{64}$ **33.** 16, $\frac{8}{3}$ **35.** 48, $y - 64 = 48(x - 4)$ **37.** $8x^7$ **39.** $\frac{3}{4}x^{-1/4}$
41. 0   **43.** $\frac{1}{5}x^{-4/5}$ **45.** 4, $\frac{1}{3}$

EXERCISES 1.4, page 75

1. No limit   **3.** 1   **5.** No limit   **7.** -5 **9.** 5   **11.** No limit   **13.** 288   **15.** 0   **17.** 3   **19.** -4 **21.** -8 **23.** $\frac{6}{7}$

25. No limit   **27.** $-\frac{2}{11}$ **29.** 6   **31.** 3   **33.** $-\frac{2}{121}$ **35.** $\dfrac{-\sqrt{3}}{6}$ **37.** 0   **39.** 3   **41.** 0   **43.** 0   **45.** 2

EXERCISES 1.5, page 81

1. No   **3.** Yes   **5.** No   **7.** No   **9.** Yes   **11.** No   **13.** Continuous, differentiable
15. Continuous, not differentiable   **17.** Continuous, not differentiable   **19.** Not continuous, not differentiable

EXERCISES 1.6, page 87

1. $3x^2 + 2x$ **3.** $2x + 3$ **5.** $5x^4 - \dfrac{1}{x^2}$ **7.** $4x^3 + 3x^2 + 1$ **9.** $6x$ **11.** $3x^2 + 14x$ **13.** $-\dfrac{8}{x^3}$ **15.** $3 + \dfrac{1}{x^2}$

17. $x^2 - x$ **19.** $\dfrac{1}{x^6}$ **21.** $\dfrac{-1}{2\sqrt{x}}$ **23.** $30(3x + 1)^9$ **25.** $\dfrac{9x^2 + 1}{2\sqrt{3x^3 + x}}$ **27.** $6(4x - 1)(2x^2 - x + 4)^5$ **29.** $\frac{1}{3} - 3x^{-2}$

31. $10(1 - 5x)^{-2}$ **33.** $4x^3(1 - x^4)^{-2}$ **35.** $-2(x^2 + x)^{-3/2}(2x + 1)$

37. $\dfrac{3}{2}\left(\dfrac{\sqrt{x}}{2} + 1\right)^{1/2}\left(\dfrac{1}{4}x^{-1/2}\right)$ or $\dfrac{3}{8\sqrt{x}}\left(\dfrac{\sqrt{x}}{2} + 1\right)^{1/2}$ 39. 4   41. 15   43. $f'(4) = 48, y = 48x - 191$

45. $f'(x) = 36x^3 + 18x^2 - 22x - 4 = 2(3x^2 + x - 2)(6x + 1)$ 47. 6 or -66

EXERCISES 1.7, page 92

1. $10t(t^2 + 1)^4$ 3. $(2t - 1)^{-1/2}$ 5. $\frac{2}{3}(T^3 + 5T)^{-1/3}(3T^2 + 5)$ 7. $6P - \frac{1}{2}$ 9. $2a^2t + b^2$

11. $f'(x) = x - 7, f''(x) = 1$ 13. $y' = \frac{1}{2}x^{-1/2}, y'' = -\frac{1}{4}x^{-3/2}$ 15. $f'(r) = 2\pi(hr + 1), f''(r) = 2\pi h$

17. $g'(x) = -5, g''(x) = 0$ 19. $f'(P) = 15(3P + 1)^4, f''(P) = 180(3P + 1)^3$ 21. 20   23. 54   25. 34

27. $8k(2P - 1)^{-3}$ 29. $f'(3) = -\frac{1}{2}, f''(3) = -\frac{1}{8}$ 31. 20   33. (a) $f'''(x) = 60x^2 - 24x$ (b) $f'''(x) = \dfrac{15}{2\sqrt{x}}$

EXERCISES 1.8, page 98

1. 13   3. 63 units/hr   5. 1 g/week   7. $-\frac{200}{27}$ units/day, $\frac{8}{3}$ units/day, increasing   9. (a) \$16 (b) extra cost = \$16.10

11. (a) \$9.01 (b) $\approx$ \$9.10   13. (a) 7 km/hr (b) 16.5 km

15. (a) 160 ft/sec (b) 96 ft/sec (c) -32 ft/sec^2 (d) $t = 10$ (e) -160 ft/sec   17. $t = 20$ 19. $(2 \pm \sqrt{2})/2$ 21. .12

23. 2.0025   25. It approximates the increase in the monthly payment per one percent increase in the interest rate.

CHAPTER 1: SUPPLEMENTARY EXERCISES, page 101

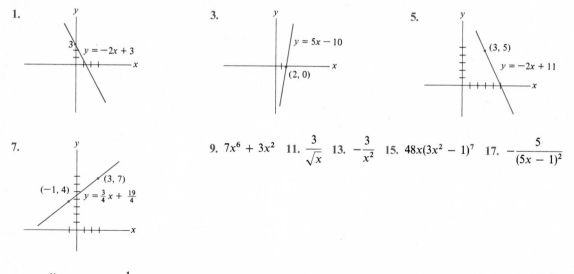

1.
$y = -2x + 3$

3.
$y = 5x - 10$
$(2, 0)$

5.
$(3, 5)$
$y = -2x + 11$

7.
$(3, 7)$
$(-1, 4)$
$y = \frac{3}{4}x + \frac{19}{4}$

9. $7x^6 + 3x^2$ 11. $\dfrac{3}{\sqrt{x}}$ 13. $-\dfrac{3}{x^2}$ 15. $48x(3x^2 - 1)^7$ 17. $-\dfrac{5}{(5x - 1)^2}$

19. $\dfrac{x}{\sqrt{x^2 + 1}}$ 21. $-\dfrac{1}{4x^{5/4}}$ 23. 0   25. $10[x^5 - (x - 1)^5]^9[5x^4 - 5(x - 1)^4]$ 27. $\frac{3}{2}t^{-1/2} + \frac{3}{2}t^{-3/2}$ 29. $\dfrac{2(9t^2 - 1)}{(t - 3t^3)^2}$

31. $\frac{9}{4}x^{1/2} - 4x^{-1/3}$ 33. 28   35. 14, 3   37. $\frac{15}{2}$ 39. 33   41. $4x^3 - 4x$ 43. $-\frac{3}{2}(1 - 3P)^{-1/2}$ 45. 29

47. $300(5x + 1)^2$ 49. -2 51. $3x^{-1/2}$ 53. Slope -4; tangent $y = -4x + 6$ 55.

$\left(\frac{3}{2}, \frac{9}{4}\right)$
$y = x^2$
$y = 3x - \frac{9}{4}$
$\frac{3}{2}$

57. $y = 2$ 59. 96 ft/sec   61. 4   63. Does not exist   65. $-\frac{1}{50}$ 67. \$12.53

CHAPTER 2

1. (a), (e), (f)   3. (b), (c), (d)

5. Decreasing for $x < -2$, minimum point at $x = -2$, minimum value $= -2$, increasing for $x > -2$, concave up, y-intercept $(0, 0)$, x-intercepts $(0, 0)$ and $(-3.6, 0)$.

7. Decreasing for $x < 0$, minimum point at $x = 0$, increasing for $0 < x < 2$, maximum point at $x = 2$, decreasing for $x > 2$, concave up for $x < 1$, concave down for $x > 1$, inflection point at $(1, 3)$, y-intercept $(0, 2)$, x-intercept $(3.4, 0)$.

9. Decreasing for $x < 2$, min. at $(2, 3)$, increasing for $x > 2$, concave up for all x, no inflection points, defined for $x > 0$, the line $y = x$ is an asymptote, the y-axis is an asymptote, minimum value $= 3$.

11. Decreasing for $1 \leq x < 3$, minimum point at $x = 3$, increasing for $x > 3$, maximum value $= 6$ (at $x = 1$), minimum value $= 1$ (at $x = 3$), inflection point at $x = 4$, concave up for $1 \leq x < 4$, concave down for $x > 4$, the line $y = 4$ is an asymptote.

13. Slope increases for all x.   15. Slope decreases for $x < 3$, increases for $x > 3$. Minimum slope occurs at $x = 3$.

17. Oxygen content decreases until time a, at which time it reaches a minimum. After a, oxygen content steadily increases. The rate of increase increases until b, and then decreases. Time b is the time when oxygen content is increasing fastest.

19.

21. The parachutist's speed levels off to 15 ft/sec.   23. (a) Yes (b) Yes

25.

27.

29.

31. Relatively low

1. (b), (c), (f)   3. (d), (e), (f)   5.

7.

9.

11. The second curve   13. The second curve

15.
(3, 4)

17.
(3, 1)

19.
(3, −2)

EXERCISES 2.3, page 129

1.
(0, −8)

3.
$(-1, -\frac{9}{2})$

5.
(3, 10)

7.
(−4, 6)

9.
(−3, 0)   (−1, −4)

11.
(−2, 16)   (2, −16)

13.
(9, 81)   (−3, −15)

15.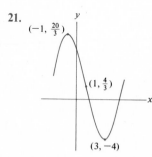
$(4, -\frac{4}{3})$ (0, −12)

17.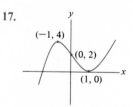
(−1, 4)   (0, 2)   (1, 0)

19.
(2, 5)   (0, 1)   (1, 3)

21.
$(-1, \frac{20}{3})$ $(1, \frac{4}{3})$ (3, −4)

23.
(−2, 64) $(\frac{1}{2}, \frac{3}{2})$ (3, −61)

25. No, $f''(x) = 2a \neq 0$ **27.** (4, 3) min   **29.** (1, 5) max   **31.** (−.1, −3.05) min

EXERCISES 2.4, page 135

1. $\left(\dfrac{3 \pm \sqrt{5}}{2}, 0\right)$ **3.** $(-2, 0), (-\frac{1}{2}, 0)$ **5.** $(\frac{1}{2}, 0)$ **7.** The derivative $x^2 - 4x + 5$ has no zeros.

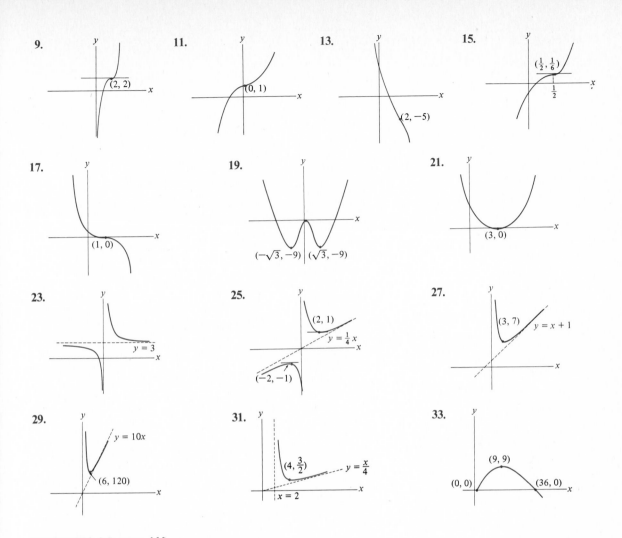

9. (2, 2)

11. (0, 1)

13. (2, −5)

15. $(\frac{1}{2}, \frac{1}{6})$, $\frac{1}{2}$

17. (1, 0)

19. $(-\sqrt{3}, -9)$, $(\sqrt{3}, -9)$

21. (3, 0)

23. $y = 3$

25. (2, 1), $y = \frac{1}{4}x$, (−2, −1)

27. (3, 7), $y = x + 1$

29. $y = 10x$, (6, 120)

31. $(4, \frac{3}{2})$, $y = \frac{x}{4}$, $x = 2$

33. (9, 9), (0, 0), (36, 0)

EXERCISES 2.5, page 144

1. 36   **3.** $t = 4$, $f(4) = 8$ **5.** 20   **7.** 150 ft   **9.** 250   **11.** $x = 50$, $h = 50$ **13.** $x = 4$ in., $h = 2$ in.   **15.** $x = \dfrac{14}{4 + \pi}$ ft

17. $x = 10$ ft, $h = 7.5$ ft   **19.** 20 cm × 20 cm × 20 cm   **21.** 5 ft × 5 ft × 6 ft

EXERCISES 2.6, page 151

1. 10 m × 10 m   **3.** $x = \dfrac{220}{\pi}$ yd   **5.** $\frac{30}{4}$ in.   **7.** 100 ft × 120 ft   **9.** $x = 6$ m, $h = 9$ m   **11.** 150   **13.** $1.10

15. $x = 40{,}000$ **17.** 400   **19.** $t = 20$ **21.** $2\sqrt{3} \times 6$

EXERCISES 2.7, page 163

1. $1   **3.** 32   **5.** 5   **7.** $x = 20$ units, $p = $133.33   **9.** 2 million tons, $156 per ton   **11.** (a) $1.00 (b) $1.15
13. (a) $x = 15 \cdot 10^5$, $p = $45. (b) No. Profit is maximized when price is increased to $50.

CHAPTER 2: SUPPLEMENTARY EXERCISES, page 166

1. Graph goes through (1, 2), increasing at $x = 1$.   3. Increasing and concave up at $x = 3$.
5. (10, 2) is a minimum point.   7. Graph goes through (5, -1), decreasing at $x = 5$.   9. (-2, 0) is a maximum point.

11.

13.

15.

17.

19.

21.

23.

25.

27.

29.

31. $f'(x) = 3x(x^2 + 2)^{1/2}, f'(0) = 0$

33. $f''(x) = -2x(1 + x^2)^{-2}, f''(x)$ is positive for $x < 0$ and negative for $x > 0$.
35. $1\frac{1}{2}$ ft wide $\times$ 3 ft long $\times$ 2 ft deep   37. $x = 25\sqrt{3}, y = 25\sqrt{6}$ 39. 2000 cans

CHAPTER 3

EXERCISES 3.1, page 174

1. $(x + 1)(3x^2 + 5) + (x^3 + 5x + 2)(1)$ or $4x^3 + 3x^2 + 10x + 7$
3. $(3x^2 - x + 2)(4x) + (2x^2 - 1)(6x - 1)$ or $24x^3 - 6x^2 + 2x + 1$
5. $(x^4 + 1)(3) + (3x + 5)(4x^3)$ or $15x^4 + 20x^3 + 3$

7. $(x^2 - 3x + 1)\left(1 - \dfrac{2}{x^2}\right) + \left(x + \dfrac{2}{x}\right)(2x - 3)$ or $3x^2 - 6x + 3 - \dfrac{2}{x^2}$

9. $(2x - 7) \cdot 5(x - 1)^4 + (x - 1)^5(2)$ or $(x - 1)^4(12x - 37)$

11. $(x^2 + 3) \cdot 10(x^2 - 3)^9(2x) + (x^2 - 3)^{10}(2x)$ or $2x(x^2 - 3)^9(11x^2 + 27)$

13. $(2x + 9)^{-2}(-3) + (5 - 3x)(-2)(2x + 9)^{-3}(2)$ or $(6x - 47)(2x + 9)^{-3}$

15. $\sqrt{x} \cdot 3(3x^2 - 1)^2(6x) + (3x^2 - 1)^3 \cdot \dfrac{1}{2\sqrt{x}}$ or $(3x^2 - 1)^2(\tfrac{39}{2}x^{3/2} - \tfrac{1}{2}x^{-1/2})$

17. $(1 - 4x)\dfrac{x}{\sqrt{x^2 - 1}} + \sqrt{x^2 - 1}(-4)$ or $(x^2 - 1)^{-1/2}(-8x^2 + x + 4)$

19. $\dfrac{(x^2 + 1)(2x) - (x^2 - 1)(2x)}{(x^2 + 1)^2}$ or $\dfrac{4x}{(x^2 + 1)^2}$

21. $\dfrac{(-1)(10x + 2)}{(5x^2 + 2x + 5)^2}$ or $-\dfrac{10x + 2}{(5x^2 + 2x + 5)^2}$

23. $\dfrac{(x + 1)\dfrac{1}{2\sqrt{x}} - \sqrt{x}(1)}{(x + 1)^2}$ or $\dfrac{1 - x}{2\sqrt{x}(x + 1)^2}$

25. $\dfrac{(x^2 + 1)^2(1) - x \cdot 2(x^2 + 1)(2x)}{(x^2 + 1)^4}$ or $\dfrac{-3x^2 + 1}{(x^2 + 1)^3}$

27. $\dfrac{(x + 2)(1) - (x - 2)(1)}{(x + 2)^2}$ or $\dfrac{4}{(x + 2)^2}$

29. $\dfrac{\sqrt{4x + x^2}(-1) - (3 - x)\left(\dfrac{2 + x}{\sqrt{4x + x^2}}\right)}{4x + x^2}$ or $-\dfrac{5x + 6}{(4x + x^2)^{3/2}}$

31. $\dfrac{(3x^2 + 1)^2 \cdot 3(x^5 + 1)^2(5x^4) - (x^5 + 1)^3 \cdot 2(3x^2 + 1)(6x)}{(3x^2 + 1)^4}$ or $\dfrac{x(x^5 + 1)^2(33x^5 + 15x^3 - 12)}{(3x^2 + 1)^3}$

33. $y - 16 = 88(x - 3)$ 35. $150; AC(150) = 35 = C'(150)$ 37. $2 \times 3 \times 1$

39. (a) $f(0) = 0, f(\tfrac{1}{2}) = \tfrac{1}{2}, f(1) = 1$ (c) Concave up for $0 < x < \tfrac{1}{2}$; concave down for $\tfrac{1}{2} < x < 1$; inflection point at $(\tfrac{1}{2}, \tfrac{1}{2})$ (d)   43. $f(x)$

EXERCISES 3.2, page 181

1. $\dfrac{1}{x^3 + 1}$ 3. $\sqrt{\dfrac{x^2}{x^2 - 1}}$ 5. x^4 7. $f(x) = \sqrt{x}, g(x) = x^2 + 3x - 1$ 9. $f(x) = \sqrt{x}, g(x) = \dfrac{x - 1}{x + 1}$

11. $f(x) = x^5, g(x) = x^3 + 9x - 2$ 13. $f(x) = \dfrac{x - 1}{x + 1}, g(x) = x^2$ 15. $f(x) = \sqrt{x + 1}, g(x) = x^{1/3}$

17. $f(x) = x^2 + 1, g(x) = x^3 + 1$ 19. $f(x) = x(x^2 + 1)^2, g(x) = x^2$ 21. $30x(x^2 + 1)^{14}$ 23. $2x^3(1 + x^4)^{-1/2}$

25. $3\left[\dfrac{2}{x} + 5(3x - 1)^4\right]^2\left[-\dfrac{2}{x^2} + 60(3x - 1)^3\right]$ 27. $8x(x^3 - 1)(x^2 + 1)^3 + 3x^2(x^2 + 1)^4$ or $x(x^2 + 1)^3(11x^3 + 3x - 8)$

29. $3(4 - x)^3(4 + x)^2 - 3(4 + x)^3(4 - x)^2$ or $-6x(16 - x^2)^2$

31. $\dfrac{1}{2}[2 + \sqrt{1 + x^2}]^{-1/2}\left[\dfrac{1}{2}(1 + x^2)^{-1/2}(2x)\right]$ or $\dfrac{x}{2}[2 + \sqrt{1 + x^2}]^{-1/2}(1 + x^2)^{-1/2}$

33. $\dfrac{\sqrt{1 + x^2} - x(\tfrac{1}{2})(1 + x^2)^{-1/2}(2x)}{1 + x^2}$ or $\dfrac{1}{(1 + x^2)^{3/2}}$

35. $\dfrac{(1 - \sqrt{2x}) \cdot \tfrac{1}{2}(2x)^{-1/2}(2) - (1 + \sqrt{2x}) \cdot (-\tfrac{1}{2})(2x)^{-1/2}(2)}{(1 - \sqrt{2x})^2}$ or $\dfrac{2(2x)^{-1/2}}{(1 - \sqrt{2x})^2}$

37.

39. $x^3 + 1$

EXERCISES 3.3, page 187

1. $-\dfrac{x}{y}$ **3.** $\dfrac{1 + 6x}{5y^4}$ **5.** $-\dfrac{y}{x}$ **7.** $\dfrac{y}{2x}$ **9.** $\dfrac{1 - y}{x + 1}$ **11.** $\dfrac{2x}{2y - 1}$ **13.** $\frac{1}{2}$ **15.** $-\frac{2}{15}$ **17.** $\frac{3}{5}$ **19.** $-\dfrac{x^3}{y^3}\dfrac{dx}{dt}$

21. $\dfrac{2x - y}{x}\dfrac{dx}{dt}$ **23.** $\dfrac{2x + 2y}{3y^2 - 2x}\dfrac{dx}{dt}$ **25.** \$331,250 per month   **27.** $5\frac{1}{3}$ thousand books per year

29. (a) $\dfrac{dr}{dt} = \dfrac{1}{4\pi r^2}\dfrac{dV}{dt}$ (b) $\dfrac{5}{4\pi}\,(\approx .40)$ cm/sec

CHAPTER 3: SUPPLEMENTARY EXERCISES, page 190

1. $\sqrt{x}(3x^2) + (x^3 + 1) \cdot \dfrac{1}{2\sqrt{x}}$ or $\dfrac{1}{2\sqrt{x}}(7x^3 + 1)$ **3.** $\dfrac{(\sqrt{x} + 1) \cdot \frac{1}{2}x^{-1/2} - x^{1/2}(\frac{1}{2}x^{-1/2})}{(\sqrt{x} + 1)^2}$ or $\dfrac{1}{2\sqrt{x}(\sqrt{x} + 1)^2}$

5. $30x^2(x^2 + 5x + 9)(x^3 - 2)^9 + (x^3 - 2)^{10}(2x + 5)$

7. $\dfrac{(x^2 + 2)[2(x + 2) - 2(x - 2)] - [(x + 2)^2 - (x - 2)^2](2x)}{(x^2 + 2)^2}$ or $\dfrac{16 - 8x^2}{(x^2 + 2)^2}$

9. $2[x + 2 + (x + 2)^2][1 + 2(x + 2)]$ or $(x^2 + 5x + 6)(10 + 4x)$ **11.** $[g(x)]^2 + g(x) + 3, 2g(x) + 1$
13. $[g(x)]^{3/2} + [g(x)]^{1/2}, \frac{3}{2}[g(x)]^{1/2} + \frac{1}{2}/g(x)^{1/2}$ **15.** $[3g(x) - 1]^5, 15[3g(x) - 1]^4$ **17.** $-15[g(x)]^{-4}g'(x)$

19. $-\frac{3}{4}[4 - g(x)]^{-1/4}g'(x)$ **21.** $-[g(x) + 1]^{-3/2}g'(x)$ **23.** $\dfrac{3x^2}{x^6 + 1}$ **25.** $\dfrac{2x}{(x^2 + 1)^2 + 1}$ **27.** $\frac{1}{2}\sqrt{1 - x}$ **29.** $\frac{3}{2}$

31. $1.89 \text{ m}^2/\text{yr}$ **33.** -3 **35.** $\frac{3}{5}$ **37.** $-\frac{1}{25}$ **39.** 200 dishwashers per month

CHAPTER 4

EXERCISES 4.1, page 196

1. $2^{2x}, 3^{(1/2)x}, 3^{-2x}$ **3.** $2^{2x}, 3^{3x}, 2^{-3x}$ **5.** $2^{-4x}, 2^{9x}, 3^{-2x}$ **7.** $2^{(1/2)x}, 3^{(4/3)x}$ **9.** $2^{-2x}, 3^x$ **11.** $3^{2x}, 2^{6x}, 3^{-x}$
13. $2^x, 3^x, 3^x$ **15.** (a) 8   (b) $\frac{1}{8}$ (c) 5.66   (d) 17.15   (e) 1.15   (f) 1.87   (g) .18   (h) .07   **17.** 1   **19.** 2   **21.** -1 **23.** $\frac{1}{5}$
25. $\frac{5}{2}$ **27.** -1 **29.** 4   **31.** 2^h **33.** $2^h - 1$ **35.** $3^x + 1$ **37.** $3^{5x} + 1$

EXERCISES 4.2, page 202

1. 1.1612, 1.105, 1.10   **3.** 1.005, 1.002, 1   **5.** $10e^{10x}$ **7.** $4e^{4x}$ **9.** $e^{2x} + e^{5x}$ **11.** e^{1+x} **13.** e^{2x} **15.** 7.3891

17. .60653   **19.** $x = 4$ **21.** $x = 4, -2$ **23.** $xe^x + e^x$ **25.** $\dfrac{e^x}{(1 + e^x)^2}$ **27.** $20e^x(1 + 5e^x)^3$

EXERCISES 4.3, page 208

1. $-e^{-x}$ **3.** $5e^x$ **5.** $2te^{t^2}$ **7.** $\dfrac{e^x + e^{-x}}{2}$ **9.** $-2(e^{-2x} + 1)$ **11.** $3(e^x + e^{-x})^2(e^x - e^{-x})$ **13.** $-\frac{2}{3}e^{3-2x}$ **15.** $3e^{3t}$

17. $\left(3x^2 + 1 + \dfrac{1}{x^2}\right)e^{x^3+x-(1/x)}$ 19. $4(2x + 1 - e^{2x+1})^3(2 - 2e^{2x+1})$ 21. $3x^2 e^{x^2} + 2x^4 e^{x^2}$ 23. $3xe^{x^3} - x^{-2}e^{x^3}$

25. $-xe^{-x+2}$ 27. $\left(-\dfrac{1}{x^2} + \dfrac{1}{x} + 3\right)e^x$ 29. $\dfrac{2e^x}{(e^x + 1)^2}$ 31. Max at $x = 1$ 33. Min at $x = \frac{11}{2}$ 35. Max at $x = 3$

37. Min at $x = 1$; max at $x = 3$ 39. Max at $x = -6$; min at $x = -5$ 41. $.02e^{-2e^{-.01x}}e^{-.01x}$ 43. $y = Ce^{-4x}$
45. $y = e^{-.5x}$ 49.

$(-\dfrac{1}{\sqrt{2}}, e^{-1/2})$ $(0, 1)$ $(\dfrac{1}{\sqrt{2}}, e^{-1/2})$

EXERCISES 4.4, page 214

1. -1 3. $-\ln 1.7$ 5. $e^{2.2}$ 7. 2 9. e 11. 1 13. $\frac{1}{2}\ln 5$ 15. $4 - e^{1/2}$ 17. $\pm e^3$ 19. $\dfrac{\ln(.5)}{-.00012}$ 21. $\frac{3}{5}$ 23. $\dfrac{e}{2}$

25. $3\ln\frac{9}{2}$ 27. $5\ln 6$ 29. $\frac{1}{5}\ln\frac{2}{5}$ 31. $-\ln\frac{3}{2}$ 33. $(-\ln 3, 3 - 3\ln 3)$, minimum 35. $(\frac{1}{2}\ln\frac{3}{2}, \frac{1}{2})$, minimum
37. 39. 109.947

$(2\ln 51, \dfrac{250}{51^{1.02}})$

$4\ln 51$

EXERCISES 4.5, page 218

1. $\dfrac{1}{x}$ 3. $\dfrac{1}{x + 5}$ 5. $-\dfrac{\ln(x + 1)}{x^2} + \dfrac{1}{x(x + 1)}$ 7. $\left(\dfrac{1}{x} + 1\right)e^{\ln x + x}$ 9. $\dfrac{1}{x}$ 11. $\dfrac{2\ln x}{x} + \dfrac{1}{x}$ 13. $\dfrac{1}{x}$ 15. $\dfrac{\ln x - 2}{(\ln x)^3}$

17. $2e^{2x}\ln x + \dfrac{e^{2x}}{x}$ 19. $\dfrac{5e^{5x}}{e^{5x} + 1}$ 21. $2(\ln 4)t$ 23. $\dfrac{6\ln t - 3(\ln t)^2}{t^2}$ 25. $y = 1$ 27. $\left(e^2, \dfrac{2}{e}\right)$, yes

29. 31. $\dfrac{1 + 3\ln 10}{100}$

$(1, \frac{1}{2})$ $(2, \ln 2)$

33. $P(x) = 300\ln(x + 1) - 2x$ and $P'(149) = 0$. Since $P''(149) < 0$, the graph of $P(x)$ is concave down at $x = 149$. So $P(x)$ has a maximum point there.

35. $\dfrac{1}{e}$

EXERCISES 4.6, page 222

1. $\ln 5x$ 3. $\ln 3$ 5. $\ln 2$ 7. x^2 9. $\ln\dfrac{x^5 z^3}{\sqrt{y}}$ 11. $3\ln x$ 13. $3\ln 3$ 15. $\dfrac{1}{x + 5} + \dfrac{2}{2x - 1} - \dfrac{1}{4 - x}$

17. $\dfrac{1}{x + 1} + \dfrac{3}{3x - 2} - \dfrac{1}{x + 2}$ 19. $\dfrac{1}{2x} - \dfrac{2x}{x^2 + 1}$ 21. $(x + 1)^3(4x - 1)^2\left[\dfrac{3}{x + 1} + \dfrac{8}{4x - 1}\right]$

23. $(x - 2)^3(x - 3)^5(x + 2)^{-7}\left[\dfrac{3}{x - 2} + \dfrac{5}{x - 3} - \dfrac{7}{x + 2}\right]$ 25. $x^x[1 + \ln x]$ 27. $e^x\sqrt{x^2 - 1}\left[1 + \dfrac{x}{x^2 - 1}\right]$

29. $x^{\ln x} \cdot \dfrac{2 \ln x}{x}$ 31. $\dfrac{\sqrt{x-1}(x-2)}{x^2-3} \cdot \left[\dfrac{1}{2} \cdot \dfrac{1}{x-1} + \dfrac{1}{x-2} - \dfrac{2x}{x^2-3} \right]$ 33. $y = cx^k$ 35. $h = 3, k = \ln 2$

CHAPTER 4: SUPPLEMENTARY EXERCISES, page 225

1. 81   3. $\frac{1}{25}$ 5. 4   7. 9   9. e^{3x^2} 11. e^{2x} 13. $e^{11x} + 7e^x$ 15. $x = 4$ 17. $x = -5$ 19. $70e^{7x}$ 21. $e^{x^2} + 2x^2e^{x^2}$

23. $e^x \cdot e^{e^x} = e^{x+e^x}$ 25. $\dfrac{(2x-1)(e^{3x}+3) - 3e^{3x}(x^2-x+5)}{(e^{3x}+3)^2}$ 27. $y = Ce^{-t}$ 29. $y = 2e^{1.5t}$

31.

33.

35. $\frac{2}{3}$ 37. 1   39. $\sqrt{5}$

41. $\frac{1}{2} \ln 5$ 43. $e^{5/2}$ 45. $e, \dfrac{1}{e}$ 47. $\dfrac{5}{5x-7}$ 49. $\dfrac{2 \ln x}{x}$ 51. $\dfrac{6x^5 + 12x^3}{x^6 + 3x^4 + 1}$ 53. $\dfrac{1}{x} + 1 - \dfrac{1}{2(1+x)}$ 55. $\dfrac{1}{x \ln x}$

57. $\ln x$ 59. $\dfrac{e^x}{x} + e^x \ln x$ 61. $(x^2+5)^6(x^3+7)^8(x^4+9)^{10} \left[\dfrac{12x}{x^2+5} + \dfrac{24x^2}{x^3+7} + \dfrac{40x^3}{x^4+9} \right]$ 63. $10^x \ln 10$

65.

67.

CHAPTER 5

EXERCISES 5.1, page 236

1. $P(t) = 400e^{.07t}$ 3. (a) 5000 (b) $t = 6.9$ 5. .017   7. (a) $P(t) = 4e^{.02t}$ (b) 6.7 billion (c) 1994   9. 71 min
11. (a) $P(t) = 30e^{-.08t}$ (b) 13.48 g   13. (a) $P(t) = 100e^{-.01t}$ (b) 74.082 g (c) 69 yr   15. (a) .13 (b) 7.7105 g
17. 58.285%   19. 8990 yr   21. $f(t) = e^{-.081t}$ 23. 184 yr   25. 20,022 yr   27. (a) $P(t) = 500e^{-.23t}$ (b) 10 months

EXERCISES 5.2, page 246

1. $1210   3. $10,000(1.02)^{12}$ 5. $2316.37   7. $616.84   9. 11.45%   11. 39% interest compounded continuously
13. 8.9 yr   15. 1991   17. $446.26   19. 7.25%   21. $5488.10   23. 29%

EXERCISES 5.3, page 254

1. 20%, 4%   3. 30%, 30%   5. 60%, 300%   7. $-25\%, -10\%$ 9. 12.5%   11. 5.8 yr   13. $p/(140 - p)$, elastic
15. $2p^2/(116 - p^2)$, inelastic   17. $p - 2$, elastic   19. (a) Inelastic (b) raised   21. (a) Inelastic (b) decrease
23. (a) 2 (b) yes

EXERCISES 5.4, page 266

1. (a) $f'(x) = 10e^{-2x} > 0$, $f(x)$ increasing; $f''(x) = -20e^{-2x} < 0$, $f(x)$ concave down

(b) As x becomes large, $e^{-2x} = \dfrac{1}{e^{2x}}$ approaches 0 (c)

3. $y' = 2e^{-x} = 2 - (2 - 2e^{-x}) = 2 - y$
5. $y' = 30e^{-10x} = 30 - (30 - 30e^{-10x}) = 30 - 10y = 10(3 - y)$, $f(0) = 3(1 - 1) = 0$ 7. 4.8 hr

CHAPTER 5: SUPPLEMENTARY EXERCISES, page 268

1. $29.92e^{-.2x}$ 3. $5488.10   5. .058   7. (a) $11.2e^{.024t}$ (b) 18.1 million (c) 1992
9. (a) $36,692.97 (b) the alternative investment is superior by $3859.   11. 400%   13. 3%, decrease   15. Increase
17. $100(1 - e^{-.083t})$

CHAPTER 6

EXERCISES 6.1, page 277

1. $\frac{1}{2}x^2 + C$ 3. $\frac{1}{3}e^{3x} + C$ 5. $3x + C$ 7. $-\frac{1}{4}$ 9. $\frac{2}{3}$ 11. -2 13. $-\frac{5}{2}$ 15. $\frac{1}{2}$ 17. -1 19. 1   21. $\frac{1}{15}$
23. $\dfrac{x^3}{3} - \dfrac{x^2}{2} - x + C$ 25. $4\sqrt{x} - 2x^{3/2} + C$ 27. $4t + e^{-5t} + \dfrac{e^{2t}}{6} + C$ 29. $\frac{2}{5}t^{5/2} + C$ 31. C 33. $\dfrac{x^2}{2} + 3$
35. $\frac{2}{3}x^{3/2} + x - \frac{28}{3}$ 37. $2\ln|x| + 2$ 39. (a) $-16t^2 + 96t + 256$ (b) 8 sec (c) 400 ft
41. $P(t) = 60t + t^2 - \frac{1}{12}t^3$ 43. $20 - 25e^{-.4t}$ °C   45. $-95 + 1.3x + .03x^2 - .0006x^3$ 47. $5875(e^{.016t} - 1)$

EXERCISES 6.2, page 285

1. 0   3. 5   5. 30   7. $\frac{4}{3}(1 - e^{-3})$ 9. 14   11. $5(e - 1)$ 13. $\ln 2$ 15. $1\frac{7}{9}$ 17. $\frac{1}{5}$ 19. $3\frac{3}{4}$ 21. $\frac{115}{6} + \ln\frac{7}{9}$
23. 323.2 km   25. 80.8   27. $1,185.75
29. The additional profit generated by raising the sales level from a to b units.   31. 10   33. $2(e^{1/2} - 1)$ 35. $6\frac{3}{5}$
37. 15   39. 9   43. 10

EXERCISES 6.3, page 294

1. $\frac{64}{3}$ 3. $\frac{52}{3}$ 5. 20   7. 18   9. $\frac{32}{3}$ 11. $\frac{32}{3}$ 13. (a) $\frac{9}{2}$ (b) $\frac{19}{3}$ (c) $\frac{79}{6}$ 15. $\frac{3}{2}$ 17. $2 + 12\ln(\frac{3}{2})$
19. 4.48 billion barrels saved

EXERCISES 6.4, page 306

1. .21875, .2421875, actual value .25   3. .23216, actual value .23254   5. 1.09286, actual value 1.09861   7. $20
9. $50   11. $200   13. $25   15. Intersection (100, 10), consumers' surplus = $100, producers' surplus = $250
17. $3236.68   19. 4.52$P$ dollars   21. 3   23. 0   25. 8   27. 55 deg   29. ≈ 82 g
31. The sum is closely approximated by $\int_0^3 (3 - x)^2\, dx$, and the value of this integral is 9.
33. The sum is closely approximated by $\int_0^1 (2x + x^3)\, dx$, and the value of this integral is $\frac{5}{4}$.

35. 100.08; they estimate the amount of oil consumed in 1976, 1977, 1978, and 1979.   37. $\frac{4}{3}\pi r^3$ 39. $\dfrac{31\pi}{5}$

41. 8π 43. $\dfrac{\pi}{2}\left(1 - \dfrac{1}{e^{2r}}\right)$

EXERCISES 6.5, page 316

1. $\frac{1}{6}(x^2 + 4)^6 + C$ 3. $\frac{1}{4}(x^2 - 5x)^4 + C$ 5. $e^{5x-3} + C$ 7. $\ln|x^3 - 1| + C$ 9. $\frac{2}{3}\sqrt{x^3 - 1} + C$ 11. $-e^{(1/x)} + C$
13. $\frac{1}{3}\ln|x^3 + 3x + 2| + C$ 15. $\frac{1}{11}(x^5 - 2x + 1)^{11} + C$ 17. $\frac{3}{2}\ln|2x - 4| + C$ 19. $-\frac{1}{2}e^{-x^2} + C$

21. $\frac{1}{5}xe^{5x} - \frac{1}{25}e^{5x} + C$ 23. $\frac{x}{10}(2x + 1)^5 - \frac{1}{120}(2x + 1)^6 + C$ 25. $\frac{2}{3}x(x + 1)^{3/2} - \frac{4}{15}(x + 1)^{5/2} + C$

27. $-3xe^{-x} - 3e^{-x} + C$ 29. $x \ln x - x + C$ 31. $\frac{1}{2}xe^{2x} - \frac{1}{4}e^{2x} + C$ 33. $\frac{1}{2}e^{x^2} + C$
35. $\frac{2}{3}(2x - 1)(3x - 3)^{1/2} - \frac{8}{27}(3x - 3)^{3/2} + C$ 37. $-\frac{1}{18}(6x^2 + 9x)^{-6} + C$ 39. $2\sqrt{x} \ln x - 4\sqrt{x} + C$
41. $\ln|\ln 5x| + C$ 43. \$738.80

EXERCISES 6.6, page 323

1. 0   3. No limit   5. $\frac{1}{4}$ 7. 2   9. 5   11. 6   13. 2   15. $\frac{1}{2}$ 17. 2   19. 1
21. Area under curve from 1 to b is $-4 + 4b^{1/4}$. This has no limit as $b \to \infty$.   23. $\frac{1}{2}$ 25. Divergent   27. $\frac{1}{6}$
29. $\frac{2}{3}$ 31. 1   33. $2e$ 35. Divergent   37. $\frac{1}{2}$ 39. 2   41. $\frac{1}{4}$ 43. 2   45. $\frac{1}{3}$ 49. \$50,000

EXERCISES 6.7, page 329

1. (a) $\frac{11}{32}$ (b) $\frac{7}{27}$ (c) $\frac{27}{32}$ (d) $\frac{27}{32}$ 3. $\frac{3}{8}$ 5. (a) .18127 (b) .45085 (c) .13534 (d) .5
7. (a) $f(x) = \frac{1}{3}, 0 \le x \le 3$ (b) $\frac{2}{3}$ (c) $\frac{1}{3}$ 9. (a) .25 (b) 5000 acres   11. .88   13. (a) .39347 (b) .36788
15. .22313

EXERCISES 6.8, page 334

1. 7; $a = 5$ 3. 2; $a = -2, b = 5$ 5. 10; $a = \sqrt{8}$ 7. $-\frac{1}{3}\ln\left|\dfrac{3 + \sqrt{9 - x^2}}{x}\right| + C$

9. $\dfrac{x}{4} - \dfrac{1}{8}\ln|4 - e^{2x}| + C$ 11. $\dfrac{2x - 20}{3}\sqrt{x + 5} + C$ 13. $\dfrac{1}{5}\ln\left|\dfrac{x - 4}{x + 1}\right| + C$ 15. $-\ln\left|\dfrac{x}{x - 1}\right| + C$

17. $\dfrac{4}{\sqrt{3}}\ln\left|\dfrac{\sqrt{3 - 2x} - \sqrt{3}}{\sqrt{3 - 2x} + \sqrt{3}}\right| + C$ 19. $\frac{1}{3}x^2 e^{3x} - \frac{2}{9}xe^{3x} + \frac{2}{27}e^{3x} + C$ 21. $\dfrac{x}{2}\sqrt{9x^2 + 1} + \dfrac{1}{6}\ln|3x + \sqrt{9x^2 + 1}| + C$

23. $\dfrac{1}{\sqrt{3}}\ln|\sqrt{3}x + \sqrt{3x^2 - 1}| + C$ 25. $\frac{1}{4}\ln|x^4 + \sqrt{1 + x^8}| + C$ 27. $\frac{5}{7}\ln|x + 5| + \frac{2}{7}\ln|x - 2| + C$

29. $\dfrac{10 \ln x - 4}{75}\sqrt{1 + 5 \ln x} + C$

CHAPTER 6: SUPPLEMENTARY EXERCISES, page 337

1. $-2e^{-x/2} + C$ 3. $-\frac{1}{3}(x^3 - 3x + 2)^{-1} + C$ 5. $\frac{1}{16}(2x + 3)^8 + C$ 7. $\frac{1}{4}(e^x + 4)^4 + C$

9. $-2(2x + 1)e^{-x/2} - 8e^{-x/2} + C$ 11. $\frac{1}{3}\ln|x^3 - 3x + 2| + C$ 13. $\dfrac{x}{16}(2x + 3)^8 - \frac{1}{288}(2x + 3)^9 + C$

15. $\frac{1}{3}(x - 5)^3 - 7$ 17. $\frac{3}{4}$ 19. $\frac{1}{5} - \frac{1}{5}e^{-2}$ 21. $\ln\left(\dfrac{e + e^{-1}}{2}\right)$ 23. $\frac{76}{3}$ 25. $\frac{125}{6}$ 27. $\frac{15}{2}$ 29. $\frac{40}{99}$, exact value: .40547

31. \$433.33   33. 15   35. $.02x^2 + 150x + 500$ dollars
39. (a) 68.5 quadrillion Btu (b) The improper integral represents the total reserves that will be used up during all years after $t = 0$ (1980). This should equal the reserves in 1980, assuming that gas is used until reserves are exhausted. In fact, $\int_0^\infty R(t)\,dt \approx 159$.
41. The total cubic centimeters of drug injected during the first 4 minutes   43. $\frac{1}{3}e^6$ 45. $\frac{2}{25}$
49. (a) $e^{-1/3} \approx .72$ (b) $r(t) = e^{-t/72}$

CHAPTER 7

1. $f(1, 0) = 1, f(0, 1) = 8, f(3, 2) = 25$ **3.** $f(0, 1) = 0, f(3, 12) = 18, f(a, b) = 3\sqrt{ab}$
5. $f(2, 3, 4) = -2, f(7, 46, 44) = \frac{7}{2}$
11. \$50. \$50 invested at 5% continuously compounded interest will yield \$100 in 13.8 years

13.

15. They correspond to the points having the same altitude above sea level.

1. $5y, 5x$ **3.** $4xe^y, 2x^2e^y$ **5.** $-\dfrac{y^2}{x^2}, \dfrac{2y}{x}$ **7.** $4(2x - y + 5), -2(2x - y + 5)$ **9.** $(2xe^{3x} + 3x^2e^{3x}) \ln y, x^2e^{3x}/y$

11. $\dfrac{2y}{(x + y)^2}, -\dfrac{2x}{(x + y)^2}$ **13.** $\dfrac{3\sqrt{K}}{2\sqrt{L}}$ **15.** $\dfrac{2xy}{z}, \dfrac{x^2}{z}, -\dfrac{1 + x^2y}{z^2}$ **17.** $ze^{yz}, xz^2e^{yz}, x(yz + 1)e^{yz}$ **19.** $1, 3$ **21.** -12

23. $\dfrac{\partial f}{\partial x} = 3x^2y + 2y^2, \dfrac{\partial^2 f}{\partial x^2} = 6xy, \dfrac{\partial f}{\partial y} = x^3 + 4xy, \dfrac{\partial^2 f}{\partial y^2} = 4x, \dfrac{\partial^2 f}{\partial y \, \partial x} = \dfrac{\partial^2 f}{\partial x \, \partial y} = 3x^2 + 4y$

25. (a) Marginal productivity of labor = 480; of capital = 40 (b) 240 fewer units produced
27. As the price of a bus ride increases, fewer people will ride the bus if the price of a train ticket remains constant. An increase in train ticket prices, coupled with constant bus fare, should cause more people to ride the bus.

29. $\dfrac{\partial f}{\partial r}, \dfrac{\partial f}{\partial m}, \dfrac{\partial f}{\partial s}$ **31.** $\dfrac{\partial V}{\partial P}(20,300) = -.06, \dfrac{\partial V}{\partial T} = .004$

33. $\dfrac{\partial^2 f}{\partial x^2} = -\frac{45}{4}x^{-5/4}y^{1/4}$. Marginal productivity of labor is decreasing.

1. $(-2, 1)$ **3.** $(26, 11)$ **5.** $(1, -3), (-1, -3)$ **7.** $(\sqrt{5}, 1)(\sqrt{5}, -1); (-\sqrt{5}, 1); (-\sqrt{5}, -1)$ **9.** $(\frac{4}{3}, \frac{4}{3})$ **11.** $(0, 0)$ min
13. $(-1, -4)$ max **15.** $(0, -1)$ min **17.** $(-1, 2)$ max; $(1, 2)$ neither max nor min
19. $(\frac{1}{4}, 2)$ min; $(\frac{1}{4}, -2)$ neither max nor min **21.** $(\frac{1}{2}, \frac{1}{6}, \frac{1}{2})$ **23.** 14 in. × 14 in. × 28 in. **25.** $x = 120, y = 80$

1. 58 at $x = 6, y = 2, \lambda = 12$ **3.** 13 at $x = 8, y = -3, \lambda = 13$ **5.** $x = \frac{1}{2}, y = 2$ **7.** $x = 2, y = 3, z = 1$

9. $x = 2, y = 2, z = 3$ **11.** $x = \dfrac{\sqrt{2}}{2}, y = \dfrac{\sqrt{2}}{2}$ **13.** $x = 4, y = 4, z = 2$ **15.** (a) $x = 81, y = 16$ (b) $\lambda = 3$

17. $x = 20, y = 60$

1. 5.009 **3.** .99 **5.** 4.95 **7.** 6.175 **9.** 75.6 **11.** Decrease in profit = \$4300 **13.** 4400π mm³ **17.** 1.5

EXERCISES 7.6, page 391

1. $y = -2x + \frac{7}{3}$ 3. $y = x + 2$ 5. $y = .4x + .9$ 7. $y = .7x - .6$ 9. $y = .5875x + 1.85$
11. (a) $y = -4.2x + 22$, (b) $8.6°C$ 13. 46

EXERCISES 7.7, page 398

1. $e^2 - 2e + 1$ 3. $2 - e^{-2} - e^2$ 5. $309\frac{3}{8}$ 7. $\frac{5}{3}$ 9. $\frac{38}{3}$ 11. $e^{-5} + e^{-2} - e^{-3} - e^{-4}$ 13. $9\frac{1}{3}$

CHAPTER 7: SUPPLEMENTARY EXERCISES, page 400

1. $2, \frac{5}{6}, 0$ 3. ≈ 20. Ten dollars increases to 20 dollars in 11.5 years. 5. $6x + y, x + 10y$ 7. $\dfrac{1}{y} e^{x/y}, -\dfrac{x}{y^2} e^{x/y}$

9. $3x^2, -z^2, -2yz$ 11. $6, 1$ 13. $20x^3 - 12xy, 6y^2, -6x^2, -6x^2$
15. $-201, 5.5$. At the level $p = 25, t = 10,000$, an increase in price of \$1 will result in a loss in sales of approximately 201 calculators, and an increase in advertising of \$1 will result in the sales of approximately 5.5 additional calculators.
17. $(3, 2)$ 19. $(0, 1), (-2, 1)$ 21. Min at $(2, 3)$ 23. Min at $(1, 4)$; neither max nor min at $(-1, 4)$
25. $20; x = 3, y = -1, \lambda = 8$ 27. $x = \frac{1}{2}, y = \frac{3}{2}, z = 2$ 29. $x = 10, y = 20$ 31. 23.6 33. 5.85 35. $y = \frac{5}{2}x - \frac{5}{3}$
37. $y = -2x + 1$ 39. 170 41. 40

Index